NOVEL NANOMATERIALS - SYNTHESIS AND APPLICATIONS

Edited by **George Z. Kyzas**
and **Athanasios C. Mitropoulos**

Novel Nanomaterials - Synthesis and Applications
http://dx.doi.org/10.5772/intechopen.70149
Edited by George Z. Kyzas and Athanasios C. Mitropoulos

Contributors

Chuan-Pei Lee, Jorge Morales-Hernandez, Ana Lilia De Jesús Lazaro, Verónica Nayelli Martínez Escobedo, Huijin Xu, Zhanbin Xing, Fuqiang Wang, Yanhuai Ding, Hui Chen, Zheng Li, Huming Ren, Xiangqiong Tang, Jiuren Yin, Ping Zhang, Nabanita Saikia, Muzamal Hussain, Muhammad Nawaz Naeem, Nabeel Z. Al-Hazeem, Marketa Vaculovicova, Vojtech Adam, Alexander Samoylov, Stanislav Ryabtsev, Vasily Popov, Petre Badica, Alexey Vereschaka, Gaik Oganian, Nikolay Sitnikov, Andre Batako, Anatoliy Aksenenko, Rodica-Mariana Ion, Sanda Maria Doncea, Daniela Turcanu-Carutiu, Young Rang Uhm, Jiatao Zhang, Sajid Muhammad, Hongpan Rong, Shuping Zhang, Lezorgia Nekabari Nwidee, Stefan Iglauer, Rajendra Sukhadeorao Dongre, Badriyah Alhalaili, Howard Mao, M. Saif Islam, Shah Kwok Wei, George Z. Kyzas, Athanasios C. Mitropoulos

First published in London, United Kingdom, 2018 by IntechOpen
IntechOpen is the global imprint of INTECHOPEN LIMITED, registered in England and Wales, registration number: 11086078, The Shard, 25th floor, 32 London Bridge Street
London, SE19SG – United Kingdom
Printed in Croatia

British Library Cataloguing-in-Publication Data
A catalogue record for this book is available from the British Library

Additional hard copies can be obtained from orders@intechopen.com

Novel Nanomaterials - Synthesis and Applications, Edited by George Z. Kyzas and Athanasios C. Mitropoulos
p. cm.
Print ISBN 978-1-78923-088-8
Online ISBN 978-1-78923-089-5

For EU product safety concerns:
IN TECH d.o.o., Prolaz Marije Krucifikse Kozulić 3, 51000 Rijeka, Croatia,
info@intechopen.com or visit our website at intechopen.com.

Meet the editors

Dr. George Z. Kyzas was born in Drama, Greece, and obtained his BSc degree in Chemistry and MSc and PhD degrees in Chemical Technology—Materials Science from the Aristotle University of Thessaloniki, Greece. His current interests include the synthesis of various adsorbent materials for the treatment of wastewaters (such as dyes, heavy metals, pharmaceuticals, and phenols). He has published over 85 significant scientific papers, authored books, edited books, and published chapters in books, teaching notes, and reports. He also acted as a guest editor in special issues of journals and presented many works in international conferences. He has been awarded with honors, grants, and fellowships for his research career/profile by the research committee of the Aristotle University of Thessaloniki, the National State Scholarships Foundation of Greece, and the Stavros Niarchos Found.

Prof. Athanasios C. Mitropoulos was born in Athens in 1957. He received his BSc degree in Chemistry from the University of Thessaloniki and MSc and PhD degrees in Physical Chemistry from the University of Bristol. In 1998, he was appointed professor at the Department of Petroleum Engineering of the Eastern Macedonia and Thrace Institute of Technology. Since 2008, he has been the president of the same institute. His expertise is in the characterization of porous media, nanoporous materials, and membranes with in-situ techniques of adsorption and small angle X ray scattering. He has published more than 100 journal papers, book chapters, and patents. He is a member of the Society of Petroleum Engineers.

Contents

Preface

The present book *Novel Nanomaterials- Synthesis and Applications* describes many points in materials of nanotechnology. Nanomaterials is one of the hottest fields in nanotechnology that studies fabrication, characterization, and analysis of materials with morphological features on the nanoscale in at least one dimension. Recent progress in synthesis and fundamental understanding of properties of nanomaterial has led to significant advancement of nanomaterial-based gas/chemical/biological sensors. The most important aspect of nanomaterial is their special properties associated with nanoscale geometries. The most fundamental characteristic of nanomaterials is their high surface area to volume ratio, which results in a number of unusual physical and chemical properties such as high molecular adsorption, large surface tension force, enhanced chemical and biological activities, large catalytic effects, and extreme mechanical strength. Another unique property of this material, which is recently the most studied, is the quantum size effect that leads to their discrete electronic band structure like those of molecules. This quantum property of nanomaterial can lead to an extraordinary high sensitivity and selectivity of biosensors and can be a benefit to the field of diagnostics. This book covers the recent progress of selected nanomaterials over a broad range of gas/chemical/biological sensing applications, and examples of nanomaterials in sensing and diagnostic application are also presented. The hottest area of nanomaterial called "carbon nanomaterial" including carbon nanotube and graphene is currently reviewed. With the extensive review of newly discovered nanomaterials used for sensors and diagnostics, this book will be interesting not only for scientists working in the field of nanomaterial-based sensor technology but also for students studying analytical chemistry, biochemistry, electrochemistry, material science, and micro- and nanotechnology. All of the above clearly indicate that the "world" of nanomaterials has various sections. Therefore, the scope of this book is wide. Specialists, researchers, and professors from more than 20 countries have published in this book their research in the nanomaterial field. We are grateful to all of the authors who have contributed their tremendous expertise to the present book. We wish to acknowledge the outstanding support of Ms. Maja Bozicevic, Publishing Process Manager, InTechOpen , who collaborated tirelessly in crafting this book.

The future of nanomaterials is indeed bright!

Dr. George Z. Kyzas (MSc, PhD) and Prof. Athanasios C. Mitropoulos (MSc, PhD)
Hephaestus Advanced Laboratory
Eastern Macedonia and Thrace Institute of Technology
Kavala, Greece

Dedicated to

The new (autumn 2017) "MPhil in Nanotechnology" (Eastern Macedonia and Thrace Institute of Technology, Kavala, Greece)

Section 1

Synthesis

Chapter 1

Introductory Chapter: Nanomaterials in the 2020s

George Z. Kyzas and Athanasios C. Mitropoulos

Additional information is available at the end of the chapter

http://dx.doi.org/10.5772/intechopen.74137

1. Introduction

This work is an introductory chapter of the book *Novel Nanomaterials - Synthesis and Applications*. Nanomaterials are a special topic of recent research and are a milestone of nanoscience and nanotechnology. Nanoscale materials are a series of substances/compounds, in which at least one dimension has smaller size than 100 nm. Nanomaterials have a broad area of development which day by day is rapidly growing. Their impact to commercial applications is huge as well as respective to academia and education. A high percentage of nanomaterials exist naturally, but the great interest is for the type of nanomaterials which can be synthesized. The latter is called as "engineered nanomaterials." These nanomaterials are formed at the molecular (nanometer) level in order to present the positive of the small/tiny size and innovative properties. Those properties are not generally observed in their common/bulk counterparts/nanoparticles. The bing bang was very crucial for the nanomaterials science given that many theories refer the existence of nanostructures after falls of early meteorites, and until now they are under examination with new nanoforms. All of the above clearly indicate that the "world" of nanomaterials is explosively growing in various sections. Therefore, the target of this book is wide. It includes many approaches of metal oxide nanostructures, nanocomposites, and polymer materials. Synthesis, characterizations, various processes, fabrications, and some promising applications are also developed and analyzed.

2. Nanomaterials science

Nanomaterials are a special topic of recent research and are a milestone of nanoscience and nanotechnology. These materials have a broad area of development which day by day is rapidly growing. Their impact to commercial applications is huge as well as the respective to academia and education.

At this point, it is mandatory to give a brief definition of nanomaterials. So, nanoscale materials are a series of substances/compounds, in which at least one dimension has smaller size than 100 nm. 1 nm is equal to 10^{-6} mm, meaning 10^5 times smaller than a possible size of a human hair (obviously measuring the diameter). Another question which must be replied is why nanomaterials have such big interest. The answer is not difficult. Since at this scale, interesting optical, attractive, electrical, and some other different properties rise. These innovative properties have the potential for extraordinary effects in various gadgets, especially in biomedicine, biochemistry, or other relatively similar fields.

Another crucial question is where nanomaterials can be found or how they are "born." A high percentage of nanomaterials exist naturally, but the great interest is for the type of nanomaterials which can be synthesized (engineered nanomaterials (EN)). This type of materials is extensively used in many processes producing even commercial products. They can be found in such things as sunscreens, cosmetics, sporting goods, stain-resistant clothing, tires, and electronics. They also exist in everyday things and especially products of medicine and biomedicine (e.g., drug delivery, imaging, or even diagnosis purposes).

Engineered nanomaterials are formed at the molecular (nanometer) level in order to present the positive of the small/tiny size and innovative properties. Those properties are not generally observed in their common/bulk counterparts/nanoparticles. Two major explanations can be given to answer why these materials at nanoscale have different properties. These materials showed high (or better increasing) relative surface area and some new types of quantum effects. Also, it is noteworthy that their conventional forms present much lower surface area to volume ratio than the new forms (EN). The latter cause higher chemical reactivity, influencing also their strength. Going to nanoscale, the importance of quantum effects is great especially regarding the determination of materials properties and characteristics; this confirms some new optical, electrical, and magnetic behaviors. Nanomaterials are already used for commercial purposes in industry. The available commercial products of nowadays belong to a broad technologic area, including stain-resistant and wrinkle-free textiles, cosmetics, sunscreens, electronics, paints, and varnishes. Nanocoatings and nanocomposites are finding uses in diverse consumer products, such as windows, sports equipment, bicycles, and automobiles. There are novel UV-blocking coatings on glass bottles which protect beverages from damage by sunlight and longer-lasting tennis balls using butyl-rubber/nano-clay composites. Nanoscale titanium dioxide, for instance, is finding applications in cosmetics, sun block creams, and self-cleaning windows, and nanoscale silica is being used as filler in a range of products, including cosmetics and dental fillings [1].

3. Historical points for nanomaterials

The big bang caused very crucial for the nanomaterials science given that many theories refer the existence of nanostructures after falls of early meteorites. After that start, the nature physically created some other nanostructure-like formations as seashells, skeletons, etc. However, the scientific/technological story and official start of nanomaterials can be pinned much later.

Michael Faraday has synthesized colloidal gold particles synthesized in 1857. In the 1930s, nanostructured catalysts were also investigated. During the 1940s, the market of the USA and Germany was filled after manufacturing with precipitated and fumed silica nanoparticles as substitutes for ultrafine carbon black for rubber reinforcements. Then, from the 1960s to the 1970s, some metallic-like nanopowders used to produce magnetic recording tapes were also developed. In 1976, Granqvist and Buhrman published a work reporting a new term; the popular inert-gas evaporation technique according to which nanocrystals were synthesized for the first time.

Nowadays, the engineering of nanophase or generally nanotechnology is rapidly developing investigating more and more structural and functional materials (inorganic and organic). Therefore, there is a possibility to change/modify or even treat some important electric, magnetic, mechanical, catalytic, optical, and electronic functions. The synthesis of materials in nanophase or the so-called cluster-assembled materials is based on the separated small cluster formation. Then, these are fused into a material (majorly in bulk form).

Author details

George Z. Kyzas* and Athanasios C. Mitropoulos

*Address all correspondence to: georgekyzas@gmail.com

Hephaestus Advanced Laboratory, Eastern Macedonia and Thrace Institute of Technology, Kavala, Greece

Reference

[1] Borm PJ, Robbins D, Haubold S, Kuhlbusch T, Fissan H, Donaldson K, Schins R, Stone V, Kreyling W, Lademann J, Krutmann J, Warheit D, Oberdorster E. The potential risks of nanomaterials: A review carried out for ECETOC. Particle and Fibre Toxicology. 2006;**3**:1 (Article ID 1)

Chapter 2

Ga_2O_3 Nanowire Synthesis and Device Applications

Badriyah Alhalaili, Howard Mao and Saif Islam

Additional information is available at the end of the chapter

http://dx.doi.org/10.5772/intechopen.72464

Abstract

In recent years, gallium oxide nanowires have been used in many scientific disciplines due to their outstanding and unique properties. Several applications have focused on incorporating gallium oxide nanowires in devices to improve their performance and efficiency. These distinctive structures bring new opportunities to several research fields and applications such as optoelectronics, electronics, and chemistry. This chapter provides a basic overview of gallium oxide's properties and the growth process of gallium oxide nanowires, with an emphasis on varied applications and future challenges.

Keywords: β-Ga_2O_3 nanowires, properties, growth, nanodevices, applications, challenges

1. Introduction

In the last few years, interest in the use of wide-bandgap materials in semiconductor devices has grown. Ideally, such materials would be abundant, inexpensive, and easy to fabricate, and have high thermal and chemical stability. However, reducing the size of these materials and the devices that utilize them down to the micro/nanoscale remains a big challenge. Furthermore, manipulating their growth to create more directional nanowires could be useful in several applications. In addition, a limited number of devices such as transistors and sensors are capable of operating at temperatures up to 500°C. Wide-bandgap materials can operate in harsh environmental conditions with high temperature, pressure, mechanical vibration and radiation. Therefore, sensors and electronic devices incorporating these materials are attractive choices for increasing the operating temperature to above 500°C while effectively integrating electronics operating even at 1000°C [1].

Preliminary studies demonstrate that some robust wide-bandgap materials can enable a number of devices and sensors that can be used in extreme environments [1, 2]. These benefits can

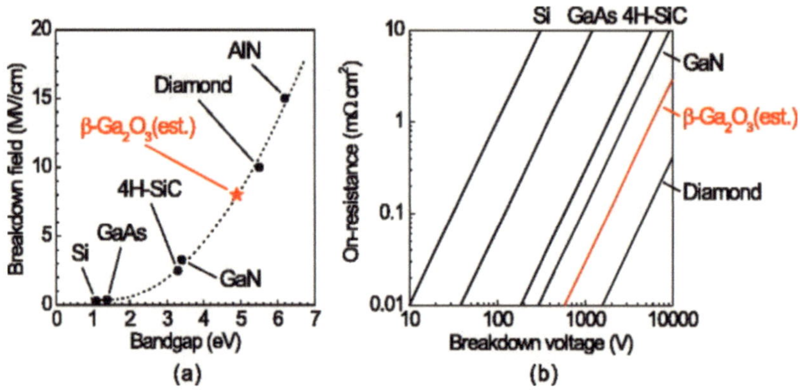

Figure 1. (a) Relationship between the bandgap and breakdown field for various semiconductor materials, (b) On-resistance as a function of breakdown voltage for major semiconductors. (Reprinted with permission from [82]. Copyright 2012, American Institute of Physics.).

be realized by exploiting the electrical, optical, thermal, physical and chemical properties of wide-bandgap semiconductors including group III nitrides such as gallium nitride (GaN), aluminum nitride (AlN) and their ternary and quaternary alloys; and silicon carbide (SiC), diamond and some wide-bandgap oxides (such as Ga_2O_3). These materials offer superior electrical, optical, mechanical, and chemical properties when compared to other widely used semiconductor materials such as Si, GaAs and InP [1, 3].

Gallium oxide has a promising future with respect to other semiconductors based on its material properties (**Figure 1a**). It has the potential to be more widely used in power and optoelectronics applications than not only Si but also SiC and GaN. This is due to its high breakdown voltage and low on-resistance (**Figure 1b**). These features have led to the evaluation of Ga_2O_3 nanowires as a novel material in semiconductor research fields; it is also less expensive and easier to fabricate than SiC and GaN. Compared to thin films, nanowires have a higher surface-to-volume ratio, which increases their sensitivity in detection. Ga_2O_3 nanowires have an established place in many applications, such as optical and sensing studies [4, 5]. For example, β-Ga_2O_3 nanowire-based devices are very attractive for use as gas sensors [6] due to their stability, moisture resistance, fast response, and long lifetime. Research interest in Ga_2O_3 nanowires has been increasing, and it has obtained recognition among other wide-bandgap materials. This chapter will mainly focus on synthesis techniques, device applications, and future challenges for β-Ga_2O_3 nanowires.

2. Fundamental properties of β-Ga_2O_3 nanowires

Gallium (III) oxide is an insulating metal oxide with five different polymorphs (α-, β- γ-, δ-, ε-). β-Ga_2O_3 has a monoclinic structure and is the most stable form both chemically and thermally. The other phases exist in a metastable state and are converted into β-phase at temperatures above 600°C. β-Ga_2O_3 has exceptional properties which have attracted the attention of scientists, including a wide bandgap of 4.9 eV, a high melting point of 1900°C, excellent electrical conductivity, and photoluminescence. Despite these wonderful properties, it is only in recent years that interest in this material has grown.

http://dx.doi.org/10.5772/intechopen.72464

Recently, researchers have showed interest in the fabrication of low-dimensional β-Ga_2O_3 nanowires because of their outstanding properties as compared to bulk structures. The main advantage of β-Ga_2O_3 nanowires is a higher surface-to-volume ratio, offering more surface states at the interface and thus greater interaction with the surroundings. β-Ga_2O_3 nanowires can be synthesized using a variety of approaches, and various parameters such as size, length and electronic properties can be controlled during their growth. Recently, a wide array of nanoscale devices such as FETs, gas sensors, photodetectors, and nanophotonic switches have been fabricated using β-Ga_2O_3 nanowires. These β-Ga_2O_3 nanowire devices are also popular for diverse applications such as displays [7], solar cell fabrication [8], ultraviolet (UV) limiters [9], and high-temperature gas sensors [10].

2.1. Crystal structure

In 1950, Ga_2O_3 was observed, and in 1960, its crystal structure was discovered by Geller [7]. β-Ga_2O_3 is part of the C2/m space group and has a base-centered arrangement with four lattice parameters (**Table 1**). The unit cell of β-Ga_2O_3 has two different Ga atoms [Ga(I) and Ga(II)] and three different O atoms [O(I), O(II) and O(III)] (**Figure 2**). The crystalline structures that make up the unit cell are octahedral (GaO_6) and tetrahedral (GaO_4). Along the b-axis, the double-chained GaO_6 octahedra that share edges are connected by single chains of GaO_4 tetrahedra that share vertices. The unit cell consists of three different oxygen ions that are juxtaposed in an irregular cubic array and are close to one another.

2.2. Electrical properties

2.2.1. Energy bandgap for Ga_2O_3

Density functional theory (DFT) has been used to calculate the electronic band structure of β-Ga_2O_3. β-Ga_2O_3 has an indirect bandgap of 4.83 eV, with a valence-band maximum (VBM) located at the M point, which is marginally less than the direct bandgap of 4.87 eV detected at the Γ-point, as shown in **Figure 3** [11]. Research has confirmed that β-Ga_2O_3 is an indirect bandgap material, but due to the weakness of the indirect transitions and the small energy difference between indirect and direct gaps, it is effectively a direct bandgap material. Peak absorption occurs around 4.9 eV. **Table 1** summarizes the electronic properties of β-Ga_2O_3.

2.2.2. Dopants for Ga_2O_3

As Ga_2O_3 is an insulator, it needs to be doped to increase its conductivity. Doping enables control of sheet resistance [12] and facilitates the formation of ohmic contacts. It can also change many properties of a material. For example, it can increase the lattice constant [13], and change crystallinity [14] and electron mobility [15]. Doping can even be used to tune the optical bandgap [16] and introduce defect levels [17]. It is typically done at the same time nanowires are grown, using methods such as chemical vapor deposition, pulsed laser deposition, and evaporation. Alternatively, dopants can be added directly into the material through ion implantation.

β-Ga_2O_3 is assumed to be an n-type semiconductor due to its shallow donor oxygen vacancies and ionization energy of 30–40 meV [9]. Its electrical conduction and free carrier concentration can be adjusted by doping it with Si. The effects of oxygen vacancies and other impurities on the electrical and optical properties of Ga_2O_3 have also been examined [11]. It was found that

Property	Attributes
Crystal structure	Monoclinic
Group of symmetry	C2/m
Lattice parameters	
a	12.214 Å
b	3.037 Å
c	5.798 Å
β	103.83 Å
Bandgap (E_g)	4.83 eV (indirect) & 4.87 (direct) at RT
Density	5.863 g/cm^3
Dopants (n/p type)	Ti/Zn, Ge, Mg
Electron affinity (χ)	3.50 eV
Electron effective mass	$0.342m_e$
Electron mobility (μ_e)	10 cm^2 V^{-1} s^{-1}
Carrier density	10^{13}–10^{16} cm^{-3}
Rel. dielectric constant (ϵ)	9.9–10.2
Melting point	1800 °C
Specific heat	0.49–0.56 Jg^{-1} K^{-1}
Thermal conductivity	
[100]	(10.9 ±1.0 W/mK)–(13 W/mK)
[−201]	13.3 ± 1.0 W/mK
[001]	14.7 ±1.5 W/mK
[110]	(27.0 ±2.0 W/mK)–(21 W/mK)
Refractive index (*n*) @ 532 nm	
‖[100]	1.9523
⊥(100)	1.9201

Table 1. Summary of the basic properties of β-Ga_2O_3 [18, 29].

it cannot be assumed that oxygen vacancies are the main cause for conduction and luminance because Si is the main impurity behind electrical conduction [18]. Furthermore, these oxygen vacancies are deep donors and hence cannot describe the n-type conductivity. Other dopants such as Si, Ge, Sn, F, and Cl behave like shallow donors. In the next sub-section, we describe some dopants and their properties. **Table 2** summarizes the materials that have been used as dopants for Ga_2O_3 and the changes they make to it.

2.2.2.1. Indium (In)

Indium is one dopant that can be used with Ga_2O_3 to improve its photoelectrical properties, especially in the context of photodetection [13]. Doping with indium narrows the bandgap, which increases the range of the photoresponse but makes the photodetector less sensitive

http://dx.doi.org/10.5772/intechopen.72464

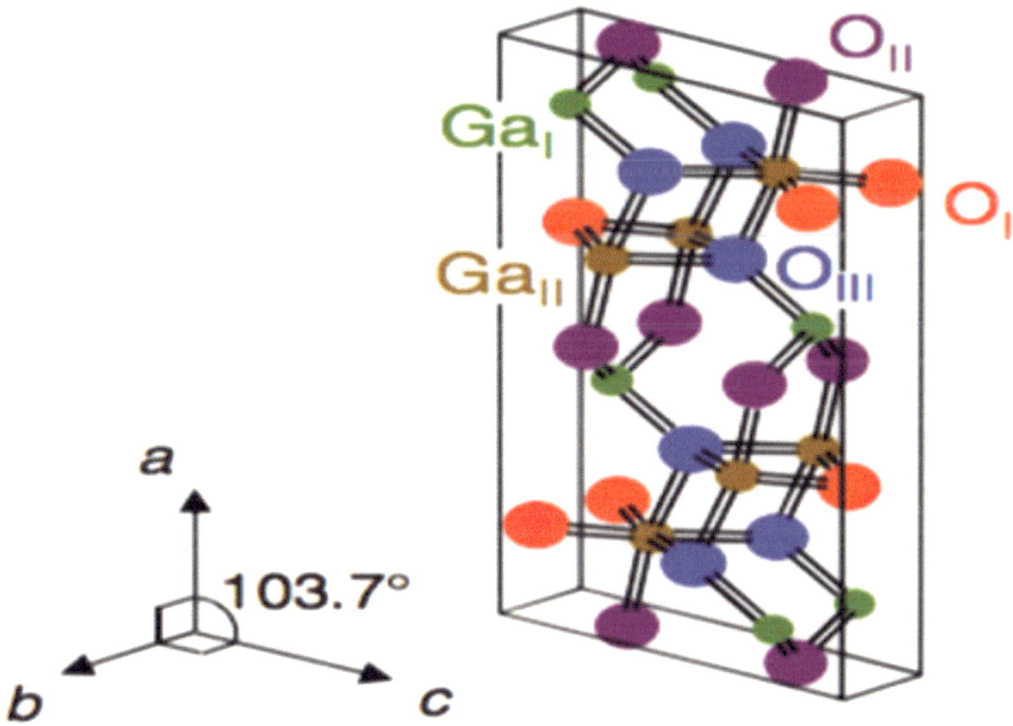

Figure 2. Unit cell of β-Ga_2O_3.

to UV wavelengths. Doping causes the photocurrent to be much larger than that of undoped devices. This is also true for the ratio of photocurrent to dark current. Indium doping can also reduce rise time and improve spectral responsivity and external quantum efficiency.

2.2.2.2. Silicon (Si)

Si has also been used to dope Ga_2O_3, significantly reducing its resistance after ion implantation of Si atoms and annealing [12]. Si doping also helps in creating an ohmic contact

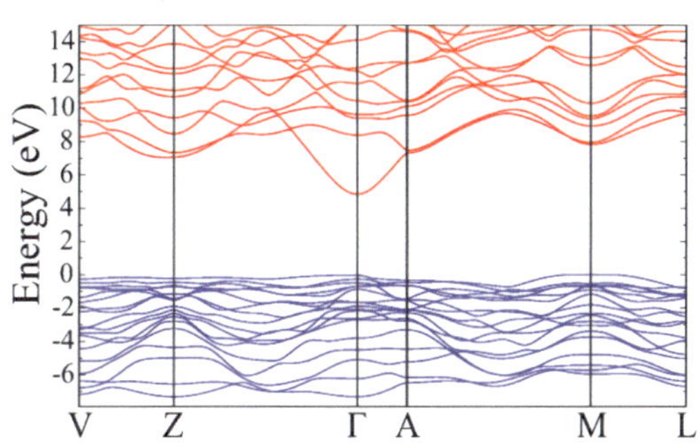

Figure 3. Electronic band structure of β-Ga_2O_3. (Reprinted with permission from Ref. [11]. Copyright 2010, American Institute of Physics.).

Materials	Properties
In	To improve the photoelectrical properties
Si	• To reduce sheet resistance of Ga_2O_3 • To support creating an ohmic contact on the surface
Sn	To increase conductivity.
Mg	A promise in making Ga_2O_3 p-type
Cu	• To decrease bandgap because CuO has a narrower bandgap than Ga_2O_3 • A potential to be a p-type dopant for Ga_2O_3
Zn	• A potential to be a p-type dopant for Ga_2O_3 • To decrease carrier concentration
Ti	• To decrease dielectric permittivity • To increase the optosensitivity • To reduce leakage current
N	Doping with N has the potential to make Ga_2O_3 p-type

Table 2. Some materials have been used as a dopant and their properties.

on the surface with Ti and Au. It was seen to increase the rejection ratio, photocurrent, responsivity, and quantum efficiency of a photodetector at high temperatures. γ-Ga_2O_3 has been doped n-type by ablation of a Ga_2O_3:Si ceramic target with laser pulses [19]. In that case, doping capability was independent of crystal structure, and carrier concentration did not vary much with temperature, meaning the Ga_2O_3 had been degenerately doped.

2.2.2.3. *Tin (Sn)*

Ga_2O_3 has also been doped with Sn to increase its conductivity [14]. Interestingly, when deposition was performed at temperatures above 410°C, doping caused the phase of the Ga_2O_3 to change from β-Ga_2O_3 to ε-Ga_2O_3. Another method of doping Ga_2O_3 with Sn is heating SnO_2 powder in a K-cell with ozone as an oxidizer [15]. However, it was observed that as the temperature of the SnO_2 powder was raised, the carrier concentration increased, which lowered electron mobility.

2.2.2.4. *Magnesium (Mg)*

Mg is one of a few materials believed to hold promise in making Ga_2O_3 p-type. Mg-doped Ga_2O_3 has been deposited into an MOCVD system with a Mg-containing source, and as the Mg concentration was increased from 1–10%, the crystalline quality degraded, with films becoming amorphous [16]. Annealing improved the crystallinity and lowered the resistivity, but the Ga_2O_3 did not become p-type as expected after doping.

2.2.2.5. Copper (Cu)

Ga_2O_3 has also been doped with Cu, which reduces the bandgap because CuO has a narrower bandgap than Ga_2O_3 [17]. The bandgap widens again after annealing. Ga^{3+} and Cu^{2+} have a similar ionic size (0.073 nm and 0.062 nm, respectively), and copper is a group IB element, so copper has the potential to be a p-type dopant for Ga_2O_3. Because Cu has fewer valence electrons than Ga does, there is a lower density of electrons between Cu and O than between Ga and O [20]. After doping, the acceptor impurity level in Ga_2O_3 is higher than the valence band.

2.2.2.6. Zinc (Zn)

Zn has potential to be used as a p-type dopant for Ga_2O_3 as it can add holes to the material after annealing and belongs to group IIB with one fewer valence electron than Ga_2O_3 [21]. While undoped β-Ga_2O_3 is clearly n-type, after doping with Zn the conductivity type could not be clearly determined because there were both electrons and holes and their concentrations were approximately equal. High transmittance was observed, indicating that crystalline quality was high. Adding Zn also caused the carrier concentration to drop from 1.4×10^{14} to 7.2×10^{11} cm^{-3}.

2.2.2.7. Titanium (Ti)

Doping Ga_2O_3 with Ti lowers its dielectric permittivity. [22]. Ti can dope about 10% of Ga_2O_3 due to their similar ionic radii (0.047 and 0.042 nm for Ga^{3+} and Ti^{4+}, respectively). This gives titanium ions high solubility in Ga_2O_3. Inclusion of Ti increases the optosensitivity of samples and reduces leakage current by increasing the activation energy and narrowing the bandgap.

2.2.2.8. Nitrogen (N)

N^{3-} has an ionic radius about the same as that of O^{2-}, so it can be expected to form shallow acceptor levels [23], which is significant because the difficulty of forming shallow acceptor levels in Ga_2O_3 is the main challenge in making it p-type. On doping Ga_2O_3 with N, the bandgap decreased and the valence band maximum and conduction band maximum met at the same point. The acceptor impurity levels were above the valence band. After Ga_2O_3 nanowires were doped with N, they exhibited p-type properties [24]. Pt has a larger work function than Ga_2O_3 and it was used for the metal contacts. When a contact metal has a work function larger than that of the semiconductor, a Schottky barrier is created when the semiconductor is n-type and an ohmic contact is exhibited for a p-type semiconductor. While an undoped Ga_2O_3 nanowire had an I-V characteristic that was rectifying, a N-doped nanowire had a linear I-V curve. This suggests that doping Ga_2O_3 with N can make it p-type.

2.3. Optical properties

2.3.1. Absorption of Ga_2O_3 nanowires

The absorption spectra of Ga_2O_3 show cutoff absorption edges at around 255–260 nm and slightly around 270 nm [18]. The band at 260 nm was due to the intrinsic band-to-band transition while the one at 270 nm was caused by Ga^{3+} vacancies in the conduction band. The absorption range (255–260 nm) can be obtained because of the transition from the valence band to the conduction band [25]. The absorption of Ga_2O_3 is influenced by polarization of incident light [26]. The difference between the absorption edge at 260 and at 270 nm was attained using the transition from the valence band.

2.3.2. Luminescence properties of Ga_2O_3 nanowires

Ga_2O_3 can produce four emissions: UV (3.2–3.6 eV), blue (2.8–3.0 eV), green (2.4 eV) [27, 28] and red [29]. Due to its wide bandgap, there are a number of defect states, which cause its different emissions from infrared to UV because its wide bandgap has a number of defect states that are responsible for them. The bandgap of Ga_2O_3 is known to be 4.8 eV, suggesting that UV emission initiated from the edge of the band is due to free electrons and recombination of self-trapped holes [30]. However, the blue and deep-blue luminescence are caused by the recombination of an electron on a defect donor state formed by oxygen vacancies and a hole on an acceptor state formed by a gallium vacancy or a gallium–oxygen vacancy pair [28, 31]. Green emission could only be achieved when the samples were doped with certain impurities such as Be, Ge, Sn, Li, Zr, and Si [18, 32]. It has been suggested that green luminescence is related to self-trapped or bound excitons [26]. Several studies have been performed on the excitation and photoluminescence of pure and Si-doped β-Ga_2O_3 single crystals [33].

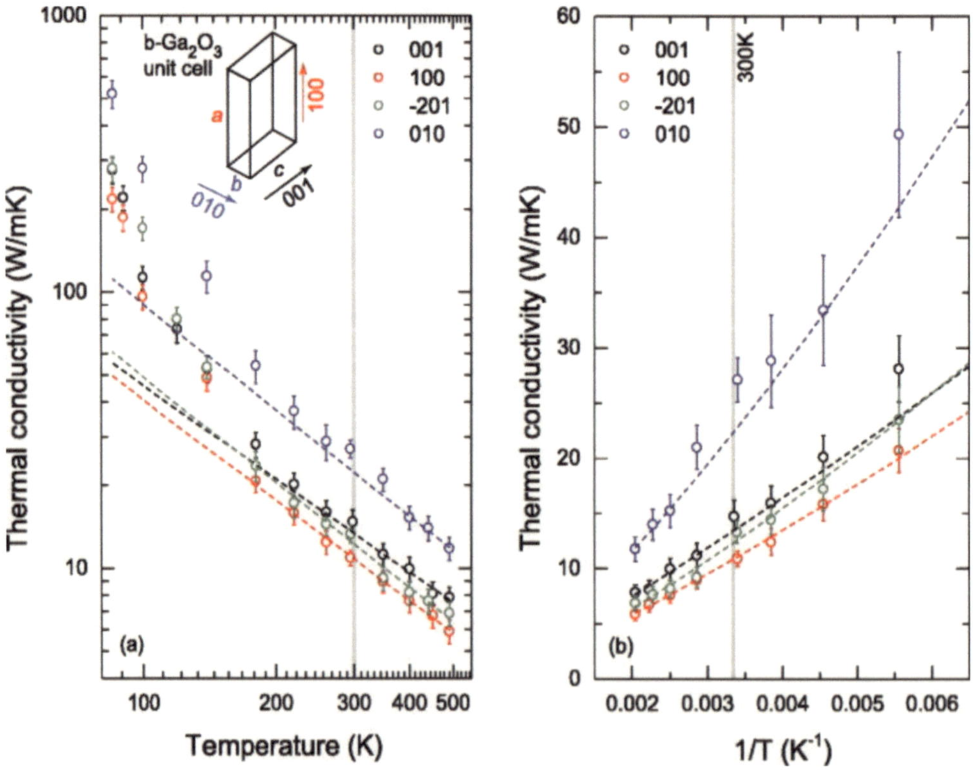

Figure 4. Thermal conductivity as a function of temperature for β-Ga_2O_3 along different crystal directions. (Reprinted with permission from Ref. [35]. Copyright 2015, American Institute of Physics.).

These studies showed that the spectra characteristic of undoped samples are independent of the excitation and emission wavelengths. However, the Si-doped sample produced a blue emission which was highly related to the excitation and emission wavelengths. Finally, it has been assumed that the red emission is caused by nitrogen impurities [32, 34].

2.3.3. Thermal properties

Ga_2O_3 is considered a weak thermal conductor compared to other semiconductors, with a thermal conductivity lower than that of other wide-bandgap materials such as SiC and GaN. Thermal conductivity in Ga_2O_3 varies depending on crystal direction due to its crystalline anisotropy. Directions with a smaller lattice constant have higher thermal conductivity., Ga_2O_3 has the highest thermal conductivity along the [010] direction and the lowest along the [100] direction. This has been measured at temperatures between 80 and 495 K (**Figure 4**). At high temperatures, thermal transport is primarily dominated by phonon scattering [35].

3. Ga_2O_3 nanowire growth mechanisms

A wide variety of methods have been used to grow Ga_2O_3 nanowires. These include thermal oxidation [36], vapor–liquid–solid mechanism [37], pulsed laser deposition [38], sputtering [39], thermal evaporation [40–42], molecular beam epitaxy [43], laser ablation [44], arc discharge [45], carbothermal reduction [46], microwave plasma [47], metalorganic chemical vapor deposition [48] and the hydrothermal method [49, 50]. **Table 3** summarizes the advantages and disadvantages of these methods. Research is ongoing to determine the best method for growing high-quality nanowires at low cost.

3.1. Thermal oxidation

Ga_2O_3 nanowires have been fabricated by oxidizing GaAs in a furnace [36]. At 1050°C and at atmospheric pressure, argon gas was used to grow Ga_2O_3 nanowires with a nanotextured surface. As the GaAs was heated, it dissociated and the arsenic evaporated, leaving behind a liquid Ga-rich surface which was then oxidized. At high temperatures, Ga_2O_3 nanowires began to grow. The roughness of the surface was believed to be caused by oxygen-deficiency defects and the difference in crystal structure between GaAs and Ga_2O_3.

When β-Ga_2O_3 nanowires are grown using this method, the material first begins as a thin film before nanowire formation if conditions are favorable. The steps in this process have been clearly described for GaAs oxidation in a furnace [51]. At high temperatures, phase separation of the GaAs occurs, and the constituents decompose. Arsenic diffuses through the substrate and evaporates from the surface, and gallium melts, forming clusters of liquid gallium on the surface. As the clusters form, the arsenic evaporates faster because it is easier for it to dissociate through the clusters than from the surface of the GaAs. Nanowire growth occurs from these clusters. Due to liquid Ga's ability to wet GaAs, the droplets are hemispherical. Oxygen in the chamber reacts with the Ga clusters to form a Ga_2O_3 thin film, with voids where the As evaporated. As the GaAs is converted into Ga_2O_3, Ga is no longer able to wet it, resulting in

Growth Mechanisms	Advantages	Disadvantages
Thermal oxidation	Simple and inexpensive	Low degree of control over growth
Vapor–liquid–solid	Good control of growth	Can interfere with doping levels
Pulsed laser deposition	High deposition rate	More defects and dislocations
RF magnetron sputtering	Non-stoichiometric deposition	• Incorporation of gas into material • It produces low quality NWs
Thermal evaporation	Inexpensive and compatible with many materials	Contamination of material
Molecular beam epitaxy	high deposition rate and low impurity levels	More surface roughness at higher temperatures
Laser ablation	High quality film	Expensive and require longer time
Arc-discharge	Faster growth	• Low quality NWs • Many defects
Carbothermal reduction	Simple process for growing a wide variety of structures	High levels of impurities
Microwave plasma	Can grow a wide variety of nanowires	Radial nonuniformity
Metalorganic chemical vapor deposition	High growth rate	Expensive and uses toxic gases
Hydrothermal method	Simple, cheap and efficient	Require precise temperature control
Sol-gel method	Simple and inexpensive	The thickness is nonuniform

Table 3. Advantages and disadvantages of β-Ga_2O_3 NWs growth techniques.

droplets with large contact angles. The droplets are unable to expand laterally and are thus forced to grow in one dimension, resulting in nanowire growth.

At oxidation temperatures above 750°C, nanowires begin to appear on the surface of Ga_2O_3 [51]. As the growth temperature is increased, the nanowires increase in length and density. The nanowires grown at 870°C (**Figure 5a**) were shorter and less dense than the ones grown at 900 and 942°C (**Figure 5b** and **c**). The samples were obtained by heating n-type (111) silicon-doped GaAs. **Figure 6** shows SEM images for nanowire growth at temperatures of 900, 1000 and

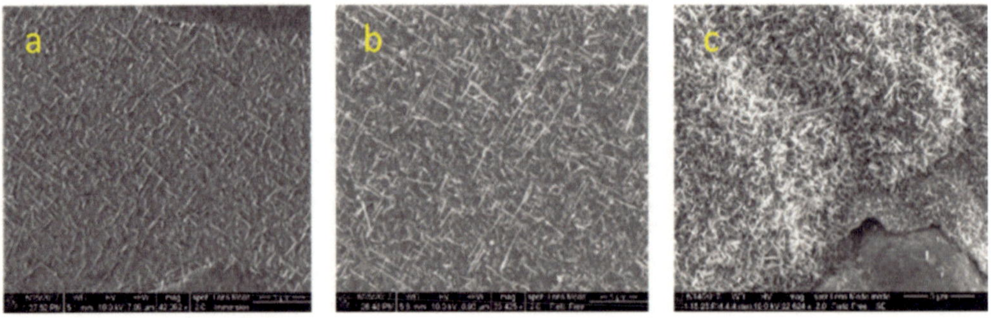

Figure 5. SEM images of n-type GaAs oxidized at (a) 870, (b) 900, (c) 942°C.

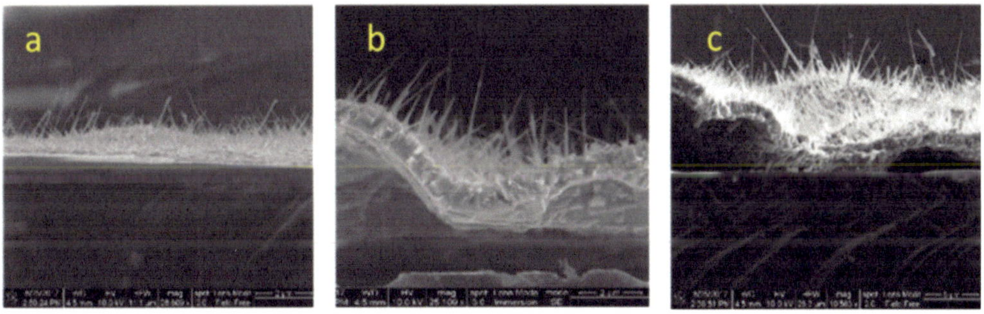

Figure 6. SEM image of cross-sections of GaAs oxidized at (a) 900, (b) 1000, (c) 1015°C.

1015°C. The average lengths of the nanowires were 1–2 μm for 900°C, 1–3 μm for 1000°C, and 2–4 μm for 1015°C. The differences in nanowire length caused by temperature are more apparent in these cross-sections. The nanowires are sharp and have apparently random orientations.

One problem seen with furnace oxidation is buckling and delamination of the Ga_2O_3 film on the surface, particularly at higher temperatures. This causes the orientation of the nanowires to be more random, as can be seen in **Figure 6**. Another problem with growing nanowires using thermal oxidation is that at some temperatures, the surface is not uniformly covered with nanowires but instead contains gallium clusters [51]. **Figure 7** shows focused ion beam (FIB) images of these bubbles on the surface that were formed at high temperatures due to the growth mechanism. If a nanowire device is used as a sensor, this means that the effective sensing surface area is reduced. The clusters could also negatively affect the performance of devices in other ways. This method is simple and inexpensive, but it does not allow precise control of nanowire growth, hence their lack of directionality.

3.2. Vapor–liquid–solid (VLS) approach

The VLS mechanism (**Figure 8**) is commonly used to grow nanowires. It requires the presence of impurities, which act as catalysts at the sites where nanowires will be grown [29]. The choice of impurity affects the diameter and growth direction of the nanowires and thus is an important growth parameter. Being impurities, the catalysts become part of the material and

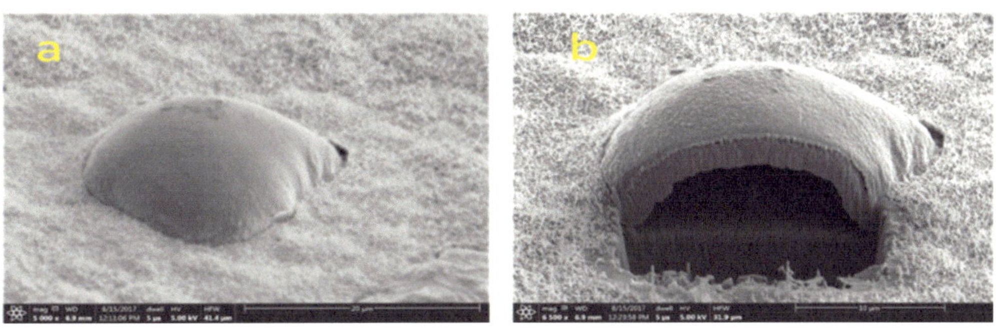

Figure 7. (a) SEM image of a Ga cluster on the surface of n-type GaAs oxidized at 942°C, (b) Closeup of the same cluster after etching with a focused ion beam, showing nanowires surrounding the cluster but none on top of it.

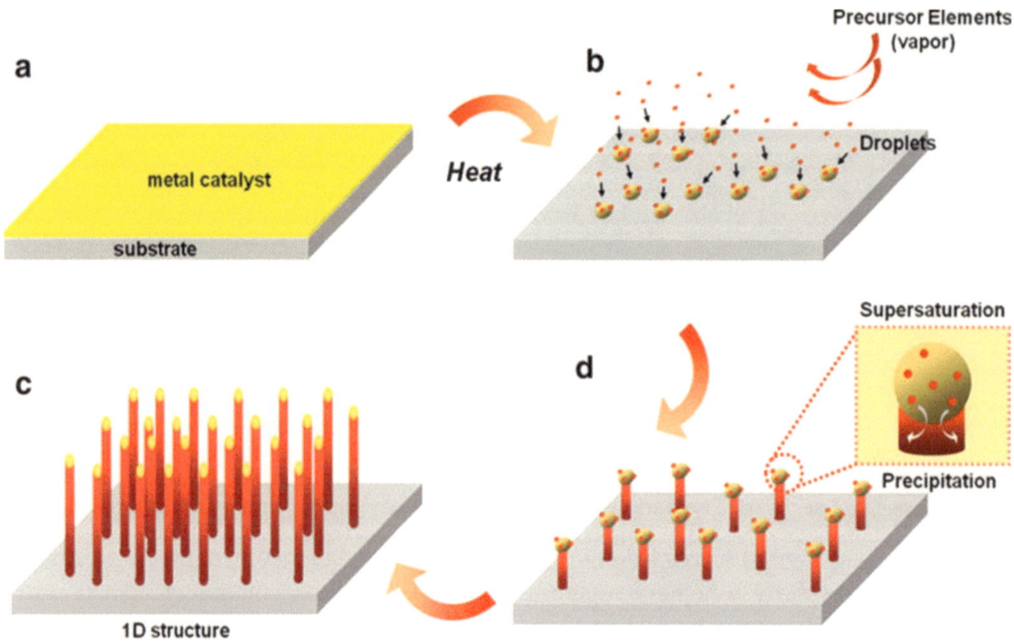

Figure 8. Vapor-liquid-solid growth of semiconductor nanowires.

can interfere with doping. Commonly used catalysts include Au, Ni, and Fe. These catalysts are commonly deposited on the substrate using the sol–gel method [37]. Nanowire growth occurs at the locations where the catalysts are deposited.

3.3. Pulsed laser deposition (PLD)

Pulsed laser deposition has been used to grow Ga_2O_3 nanostructures on α-Al_2O_3 at 900°C. The nanostructure was doped with Cr by alternating laser pulses to deposit both Ga_2O_3 and Cr [38]. This resulted in nanowires with a controlled length and diameter. The nanowires had tips larger than their bodies. PLD has a high deposition rate and is a low-cost growth method. There are many parameters involved in optimizing it for a specific process, such as the thickness of the catalyst, sample orientation, pulse intensity and number of pulses, and distance between laser and material, giving good control over the process. PLD produces more defects and dislocations due to the high energies used.

3.4. RF (radio frequency) magnetron sputtering

RF magnetron sputtering is another way of growing Ga_2O_3 nanowires [39]. If doping is needed, Ga_2O_3 powder can be mixed with the dopant (also in powder form) to form a sputter target. The fact that it produces non-stoichiometric films in an oxygen-deficient atmosphere is important in growing nanowires using Ga seeds. Growth time affects film thickness, which in turn affects the nanowires. The dopant only appears at the tip of the nanowires. Sputtering can be performed at room temperature, is low-cost, and provides good step coverage, but the films produced are not very high-quality because of damage caused during the process.

3.5. Thermal evaporation

Thermal evaporation has been used to grow Ga_2O_3 nanowires, mainly for gas-sensing applications [41, 42] at lower temperatures. In both cases, the nanowires were grown using a two-step process involving thermal evaporation and atomic layer deposition. This resulted in a Ga_2O_3 core and a different material (ZnO and SnO_2) as the shell on the outside, which acted as the sensor. The entire structure consisted of Ga_2O_3 nanowires that were coated with another material on the outside. The addition of the second material improves the performance of the gas sensors as compared to simply using Ga_2O_3 nanowires, mainly by lowering the temperature at which the sensor can operate. This method of growth is inexpensive and compatible with a wide variety of materials, but the evaporation process often leads to contamination by other substances present in the chamber.

3.6. Molecular beam epitaxy (MBE)

Molecular beam epitaxy is another option for growing Ga_2O_3. It has used to grow Ga_2O_3 on sapphire using both a Ga_2O_3 source and an elemental Ga source [43]. The pressure was less than 5×10^{-10} torr, and oxygen gas was used to protect the substrate from ions. For the Ga source, Ga was evaporated from an effusion cell through an oxygen plasma, while for the Ga_2O_3 source, Ga_2O_3 was evaporated from an iridium crucible with an oxygen plasma to decompose it into Ga_2O and O_2. Growth using elemental Ga required high temperatures to completely oxidize the Ga atoms, while with the Ga_2O_3 source, at lower temperatures, the oxide layer was thicker. Molecular beam epitaxy has also been used to grow Ga_2O_3 on β-Ga_2O_3 substrates [15]. The rate of growth was highly dependent on the substrate orientation, with the quickest growth occurring on the (010) and (310) planes and the growth on the (100) plane being very slow. At lower temperatures, surface roughness decreased. Molecular beam epitaxy has a high deposition rate and gives low impurity levels, but requires high temperatures and high vacuum conditions.

3.7. Laser ablation

Laser ablation involves the use of a laser on a Ga_2O_3 target to extract its contents and deposit them onto a substrate. Parameters such as pulse width, pulse power, distance between target and substrate, and the ablation target can all be controlled, giving this method high versatility and control over the material grown. The nanostructures it produces are high-quality but it takes a long time to grow them and it is expensive. For example, it took 5 hours to grow nanowires with diameters between 15 and 50 nm and lengths of several micrometers [52]. When a single nanowire was examined, it was found to be free of dislocations and stacking faults, attesting to the high quality of material that can be grown.

Laser ablation can also be used as part of a two-step process along with solution refluxing to grow Ga_2O_3 nanowires [53]. This approach allows nanowires to grow at the low temperature of 121°C. Rather than a Ga_2O_3 target, a high-purity Ga plate was used. While the time spent using the laser was only 20 minutes, the reflux process took a day, and calcination to convert the material grown into Ga_2O_3 took 18 hours, so this method requires more time to fabricate Ga_2O_3 nanowires.

3.8. Arc discharge

Arc discharge involves applying a direct-current arc voltage across an anode and cathode, both usually made of graphite, in an inert gas. The anode usually has a hole drilled into it to hold powders, which are used to grow the material desired, and the energy of the plasma generated is used to create nanostructures. In one study, the anode was a graphite electrode filled with GaN, graphite, and Ni powders while the cathode was a slightly larger graphite rod [45]. The growth took place in a helium environment. The Ni powder acted as a catalyst for the reaction. Nanostructures were deposited on the cathode, and most of them were nanorods. The nanorods were most commonly oriented in the [111] and [200] directions. Defects such as microtwins and stacking faults were found on some of the nanorods, suggesting that this method of growth does not produce very high-quality nanostructures.

A similar study has been carried out using a mix of Ni and Co powder as the catalyst inside the graphite anode and using a mixture of Ar and O_2 instead of He [54]. The pressure was varied, and it was found that no nanowires were present under 450 torr, suggesting that oxygen gas was necessary for their formation. Furthermore, without the Ni and Co powder there was no nanowire growth either. The nanowires obtained showed twin defects and rough step edges. The surfaces of the nanowires were covered with amorphous layers, and jog defects were also seen. Arc discharge can grow nanowires very quickly, but it produces nanowires that contain many defects.

3.9. Carbothermal reduction (CTR)

Carbothermal reduction is a simple method for growing Ga_2O_3 nanowires that involves mixing Ga_2O_3 and graphite powder and heating them in a furnace in the presence of a substrate [46]. This process can result in cactus-like nanostructures on the surface of the substrate. These nanostructures were very thin and had small spherical structures at their tips. Dense nanowire growth with a radial distribution was observed all over the surface of the samples. The cactus-like structures varied in diameter from a few micrometers to several tens of micrometers, which is much larger than conventional nanostructures. Nanowires with a vertical orientation were also seen, but their tips were often bent and a few had multiple nanorods with a random orientation at their tips. They were also tapered and had rough surfaces on closer examination. Growth time was correlated with nanowire density, and nanospheres were the first structures to grow on the surface, on top of which nanowires subsequently grew.

In another study, nanowires with lengths of tens of micrometers to hundreds of micrometers were grown using a similar process [29]. Different nanostructures were associated with different growth temperatures and substrates, with nanowires found at 900°C on Si substrates and nanosheets found at 800°C on quartz substrates. These nanosheets were large, with areas in the order of several tens of square micrometers. Their diffraction patterns were similar to that of thin films. When Si substrates were used at 800°C, Ga_2O_3 nanoribbons were observed. At high temperatures, the nanowires had sawtooth-like structures, which was believed to be due to oxygen deficiencies during growth. Although carbothermal reduction is a simple method for growing Ga_2O_3 nanostructures and can be used to create a wide variety of structures, there are high levels of carbon impurities in the final material.

3.10. Microwave plasma

The microwave plasma method of growing nanowires involves using low-melting metals as a solvent medium for the growth of nanowires [47]. Droplets or thin films of gallium on the substrate cause nucleation of nanowires with high density ($>10^{11}$ cm^{-2}). This technique was successfully used to grow unique "nanopaintbrushes." A variety of substrates were used with molten gallium and exposed to plasma in a microwave plasma reactor with H_2, CH_4, and O_2. After thin films developed, gallium droplets were added to grow nanowires. Stacking faults were not observed on the nanostructures. Four different nanostructures were seen on different parts of the substrate—nanowires, nanotubes, nanorods, and nanopaintbrushes—and it was suggested that the type of nanostructure formed depended on the initial state of the nuclei on the surface. Hydrogen appeared to etch the nuclei on the surface, preventing them from growing laterally and agglomerating with other nuclei, thus promoting vertical nanowire growth. This method of nanowire growth can produce a wide variety of nanostructures, but use of a plasma is associated with radial nonuniformity.

3.11. Metalorganic chemical vapor deposition (MOCVD)

Metalorganic chemical vapor deposition has been used to grow Ga_2O_3 nanostructures on Si/SiO_2 and Al_2O_3 [48]. As in furnace oxidation, nanostructures are only observed above a certain temperature (450°C in this case). The growth process consists of alternating pulses of a Ga precursor and H_2O and N_2 to purge the chamber. By changing the pulses of H_2O and using O_2, the morphology of the nanostructures can be controlled. With a longer pulse of H_2O, the nanowires were about a micron long and had large particles at their tips. If the water pulse was shortened, the wires would taper from the tip down, being widest at the tip and becoming thinner close to the substrate. At higher deposition temperatures, more nanostructures were observed on the surface. Metalorganic chemical vapor deposition provides relatively fast growth and high-quality films, but it is expensive and requires the use of highly toxic gases.

3.12. Hydrothermal Method

Ga_2O_3 nanowires can be obtained by calcination of gallium oxide hydroxide (GaOOH). This is considered the simplest method for growing β-Ga_2O_3 [49, 50]. The crystal of GaOOH is obtained when GaN reacts with different alkalis such as NaOH, NH_4OH, KOH, and Na_2CO_3 [49]. The GaOOH morphologies were changed into two different crystal structures: rhombohedral crystal (α-Ga_2O_3) and monoclinic crystal (β-Ga_2O_3). The morphologies of the gallium oxide were influenced by changing the temperatures [50]. **Figure 9** shows FE-SEM images of the GaOOH nanostructures grown by the hydrothermal method. Different growth temperatures were tried, such as room temperature, 50, 75, and 95°C. Temperature has a significant impact on the growth of gallium oxide. The number of shapes seen and the size of the nanowires increased as the temperature was increased. This growth method can be used at low temperatures.

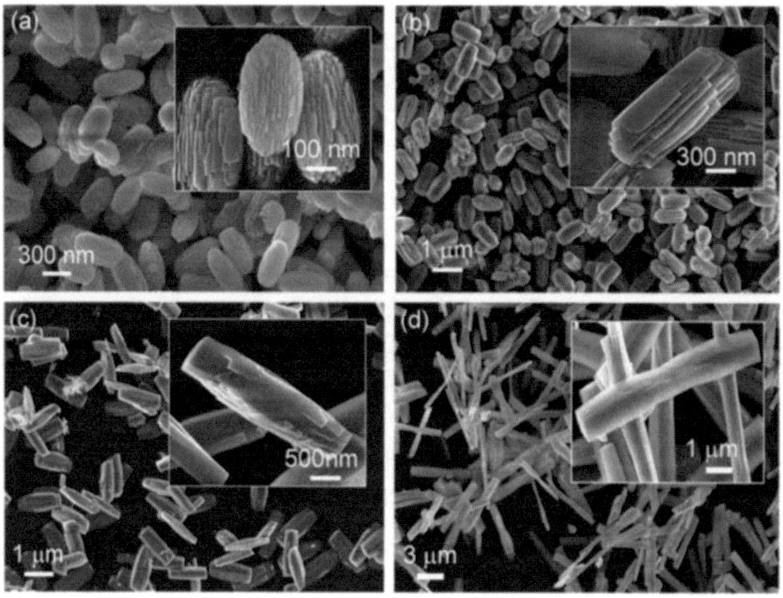

Figure 9. Effect of reaction temperature on the morphology of GaOOH nanostructures. Ref. [50] is licensed CC BY 4.0.

3.13. Ga2O3 growth substrates

Different substrates have been used to examine and produce low-cost and high-quality nanowires. The next sub-sections provide an overview of the substrates that have been utilized recently for the growth of gallium oxide nanowires. **Table 4** describes some advantages and disadvantages of different substrates on which β-Ga_2O_3 nanowires are grown.

Materials	Advantages	Disadvantages
Al_2O_3	Inexpensive and available	Lattice mismatch
GaN	Common and well-matched	Few defects observed
SiC	Large-scale fabrication processes	Lattice mismatch leads to high morphological and structural disorder
Ga_2O_3	Lack of lattice mismatch	Expensive and requires atmosphere with oxygen radicals to obtain a smooth surface.
$MgAl_6O_{10}$	Thermally stable Small lattice mismatch	Point defects: Oxygen vacancies, interstitial Ga, Ga vacancies and Ga-O vacancy pairs
MgO	It has wide bandgap and can shift the absorption edge to shorter wavelengths	Require annealing to improve crystallinity Oxygen vacancies
Si	Abundant & inexpensive	Lattice mismatch

Table 4. An overview of some substrates advantages and disadvantages were applied of β-Ga_2O_3 NWs growth techniques.

3.13.1. *Aluminum oxide (Al_2O_3)*

Al_2O_3 is one of the most commonly used materials for Ga_2O_3 growth. As sapphire is an insulator, this provides electrical isolation for Ga_2O_3. However, there is still lattice mismatch between the two materials. Lattice mismatches of 4.2% and 10.7% have been reported [44]. Sapphire is fairly inexpensive and available in different-sized wafers but there are always a few layers of α-Ga_2O_3 between a sapphire substrate and β-Ga_2O_3 [55]. It has also been reported that there is dissimilarity between the corundum crystal structure of sapphire and the monoclinic structure of β- Ga_2O_3 [56], but α-Ga_2O_3 has a corundum structure, so it is better-matched [57].

3.13.2. *Gallium Nitride (GaN)*

The lattices of GaN and Ga_2O_3 are well matched, and thus Ga_2O_3 is also commonly grown on GaN. ε-Ga_2O_3 keeps its morphology and has a low defect density when grown on GaN [55]. Furthermore, GaN has a wide bandgap, so using it as a substrate means that a photodetector will be less sensitive to longer wavelengths [58]. Both materials also have electron affinities of 4.0 eV, so there is almost no offset in the conduction band when Ga_2O_3 is grown on top of GaN.

3.13.3. *Silicon carbide (SiC)*

Depending on the orientation used, lattice mismatch between SiC and Ga_2O_3 can be fairly high [55]. When ε-Ga_2O_3 was grown on (001) SiC, morphological and structural disorder was very high, due to the difference between the crystal structures of the two materials, and the Ga_2O_3 was polycrystalline. Interestingly, in the [111] direction, high-quality Ga_2O_3 growth was observed. Because it is fairly easy to dope SiC, a p-n junction can be created using SiC as a substrate [59]. Using SiC as a substrate also enables fabrication of vertically structured devices because it is semiconducting.

3.13.4. *Gallium oxide (Ga_2O_3)*

Gallium oxide is also an occasionally used substrate for Ga_2O_3 growth. Due to the lack of lattice mismatch, high-quality crystal growth is possible [15]. To fully oxidize the gallium sub-oxides during pulsed laser deposition, an oxygen-radical-rich atmosphere is needed [60]. Without oxygen radicals, gallium tends to sublimate and its surfaces tend to be rough. Also, the high cost of Ga_2O_3 wafers makes them less attractive as a growth substrate.

3.13.5. *Magnesium aluminum oxide ($MgAl_6O_{10}$)*

As an alternative to using Al_2O_3, it is also possible to grow Ga_2O_3 on $MgAl_6O_{10}$. $MgAl_6O_{10}$ has high thermal stability and a lattice mismatch of only 2.9%, less than that of Al_2O_3 [44]. Chemically, the material is not very different from sapphire. When using $MgAl_6O_{10}$, several types of point defects in the Ga_2O_3 were observed, such as oxygen vacancies, interstitial Ga, Ga vacancies, and gallium-oxygen vacancy pairs. When $MgAl_6O_{10}$ is used, the donor band is formed by pairs of gallium vacancies and gallium-oxygen vacancies [61].

3.13.6. Magnesium oxide (MgO)

MgO has also been used as a substrate for growing Ga_2O_3. It tends to absorb water from the air, which affects the structural properties of the Ga_2O_3 on top [57]. When Ga_2O_3 is grown on MgO, it is initially amorphous and requires annealing to achieve good crystallinity [16]. However, this also decreases the bandgap, making the material less sensitive to UV light. The Ga_2O_3 film is also highly resistive, although it remains n-type due to oxygen vacancies.

3.13.7. Silicon (Si)

Ga_2O_3 can also be grown on silicon. It has been deposited on (100) Si [57]. A minimum temperature of 600°C was required for crystallization to occur. Because silicon, like silicon carbide, can be easily doped p-type, growing Ga_2O_3 on Si is an option to make a p-n junction. Ga_2O_3 grown on (100) Si using molecular beam epitaxy has good crystallinity and produces smooth surfaces [62].

4. Contacts for Ga_2O_3

Contacts to Ga_2O_3 can be either Schottky or ohmic. To make an ohmic contact, the barrier height of the metal should be low. Gold and titanium are the most commonly used metals for contacts. Metal contacts can be deposited using shadow masks, which involve depositing metal onto the substrate while it is covered by a thin plate with gaps in the desired shape.

4.1. Titanium/gold (Ti/Au)

Ti/Au interdigitated electrodes have been deposited on both β-Ga_2O_3 [63] and chromium-doped Ga_2O_3 nanowire thin films [38]. In both cases, a shadow mask was used and the contacts exhibited ohmic behavior. It was suggested that this was due to the abundance of surface states in Ga_2O_3, which makes it easy for carriers to tunnel.

4.2. Gold (Au)

A shadow mask has been used to deposit Au onto the back of a β-Ga_2O_3/Si heterojunction photodetector [62]. The deposition method was radio frequency magnetron sputtering. A Ti/Au contact was also sputtered on top of the β-Ga_2O_3. The contacts formed were ohmic. Standard photolithography techniques and liftoff have been used to deposit interdigitated Au electrodes on β-Ga_2O_3 [64]. Carrier multiplication that occurred in the area under the electrodes resulted in high responsivity and quantum efficiency.

4.3. Indium (In)

Thermal evaporation was used to deposit an In electrode on the SnO_2 part of a β-Ga_2O_3/SnO_2 heterojunction photodetector [65]. A bilayer of Ti and Au was also deposited on the β-Ga_2O_3 thin film using evaporation. The I-V curve of the device showed that the contacts were rectifying.

4.4. Titanium/aluminum (Ti/Al)

Electron beam evaporation was used to deposit Ti and Al onto the back side of a Ga_2O_3 photodiode [66]. On the other side of the diode, Schottky contacts of Au were also deposited. These contacts were rectifying. A contact consisting of alternating layers, Ti/Al/Ti/Al, was deposited through an interdigitated shadow mask onto a Ga_2O_3 photodetector [4]. The contacts were found to be non-ohmic, which was the goal.

4.5. Graphene

Graphene electrodes have also been deposited on a β-Ga_2O_3/SiC photodetector [67]. The graphene was grown by chemical vapor deposition on Cu and then transferred to the substrate. The graphene was transparent and thus allowed most incident light to reach the contact area. It also acted as a channel when the electrons and the holes were separated. On the back of the substrate, a Au/Ti electrode was deposited. A device with a gold electrode on top was also fabricated. The device with the graphene electrode showed a larger photoresponse, higher responsivity, larger I_{light}/I_{dark} ratio, and a shorter rise time.

5. Ga_2O_3 nanowire applications

Ga_2O_3 nanowires, nanobelts and nanorods have been extensively studied in recent years. These Ga_2O_3 nanostructures provide a path to a new generation of devices, but most Ga_2O_3 nanostructure-based devices are limited to a single nanowire, and the integration of individual devices on a single chip is still a real challenge. Another crucial issue is controlled doping in nanostructures and the formation of high-quality ohmic contacts between nanostructures and electrodes. Therefore, the construction of a device via a simple and cost-effective method is still a great challenge. In spite of these difficulties, various Ga_2O_3 nanowire applications such as Ga_2O_3 nanowire-based FETs, gas sensors and UV photodetectors are reported.

5.1. UV photodetectors

The spectrum of radiation between 200 and 280 nm is called solar-blind due to the absorption of solar radiation in that spectrum by the ozone layer. Because of this lack of background noise, it is possible to detect very weak signals in this spectrum. In recent years, the need for solar-blind UV photodetectors has increased, due to their potential in several applications such as flame detection and missile warning [68]. When photons strike the surface of a UV photodetector, they generate electron–hole pairs. These electron–hole pairs change the conductivity of the material, which shows that light has been detected. For a device to detect signals in the UV spectrum, a material with a wide bandgap should be used. It is important that the material detect UV light but is not affected by light from other parts of the optical spectrum.

Photomultiplier tubes are used for UV detection in the solar-blind region, but they have several problems. They are bulky, require high bias voltages, and have high leakage current. Silicon-based photodiodes are also used, but due to their narrow bandgap, they require additional filters to block light from unwanted parts of the optical spectrum. Wide-bandgap materials are preferred for use in solar-blind UV photodetectors because they are transparent to deep UV. Of the different materials available, Ga_2O_3 is particularly promising because of its wide bandgap of 4.9 eV, high melting point, and chemical stability.

5.2. Types of photodetector structures

Many different types of photodetectors can be made. Photoconductors, MSM photodiodes, Schottky diodes, p-n junctions, p-i-n junctions, and avalanche photodiodes have all been fabricated with Ga_2O_3.

5.3. Photoconductors

A photoconductor is basically a radiation-sensitive resistor. When a photon with more energy than the bandgap of the material is absorbed, an electron–hole pair is formed, which makes the material more conductive [69]. Photoconductors have a few key advantages, such as high internal gain at room temperature, large photoresponsivity, and lack of a need for amplifying equipment. They are also compatible with planar IC technologies [70], so it is not difficult to integrate them with other devices on a chip. However, a photoconductor has strong persistent photoconductivity, meaning that photocurrent persists even after removal of illumination [71]. In addition, the response speed is very slow, and the responsivity depends on the amount of time it is kept in the dark. It is not possible for a photoconductor to operate at zero bias, meaning that it consumes more power. Photoconductors also exhibit sublinear behavior for incident power [72] and poor contrast between UV and visible light.

5.4. MSM photodiode

MSM (metal–semiconductor–metal) photodiodes are a commonly used structure for Ga_2O_3 photodetectors. They consist of two back-to-back Schottky diodes with an interdigitated electrode on top of an active light-collecting region [69]. An MSM photodiode has fast operation due to a low capacitance per unit area. The speed of the device is limited by transit time rather than by RC time constant. Using electron beam lithography, electrode width and spacing can be made very small, improving the speed of the device. The device also has a simple structure that is easy to fabricate and integrate. However, there is intrinsic low responsivity due to the interdigitated electrodes covering the active region.

MSM photodiodes have high gain and are easy to integrate with read-out circuitry [73]. The interdigitated electrodes need to be close together to maintain device performance, but this lowers responsivity because it blocks part of the incoming light. It is also possible to illuminate the device from the back, but this makes fabrication much harder. Fabrication of an MSM photodiode requires only a single photolithography step [74] because it only needs a single active layer of dopants. The fabrication process is compatible with that of FETs, but fine feature sizes are required. It is difficult to reliably control the metal–semiconductor interface, and the reflection of light from surface metals is a problem. It is possible to make the dark current in an MSM photodiode very low [72]. The photoresponse is linear with optical power. There is good contrast between the visible and the UV spectrum, and the bandwidth is wide. However, noise is a significant problem [71] and the material has high resistivity [75].

5.5. Schottky photodiode

A Schottky diode consists of a metal layer in contact with a semiconductor layer that exhibits rectifying behavior due to the difference in the work function between the two layers [69]. It has high quantum efficiency, high response speed, and low dark current and exhibits good contrast between UV and visible light. A Schottky diode can operate at zero bias, and doping levels are important for controlling barrier height. Unlike photoconductors, Schottky diodes have a very low persistent photoconductivity [71]. However, a lack of uniformity in metal films leads to large leakage currents, and external quantum efficiency is reduced by transparent Schottky contacts. A back-illuminated Schottky contact does not have this problem, but those devices are more difficult to fabricate. Schottky diodes have a constant responsivity for excitations above the bandgap, independent of power and temperature [72], and the time response is limited by the RC time constant. They have some limitations. For example, the maximum responsivity is limited by reflection in the transparent top contact [75]. The leakage current of the contacts is strongly dependent on the Schottky barrier height, with a higher barrier corresponding to lower leakage current. Defects will also result in a large leakage current.

5.6. p-n photodiode

A p-n photodiode is a p-n diode made with materials that allow light to penetrate the p-n junction [69]. It has a fast response speed and low dark current, and can operate without an applied bias, so it consumes less power. The photoresponse is also linear with optical power [72]. For Ga_2O_3 devices, there is good rejection of light from the visible spectrum. The time response is limited by p-doping. A p-n photodiode can be either a homojunction, with both sides made of the same material, or a heterojunction, with different materials. Due to the difficulty of making p-type Ga_2O_3, only heterojunction photodiodes have been made, with materials such as Si [62] and SiC [59] used on the p side. With heterojunctions, it is possible to tune the bandgap of each material.

5.7. p-i-n photodiode

p-i-n photodiodes are similar to p-n photodiodes, except for the addition of an intrinsic layer between the p and n materials. In a p-i-n junction photodiode, absorbed photons generate electron–hole pairs, which are collected by the n and p layers due to the reverse bias. Carriers generated in the junction experience a high electric field and are separated rapidly, giving the detector a fast response [75]. The addition of an intrinsic layer improves absorption and increases the quantum efficiency. To reduce the capacitance, the thickness of the intrinsic layer can be increased. However, a thicker intrinsic layer increases the transit time [74]. The junctions are critical to device performance, because if recombination occurs in the junctions, device performance is degraded. There is a good UV-to-visible rejection ratio, sharp spectral responsivity cutoff, and fast response time.

5.8. Avalanche photodiode

An avalanche photodiode operates at high speeds and multiplies photocurrent internally, resulting in high sensitivity. Above the breakdown voltage, electron–hole pairs are accelerated by a large applied field, causing impact ionization. The avalanche effect also amplifies noise. The gain changes with temperature as well as applied bias. Because large voltages are applied, the device consumes large amounts of power. There is no persistent photoconductivity [65] and the quantum efficiency is high. The detectivity and selectivity are also high, and the response and decay times are fast. An avalanche photodiode requires less chip area compared to other type of photodetectors [70], has high gain, and can operate with a high bandwidth. Its fabrication is also compatible with IC fabrication technology. There are uniform junction regions to handle high applied fields [75]. The thickness of the multiplication layer affects the electric field profile and spectral response.

5.9. Ga_2O_3 nanowire-based field ionization

The high aspect ratio of Ga_2O_3 nanowires gives them potential to be used as field emission devices [76, 77]. For this application, it is desirable for the material to exhibit emission at a low electric field and remain stable at high current densities [51]. The field enhancement factor is also an important figure of merit for determining whether a material is suitable for field emission. If a nanowire with a sharp tip and a high aspect ratio is used, there is a large electric field at the tip, which reduces the potential barrier for field emission and increases the field emission current. Ga_2O_3 nanowires with ultra-sharp tips 3.5 nm in radius (**Figure 10**) have been fabricated [51], demonstrating that Ga_2O_3 has potential in this area. In fact, the field emission characteristics of turn-on field, threshold electric field, and geometrical field enhancement factor of these nanowires were comparable to those of diamond nanostructures and single-wall carbon nanotubes.

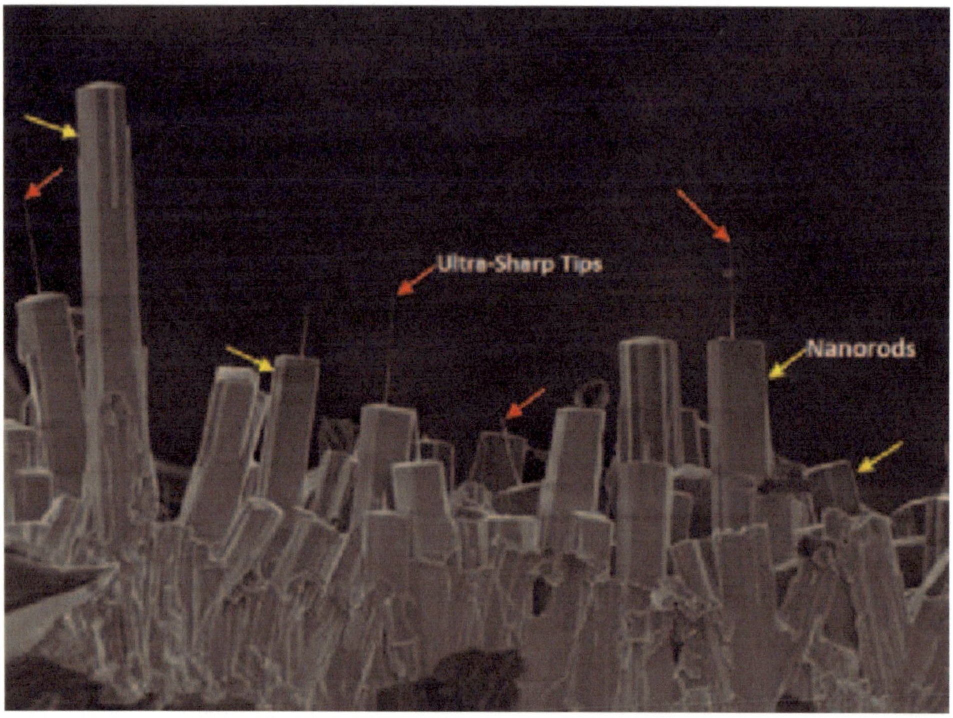

Figure 10. Transmission electron microscopy (TEM) image of the sharp tip. An ensemble of Ga or GaxO droplets is visible on the tip.

6. Ga_2O_3 FETs

A single Ga_2O_3 nanowire-based FET has been fabricated to measure the electrical properties of the as-grown nanowires [77, 78]. The device was fabricated on a silicon dioxide (SiO_2) film on top of an n-type silicon (Si) wafer with parallel pairs of Au contacts. These parallel Au electrodes acted as the source and drain, and an n-type Si layer served as the back gate. The Ga_2O_3 nanowires were first dispersed in ethanol, and then dried onto an n-type silicon chip. It was found that the conductance increased as the back-gate voltage increased. This suggests that the Ga_2O_3 nanowires had n-type characteristics, likely due to the presence of oxygen vacancies and extra gallium atoms in the lattice.

6.1. Ga_2O_3 nanowire-based temperature/gas sensor

β-Ga_2O_3 nanowire-based temperature sensors have unique conductivity behavior up to high temperatures. This property distinguishes this material from other semiconductors for nanodevice applications. Many studies have shown different methods for fabricating β-Ga_2O_3 nanowire-based gas sensors [5, 79, 80]. β-Ga_2O_3 appears to be an insulating material at room temperature; however, at high temperatures, it behaves as an n-type material and its conductance is greatly affected by its surroundings. To fabricate a gas sensor, Ga_2O_3 nanowires have been grown using the VLS method [80]. The developed sensor was cheap, easily fabricated and able to detect various chemical constituents at room temperature. As the target chemical substance approaches the nanowires, the gas may be physically adsorbed onto defects on the surface of the nanowire. This changes the dielectric constant of the nanowires, and the device detects the change in capacitance. As the concentration of the gas increases, the capacitance also increases. Because this sensor does not require an external heat source to recover quickly, it can operate at low power.

6.2. Photoelectrical generation of hydrogen

Ga_2O_3 nanowires show promise as a photocatalyst for splitting water into hydrogen and oxygen [81]. Efficiency was measured as 0.906%, much higher than the 0.581% efficiency of GaN thin films used for the same purpose. Ga_2O_3 is particularly attractive because of the tunability of its optoelectronic properties through doping and alloying, and enhancement of photoelectrochemical efficiency due to the presence of defect bands. However, when the nanowires were grown using a GaN substrate at high temperatures, an interfacial layer was observed between the nanowires and the GaN. This interfacial layer degraded the photoelectrochemical performance of the nanowires. Because the coverage of the substrate with nanowires is not complete, the substrate also contributes to the photoelectrochemical process.

7. Challenges & future perspectives

β-Ga_2O_3 nanowires have unique properties that distinguish them from other semiconductors. For instance, these nanowires have a wide bandgap and high chemical and thermal stability. Thus, Ga_2O_3 could be excellent for nanodevice applications in the future. However, a number of challenges associated with Ga_2O_3 still exist and need to be addressed.

7.1. Dopants

One issue is the introduction of specific dopants to enhance the electrical properties of nanowires. Specifically, a dopant that can reliably make Ga_2O_3 p-type needs to be found. More investigation is also required to evaluate the unintentional doping of Ga_2O_3 nanowires during their growth. The influence of these impurities on device performance also requires further study. Some procedural challenges for the growth of Ga_2O_3 nanowires also need to be solved. For instance, more directional Ga_2O_3 nanowire growth is highly desirable and the nanowires should also be more uniform in size and length. Finally, techniques to pattern and etch the Ga_2O_3 nanowires would be highly valuable.

7.2. Contacts

Different metals have been tested as contacts in Ga_2O_3 devices including titanium/gold, gold, indium, and graphene. These contacts are a key component in the performance of nanowire devices. Thus, more investigation into the ideal materials for ohmic and Schottky contacts in Ga_2O_3 nanowire devices is required. Integrating particular components of these nanowire devices to build a complete system at low cost is still a challenge. Research should also be performed on ways to improve carrier mobility and reduce contact resistance at the device level, and to lower costs to enable large-scale production.

7.3. Growth techniques

More research is required to identify the different factors that influence Ga_2O_3 nanowire growth and how these affect their optical, structural, thermal and electrical properties. Lattice mismatch needs to be taken into account when selecting a substrate for nanowire growth. High-quality fabrication at low cost for Ga_2O_3 nanowire growth is needed to make various nanowire devices such as FETs and UV photodetectors more attractive. The growth techniques for Ga_2O_3 nanowires should also be more closely examined to see if they can be improved on. Currently, different techniques such as molecular bean epitaxy (MBE), pulsed laser deposition (PLD), metal–organic chemical vapor deposition (MOCVD) and the hydrothermal method show promise in growing vertical Ga_2O_3 nanowires with high precision and control. Because Ga_2O_3 nanowires are a new material that has only recently been used in devices, integration of these devices with other devices on a chip is still necessary.

8. Conclusions

Ga_2O_3 nanowires exhibit unique electrical, thermal and optical properties that make them particularly attractive for use in future devices in sensing and optical applications. Using nanowires improves on the performance of thin films. These nanowires have a high surface-to-volume ratio, for high detection sensitivity. From a materials point of view, the wide bandgap of Ga_2O_3 also makes it less sensitive to visible light than other wide-bandgap materials such as GaN and SiC and allows it to handle higher power and temperatures. Thus, Ga_2O_3 nanowires are expected to play a key role in devices of the next generation.

Author details

Badriyah Alhalaili*, Howard Mao and Saif Islam

*Address all correspondence to: balhalaili@ucdavis.edu

Department of Electrical and Computer Engineering, University of California, Davis, USA

References

[1] Maier D, Alomari M, Grandjean N, Carlin JF, Diforte-Poisson MA, Dua C, et al. InAlN/GaN HEMTs for operation in the 1000 degrees C regime: A first experiment. IEEE Electron Device Letters. Jul 2012;**33**:985-987

[2] Tanaka A, Chen RJ, Jungjohann KL, Dayeh SA. Strong geometrical effects in submillimeter selective area growth and light extraction of GaN light emitting diodes on sapphire. Scientific Reports. Nov 27 2015;**5**:2

[3] Pearton SJ. GaN and Related Materials II. Australia: Gordon and Breach Science Publishers; 2000

[4] Weng WY, Hsueh TJ, Chang SJ, Huang GJ, Hsueh HT. A beta-Ga_2O_3 solar-blind photodetector prepared by furnace oxidization of GaN thin film. IEEE Sensors Journal. Apr 2011; **11**:999-1003

[5] Mazeina L, Perkins FK, Bermudez VM, Arnold SP, Prokes SM. Functionalized Ga_2O_3 nanowires as active material in room temperature capacitance-based gas sensors. Langmuir. Aug 17 2010;**26**:13722-13726

[6] Wang ZL. Nanowires and Nanobelts: Materials, Properties, and Devices. Boston: Kluwer Academic Publishers; 2003

[7] Geller S. Structure of beta-Ga_2O_3. Journal of Solid State Chemistry. 1977;**20**:209-210

[8] Ahman J, Svensson G, Albertsson J. A reinvestigation of beta-gallium oxide. Acta Crystallographica, Section C: Crystal Structure Communications. Jun 15 1996;**52**:1336-1338

[9] Lorenz MR, Woods JF, Gambino RJ. Some electrical properties of semiconductor beta-Ga_2O_3. Journal of Physics and Chemistry of Solids. 1967;**28**:403-&

[10] Villora EG, Shimamura K, Yoshikawa Y, Ujiie T, Aoki K. Electrical conductivity and carrier concentration control in beta-Ga(2)O(3) by Si doping. Applied Physics Letters. May 19 2008;**92**

[11] Varley JB, Weber JR, Janotti A, Van de Walle CG. Oxygen vacancies and donor impurities in beta-Ga_2O_3. Applied Physics Letters. Oct 4 2010;**97**(14):4-6

[12] Ahn S, Ren F, Oh S, Jung Y, Kim J, Mastro MA, et al. Elevated temperature performance of Si-implanted solar-blind beta-Ga_2O_3 photodetectors. Journal of Vacuum Science & Technology B. Jul 2016;**34**:4-7

[13] Tian W, Zhi CY, Zhai TY, Chen SM, Wang X, Liao MY, et al. In-doped Ga_2O_3 nanobelt based photodetector with high sensitivity and wide-range photoresponse. Journal of Materials Chemistry. 2012;**22**:17984-17991

[14] Matsuzaki K, Hiramatsu H, Nomura K, Yanagi H, Kamiya T, Hirano M, et al. Growth, structure and carrier transport properties of Ga_2O_3 epitaxial film examined for transparent field-effect transistor. Thin Solid Films. Feb 1 2006;**496**:37-41

[15] Higashiwaki M, Sasaki K, Kuramata A, Masui T, Yamakoshi S. Gallium oxide (Ga_2O_3) metal-semiconductor field-effect transistors on single-crystal beta-Ga_2O_3 (010) substrates. Applied Physics Letters. Jan 2012;**100**(1)

[16] Feng XJ, Li Z, Mi W, Luo Y, Ma J. Mg-doped beta-Ga_2O_3 films with tunable optical band gap prepared on MgO (110) substrates by metal-organic chemical vapor deposition. Materials Science in Semiconductor Processing. Jun 2015;**34**:52-57

[17] Cheng Y, Liang HW, Liu Y, Xia XC, Shen RS, Song SW, et al. Influence of N-2 and O-2 annealing treatment on the optical bandgap of polycrystalline Ga_2O_3:Cu films. Materials Science in Semiconductor Processing. Oct 2013;**16**:1303-1307

[18] Stepanov SI, Nikolaev VI, Bougrov VE, Romanov AE. Gallium oxide: Properties and applications – A review. Reviews on Advanced Materials Science. 2016;**44**:63-86

[19] Oshima T, Matsuyama K, Yoshimatsu K, Ohtomo A. Conducting Si-doped gamma-Ga_2O_3 epitaxial films grown by pulsed-laser deposition. Journal of Crystal Growth. Jul 1 2015;**421**:23-26

[20] Yan HY, Guo YR, Song QG, Chen YF. First-principles study on electronic structure and optical properties of Cu-doped beta-Ga_2O_3. Physica B: Condensed Matter. Feb 1 2014;**434**:181-184

[21] Wang XH, Zhang FB, Saito K, Tanaka T, Nishio M, Guo QX. Electrical properties and emission mechanisms of Zn-doped beta-Ga_2O_3 films. Journal of Physics and Chemistry of Solids. Nov 2014;**75**:1201-1204

[22] Dakhel AA, Alnaser WE. Experimental analysis of Ga_2O_3: Ti films grown on Si and glass substrates. Microelectronics Reliability. May 2013;**53**:676-680

[23] Zhang LY, Yan JL, Zhang YJ, Li T. Effects of N-doping concentration on the electronic structure and optical properties of N-doped beta-Ga_2O_3. Chinese Physics B. Jun 2012;**21**:8

[24] Liu LL, Li MK, Yu DQ, Zhang J, Zhang H, Qian C, et al. Fabrication and characteristics of N-doped beta-Ga_2O_3 nanowires. Applied Physics A: Materials Science and Processing. Mar 2010;**98**:831-835

[25] Zhang JG, Li B, Xia CT, Pei GQ, Deng Q, Yang ZH, et al. Growth and spectral characterization of beta-Ga_2O_3 single crystals. Journal of Physics and Chemistry of Solids. Dec 2006;**67**:2448-2451

[26] Villora EG, Yamaga M, Inoue T, Yabasi S, Masui Y, Sugawara T, et al. Optical spectroscopy study on beta-Ga_2O_3. Japanese Journal of Applied Physics Part 2-Letters. Jun 1 2002; **41**:L622-L625

[27] Harwig T, Kellendonk F, Slappendel S. Ultraviolet luminescence of beta-galliumsesquioxide. Journal of Physics and Chemistry of Solids. 1978;**39**:675-680

[28] Harwig T, Kellendonk F. Some observations on photo-luminescence of doped beta-galliumsesquioxide. Journal of Solid State Chemistry. 1978;**24**:255-263

[29] Kumar S, Singh R. Nanofunctional gallium oxide (Ga_2O_3) nanowires/nanostructures and their applications in nanodevices. Physica Status Solidi Rapid Research Letters. Oct 2013;**7**:781-792

[30] Geller S. Crystal structure of beta- Ga_2O_3. Journal of Chemical Physics. 1960;**33**:676-684

[31] Binet L, Gourier D. Origin of the blue luminescence of beta- Ga_2O_3. Journal of Physics and Chemistry of Solids. Aug 1998;**59**:1241-1249

[32] Song YP, Zhang HZ, Lin C, Zhu YW, Li GH, Yang FH, et al. Luminescence emission originating from nitrogen doping of beta-Ga_2O_3 nanowires. Physical Review B. Feb 2004;**69**:9

[33] Shimamura K, Villora EG, Ujiie T, Aoki K. Excitation and photoluminescence of pure and Si-doped beta-Ga(2)O(3) single crystals. Applied Physics Letters. May 19 2008;**92**:9

[34] Chang LW, Lu TY, Chen YL, Yeh JW, Shih HC. Effect of the doped nitrogen on the optical properties of beta-Ga_2O_3 nanowires. Materials Letters. Jul 31 2011;**65**:2281-2283

[35] Guo Z, Verma A, Wu XF, Sun FY, Hickman A, Masui T, et al. Anisotropic thermal conductivity in single crystal beta-gallium oxide. Applied Physics Letters. Mar 16 2015;**106**(11): 9-10

[36] Patil-Chaudhari D, Ombaba M, Oh JY, Mao H, Montgomery KH, Lange A, et al. Solar blind photodetectors enabled by nanotextured beta-Ga_2O_3 films grown via oxidation of GaAs substrates. IEEE Photonics Journal. Apr 2017;**9**:10-11

[37] Nguyen TD, Kim ET, Dao KA. Ag nanoparticle catalyst based on Ga_2O_3/GaAs semiconductor nanowire growth by VLS method. Journal of Materials Science: Materials in Electronics. Nov 2015;**26**:8747-8752

[38] Guo DY, Wu ZP, Li PG, Wang QJ, Lei M, Li LH, et al. Magnetic anisotropy and deep ultraviolet photoresponse characteristics in Ga_2O_3: Cr vermicular nanowire thin film nanostructure. RSC Advances. 2015;**5**:12894-12898

[39] Lee SY, Choi KH, Kang HC. Growth mechanism of in-doped beta-Ga_2O_3 nanowires deposited by radio frequency powder sputtering. Materials Letters. Aug 1 2016;**176**:213-218

[40] Choi KH, Cho KK, Kim KW, Cho GB, Ahn HJ, Nam TH. Catalytic growth and structural characterization of semiconducting beta- Ga_2O_3 nanowires. Journal of Nanoscience and Nanotechnology. Jun 2009;**9**:3728-3733

[41] Park S, Sun GJ, Lee C. UV-assisted room temperature-gas sensing of Ga_2O_3-core/ZnO-shell nanowires. Journal of Ceramic Processing Research. Aug 2015;**16**:367-371

[42] Jang YG, Kim WS, Kim DH, Hong SH. Fabrication of Ga_2O_3/SnO_2 core-shell nanowires and their ethanol gas sensing properties. Journal of Materials Research. Sep 2011;**26**: 2322-2327

[43] Ghose S, Rahman MS, Rojas-Ramirez JS, Caro M, Droopad R, Arias A, et al. Structural and optical properties of beta- Ga_2O_3 thin films grown by plasma-assisted molecular beam epitaxy. Journal of Vacuum Science & Technology B. Mar 2016;**34**:10-14

[44] Feng Q, Li FG, Dai B, Jia ZT, Xie WL, Xu T, et al. The properties of gallium oxide thin film grown by pulsed laser deposition. Applied Surface Science. Dec 30 2015;**359**:847-852

[45] Han WQ, Kohler-Redlich P, Ernst F, Ruhle M. Growth and microstructure of Ga_2O_3 nanorods. Solid State Communications. 2000;**115**:527-529

[46] Cao CB, Chen Z, An XQ, Zhu HS. Growth and field emission properties of cactus-like gallium oxide nanostructures. Journal of Physical Chemistry C. Jan 10 2008;**112**:95-98

[47] Sharma S, Sunkara MK. Direct synthesis of gallium oxide tubes, nanowires, and nano-paintbrushes. Journal of the American Chemical Society. Oct 16 2002;**124**:12288-12293

[48] Pallister PJ, Buttera SC, Barry ST. Self-seeding gallium oxide nanowire growth by pulsed chemical vapor deposition. Physica Status Solidi A-Applications and Materials Science. Jul 2015;**212**:1514-1518

[49] Zhao YY, Frost RL, Yang J, Martens WN. Size and morphology control of gallium oxide hydroxide GaO(OH), nano- to micro-sized particles by soft-chemistry route without surfactant. Journal of Physical Chemistry C. Mar 13 2008;**112**:3568-3579

[50] Reddy LS, Ko YH, Yu JS. Hydrothermal synthesis and photocatalytic property of beta-Ga_2O_3 nanorods. Nanoscale Research Letters. Sep 16 2015;**10**:10-16

[51] Bayam Y, Logeeswaran VJ, Katzenmeyer AM, Sadeghian RB, Chacon RJ, Wong MC, et al. Synthesis of Ga_2O_3 nanorods with ultra-sharp tips for high-performance field emission devices. Science of Advanced Materials. Feb 2015;**7**(2):211-218

[52] Hu JQ, Li Q, Meng XM, Lee CS, Lee ST. Synthesis of beta-Ga_2O_3 nanowires by laser ablation. Journal of Physical Chemistry B. Sep 19 2002;**106**:9536-9539

[53] Huang CC, Yeh CS. GaOOH, and beta- and gamma-Ga_2O_3 nanowires: Preparation and photoluminescence. New Journal of Chemistry. 2010;**34**:103-107

[54] Park GS, Choi WB, Kim JM, Choi YC, Lee YH, Lim CB. Structural investigation of gallium oxide (beta- Ga_2O_3) nanowires grown by arc-discharge. Journal of Crystal Growth. Dec 2000;**220**:494-500

[55] Boschi F, Bosi M, Berzina T, Buffagni E, Ferrari C, Fornari R. Hetero-epitaxy of epsilon-Ga_2O_3 layers by MOCVD and ALD. Journal of Crystal Growth. Jun 1 2016;**443**:25-30

[56] Liu XZ, Guo P, Sheng T, Qian LX, Zhang WL, Li YR. Beta- Ga_2O_3 thin films on sapphire pre-seeded by homo-self-templated buffer layer for solar-blind UV photodetector. Optical Materials. Jan 2016;**51**:203-207

[57] Akazawa H. Formation of various phases of gallium oxide films depending on substrate planes and deposition gases. Vacuum. Jan 2016;**123**:8-16

[58] Nakagomi S, Sato T, Takahashi Y, Kokubun Y. Deep ultraviolet photodiodes based on the beta- Ga_2O_3/GaN heterojunction. Sensors and Actuators A-Physical. Aug 1 2015; **232**:208-213

[59] Nakagomi S, Momo T, Takahashi S, Kokubun Y. Deep ultraviolet photodiodes based on beta-Ga_2O_3/SiC heterojunction. Applied Physics Letters. Aug 12 2013;**103**.18-21

[60] Wakabayashi R, Oshima T, Hattori M, Sasaki K, Masui T, Kuramata A, et al. Oxygen-radical-assisted pulsed-laser deposition of beta-Ga_2O_3 and beta-(AlxGa1-x)(2)O-3 films. Journal of Crystal Growth. Aug 15 2015;**424**:77-79

[61] Mi W, Luan CN, Li Z, Zhao CS, Feng XJ, Ma J. Ultraviolet-green photoluminescence of beta- Ga_2O_3 films deposited on MgAl6O10 (100) substrate. Optical Materials. Oct 2013;**35**: 2624-2628

[62] Guo XC, Hao NH, Guo DY, Wu ZP, An YH, Chu XL, et al. Beta- Ga_2O_3/p-Si heterojunction solar-blind ultraviolet photodetector with enhanced photoelectric responsivity. Journal of Alloys and Compounds. Mar 5 2016;**660**:136-140

[63] Guo DY, Wu ZP, Li PG, An YH, Liu H, Guo XC, et al. Fabrication of beta-Ga_2O_3 thin films and solar-blind photodetectors by laser MBE technology. Optical Materials Express. May 1 2014;**4**:1067-1076

[64] Hu GC, Shan CX, Zhang N, Jiang MM, Wang SP, Shen DZ. High gain Ga_2O_3 solar-blind photodetectors realized via a carrier multiplication process. Optics Express. May 18 2015;**23**:13554-13561

[65] Mahmoud WE. Solar blind avalanche photodetector based on the cation exchange growth of beta-Ga_2O_3/SnO_2 bilayer heterostructure thin film. Solar Energy Materials and Solar Cells. Aug 2016;**152**:65-72

[66] Suzuki R, Nakagomi S, Kokubun Y, Arai N, Ohira S. Enhancement of responsivity in solar-blind beta- Ga_2O_3 photodiodes with a au Schottky contact fabricated on single crystal substrates by annealing. Applied Physics Letters. Jun 1 2009;**94**:19

[67] Qu YY, Wu ZP, Ai ML, Guo DY, An YH, Yang HJ, et al. Enhanced Ga_2O_3/SiC ultraviolet photodetector with graphene top electrodes. Journal of Alloys and Compounds. Sep 25 2016;**680**:247-251

[68] Zou RJ, Zhang ZY, Hu JQ, Sang LW, Koide Y, Liao MY. High-detectivity nanowire photodetectors governed by bulk photocurrent dynamics with thermally stable carbide contacts. Nanotechnology. Dec 13 2013;**24**:20

[69] Liu KW, Sakurai M, Aono M. ZnO-based ultraviolet Photodetectors. Sensors. Sep 2010;**10**:8604-8634

[70] Brennan KF, Haralson J, Parks JW, Salem A. Review of reliability issues of metal-semiconductor-metal and avalanche photodiode photonic detectors. Microelectronics Reliability. Dec 1999;**39**:1873-1883

[71] Sang LW, Liao MY, Sumiya M. A comprehensive review of semiconductor ultraviolet photodetectors: From thin film to one-dimensional nanostructures. Sensors. Aug 2013;**13**:10482-10518

[72] Monroy E, Calle F, Pau JL, Munoz E, Omnes F, Beaumont B, et al. AlGaN-based UV photodetectors. Journal of Crystal Growth. Sep 2001;**230**:537-543

[73] Qian LX, Liu XZ, Sheng T, Zhang WL, Li YR, Lai PT. Beta-Ga_2O_3 solar-blind deep-ultraviolet photodetector based on a four-terminal structure with or without zener diodes. AIP Advances. Apr 2016;**6**:21

[74] Berger PR. MSM photodiodes. IEEE Potentials. May 1996;**15**:25-29

[75] Alaie Z, Nejad SM, Yousefi MH. Recent advances in ultraviolet photodetectors. Materials Science in Semiconductor Processing. Jan 2015;**29**:16-55

[76] Huang Y, Wang ZL, Wang Q, Gu CZ, Tang CC, Bando Y, et al. Quasi-aligned Ga_2O_3 nanowires grown on brass wire meshes and their electrical and field-emission properties. Journal of Physical Chemistry C. Feb 5 2009;**113**:1980-1983

[77] Lin J, Huang Y, Bando Y, Tang CC, Li C, Golberg D. Synthesis of In_2O_3 nanowire-decorated Ga_2O_3 nanobelt heterostructures and their electrical and field-emission properties. ACS Nano. Apr 2010;**4**:2452-2458

[78] Li ZJ, Zhao B, Liu P, Zhang YF. Synthesis of gallium oxide nanowires and their electrical properties. Microelectronic Engineering. Jul 2008;**85**:1613-1615

[79] Jangir R, Porwal S, Tiwari P, Mondal P, Rai SK, Ganguli T, et al. Photoluminescence study of β- Ga_2O_3 nanostructures annealed in different environments. Journal of Applied Physics. 2012;**112**:034307

[80] Arnold SP, Prokes SM, Perkins FK, Zaghloul ME. Design and performance of a simple, room-temperature Ga_2O_3 nanowire gas sensor. Applied Physics Letters. Sep 7 2009;**95**:23

[81] Lam KT, Wu YL, Chang SJ. Photoelectrochemical hydrogen generation by Ga_2O_3 nanowires. Science of Advanced Materials. May 2017;**9**:810-814

[82] Higashiwaki M, Sasaki K, Kuramata A, Masui T, Yamakoshi S. Applied Physics Letters. 2012;**100**(1):3

Chapter 3

Synthesis of TiB_2-Ni_3B Nanocomposite Powders by Mechanical Alloying

Jorge Morales Hernández,
Verónica N. Martínez Escobedo,
Héctor Herrera Hernández,
José M. Juárez García and Joel Moreno Palmerin

Additional information is available at the end of the chapter

http://dx.doi.org/10.5772/intechopen.73593

Abstract

Combination of good oxidation resistance, thermal stability, hardness and high strength are great interest properties in engineering and, that are possible to obtain with the Ni-Ti-B ternary system. Mechanical alloying (MA) is an alternative method and cheapest for the synthesis of this kind of metal-ceramic materials with respect to the traditional melt and quench process. The transformation sequence of all the mixtures reported the formation of (ɣ Ni) phase with a nodular morphology and identified the additional presence of the TiB_2 phase (needle morphology), which was more evident with the increase of titanium content (M2 and M3 mixtures) after 24 h of milling. Thermal activation of the milled powders showed the nucleation and growth of the Ni_3B (O boride) and TiB_2 (Hex) as the main phases after heat treatment, where the TiB_2 phase (thin flakes morphology) was nucleated onto Ni_3B matrix. Ternary alloy by MA took place under a metastable equilibrium, offering the possibility to form glassy alloys for compositions, which are not accessible by melting or quenching techniques.

Keywords: mechanical alloying, metal-ceramic materials, Ni_3B and TiB_2 phases, composite powder, high density

1. Introduction

Metal-ceramic materials (glassy metals) from ternary-eutectic alloys experiment several crystalline transformations where more than two phases or two stages of crystallization have been reported because they are a complex transformation. Different ternary alloys prepared by melt spinning with a high content of Ni and B (78 and 18.2 at.% respectively, Ti as complement) have reported the formation of the orthorhombic Ni_3B (O boride), complex cubic boride $(NiTi)_{23}B_6$ (ɣ phase), and f.c.c. nickel (ɣ Ni) nucleates onto it, where the variation in the number of crystals and eutectic colonies by nucleation and growth processes were activated thermally [1].

B-Ni-Ti system is of great interest for scientist and technologist due to the combination of good physical properties of TiB_2 such as low density, high hardness, good electrical and thermal conductivity, high strength and good oxidation resistance. TiB2 is a reinforced second phase with a great potential in the development of cutting tools that request high speed and the minimum dimensional variation (tolerances) and associate with their thermal stability. In combination with the Ni_3B phase in the nickel matrix, this metal matrix composite was studied to work out the requirements of the heat-resistant alloys. Physical and chemical stability at high temperature of this combination suggests the use of this compound as a corrosion and wear-resistant material under aggressive environments between other industrial applications.

Eutectic phase T,$Ti_3Ni_{20}B_6$, at high temperature transform in the compounds of (Ni)-Ni_3B, (Ni)-TiB_2 and (Ni)-$TiNi_3$, which are surrounding to the T phase during the solidification as shown in **Figure 1** [2]. Investigations since 1958 by different methods, such as sintering of

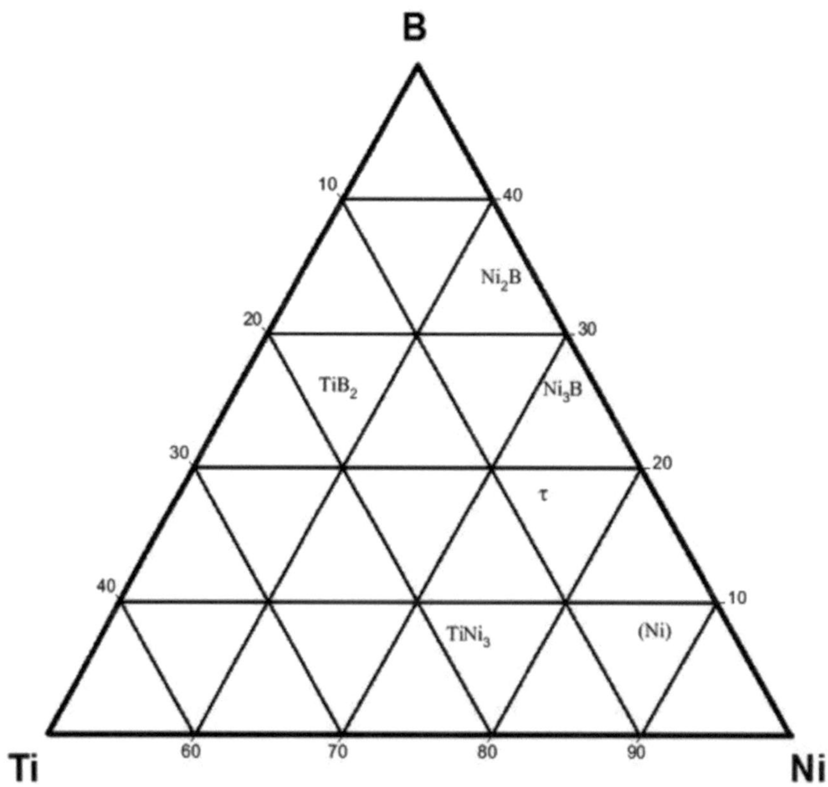

Figure 1. Partial liquidus surface projection of Ni-Ti-B in at.% [2].

powders, induction melting, arc melting, and milling of commercial powders, have reported the TiB_2 phase at 1073 K (800°C) [2]. Because the B solubility in the Ni-Ti phases is very small at temperature lower than 1073 K, making it difficult to obtain the T phase as a single phase at high temperature. Transformation sequence from the liquid phase reported four invariant equilibria in the Ni-Ti-B system where the T phase is present in all time, such equilibria are:

$$L \leftrightarrow \tau + (Ni), L \leftrightarrow \tau + TiNi_3, L \leftrightarrow \tau + TiB_2 \text{ and } L \leftrightarrow \tau + Ni_3 B \quad (1)$$

Some authors have determined the possible phases theoretically calculated in the Ni-Ti-B system, where the results shows the transformation sequence according the most negative value of free energy ΔG°, as follows: (Ni), TiB_2, NiB, Ni_4B_3, Ni_3B, and Ni_3Ti, observing that (Ni) and TiB_2 are easiest to occur thermodynamically than the other phases [2].

The self-propagating high temperature synthesis (SHS) reaction has been used for the formation of TiB_2 compound. Due to the high exothermic energy, some difficulties in the growth control during the heat transfer and solidification are associated with the concentration of strong internal stresses and the subsequent cracks formation. For these reasons, the addition of a metallic element like Ni as diluent of the Ti-B system was necessary to reduce the heat evolution during SHS reaction and synthesize finer TiB_2 particles (4–6 μm) [2]. Studies in the Ni-Ti-B system by SHS reported the formation of TiB_2 and (Ni) as main products, with transient phases such as Ni_4B_3, Ni_3B, NiB and Ni_3Ti with a Ni content in the range from 30 to 70 wt.%, showing a minimum particle size of 0.6 μm with 70 wt.% Ni [3].

Ni_3Ti-TiB_2 obtained by SHS showed that during the combustion reaction, a eutectic liquid corresponding with Ni-76 at.% Ti transforms into NiTi and Ni_3Ti at 1215 K (940°C) with the characteristic that NiTi phase was observed at the combustion front during the quenching. TiB_2 phase was formed by reducing the NiTi with boron. So, the reaction sequence from the eutectic liquid at high titanium concentration can be indicated as [4]:

$$4Ni + 2Ti \rightarrow Ni_3 Ti + NiTi \quad (2)$$

$$3NiTi + 4B \rightarrow Ni_3 Ti + 2\, TiB_2 \quad (3)$$

The glassy metallic alloys obtained by rapid quenching from the liquid phase have special features and different from those of crystalline alloys. Some problems in the production of bulk metallic glasses by casting techniques, are the changes in composition by the occurrence of many eutectic reactions in equilibrium during the rapid quenching, which affect the glass forming ability [5]. Metallic glass formation is not restricted to quenched process because a variety of methods based on the disordering process result in a crystal-to-glass transition such as the mechanical alloying (MA) of elemental powders, where a crystalline phase can be obtained from a disordered amorphous state to transform after that, in a crystalline stable phase activated thermally.

Through high temperature melting processes has been reported the formation of (Ni)-TiB_2 and (Ni)-Ni_3B composites with low titanium content. (Ni)-TiB_2 and (Ni)-Ni_3Ti composites were identified at high titanium content. The synthesis of the Ni-Ti-B system by high temperature processes implies some problems during the solidification (out of equilibria), due to the changes in composition by the occurrence of a lot of eutectic reactions during the rapid quenching, that affect the glass forming ability [5]. The glassy metallic alloys obtained by rapid quenching or melting process demand more energy consumption and elevated cost; for this reason, the mechanical alloying process was explored as an alternative method for the synthesis of this important glassy material.

2. Mechanical alloying of boride compounds

Since their beginning in the 1970s, mechanical alloying (MA) was used in the synthesis of base nickel alloys, expanding its applications to various alloy systems. MA is a non-equilibrium process similar to rapid solidification and tempering, where is possible to synthesize amorphous phases, intermetallic compounds and hardened materials by second phases, impossible to obtain with other techniques.

Conventional methods for preparing refractory compounds like boride, carbide and nitride phases are expensive with a long time-consuming. Mechanical alloying (MA) has been a one-step successful route in synthesizing these compounds, like an inexpensive and faster way [6].

Thermodynamic studies have demonstrated that MA may reproduce crystalline, quasicrystalline, nanocrystalline and amorphous alloys [6]. High melting point materials have been obtained with high-energy mills, under a strict atmosphere control and trying to reduce the wear at the mill wall and steel balls. Titanium diboride (TiB_2) is one example of this kind of compound of high melting point (2790 °C) synthesized by MA, with a wide use by their excellent physical stability at high temperature.

Some researchers have reported the synthesis of TiB_2 by the mechanochemical reaction between Al-TiO_2-B_2O_3 and Al-B_2O_3-Ti, where the reduction of the oxide compounds catalyzes the formation of TiB_2 [7]. This possibility has been proposed for the manufacturing of nanocomposites materials in the solid state, with a ceramic reinforcement (TiB_2) dispersed in an oxide matrix that could be Al_2O_3, TiO_2 or SiO_2 [8] .

Phase transformation in the Ti-B system synthesized by mechanical alloying (MA) have reported the formation of TiB_2 by gradual diffusion reaction (GDR) mechanism [9]. Ni-B system assumes some solubility of B in the Ni matrix with an increase in their lattice parameter to transform in (Ni) + Ni_3B, being more stable than the NiB phase, which need more dissolution of B in their structure [6].

Composition of the mixture, process time, steel ball-to-powder ratio, balls diameter, atmosphere and the process control agent are some important variables in the MA process that determinates the result; so that, more studies about the synthesis of TiB_2 and Ni_3B in the solid state are necessary. In this document, the effect of titanium content in the development of titanium and nickel borides by mechanical alloying was analyzed.

http://dx.doi.org/10.5772/intechopen.73593

3. Experimental

Elemental powders of Ni, Ti and B with high purity (99.7, 99.5 and 99.3%, respectively) were used as started materials. To evaluate the effect of Ti content in the formation of TiB_2 and Ni_3B compounds, three mixtures were synthetized by mechanical alloying according to the compositions shown in **Table 1**. Stainless steel vial with the powder mixture and hardened balls of 4.76 and 12.7 mm in diameter was loaded into a glove box, evacuated during 1200 s (20 min) with a mechanical vacuum pump and after filled with argon gas to prevent the powder oxidation. Ethyl alcohol (2 c.c.) was used as process control agent (PCA) to maintain the balance between fracture and cold welding of the powders. High-energy equipment SPEX mill 80 was used with a ball-to-powder weight ratio of 10:1 until a maximum milling time of 86.4 ks (24 h). Structural evolution of the powders in function of milling time (each 21.6 ks) was characterized by X-ray diffraction (XRD) in a Bruker D8 advanced diffractometer with Cu-Kα (λ = 1.542 Å). Morphology of the powders at the maximum milling time (86.4 ks) was characterized by scanning electron microscope (Jeol 2000). Particle size was validated with a Zeta Sizer equipment, model Malvern MPT-2. Determination of powder's density was evaluated at the end of the milling and after heat treatment at 1173 K (900°C), during 7.2 ks (2 h) under argon atmosphere.

Isothermal section at 1073 K (800°C) in **Figure 2** shows the position in the ternary diagram of each synthesized composition, indicating the equilibrium phases to obtain. Yellow circle in the figure shows the composition of the TiB_2 phase under equilibrium condition. Each line concentrated at this point is projected to the main phases from the Ni-B and Ni-Ti binary alloys to indicate the equilibrium phases according to the eutectic reactions. The equilibrium phases for each composition used are indicated in **Table 2**, where the TiB_2 phase was reported in the three mixtures; however, the Ni_3B phase was not observed with the maximum content of titanium (24.36 at.%), indicating the formation of a phase between titanium and nickel (TiN_3).

Mixture	Ni content		Ti content		B content	
	wt. %	at. %	wt. %	at. %	wt. %	at. %
M1	80	54.77	10	8.29	10	36.93
M2	70	47.25	20	16.34	10	36.47
M3	60	39.82	30	24.36	10	35.80

Table 1. Composition of the mixtures obtained by mechanical alloying.

Mixture	Equilibrium phases
M1	Ni_3B, Ni_2B, TiB_2
M2	T, TiB_2, Ni_3B
M3	TiN_3, TiB_2, T

Table 2. Equilibrium phases according to the isothermal section at 1073 K (800°C).

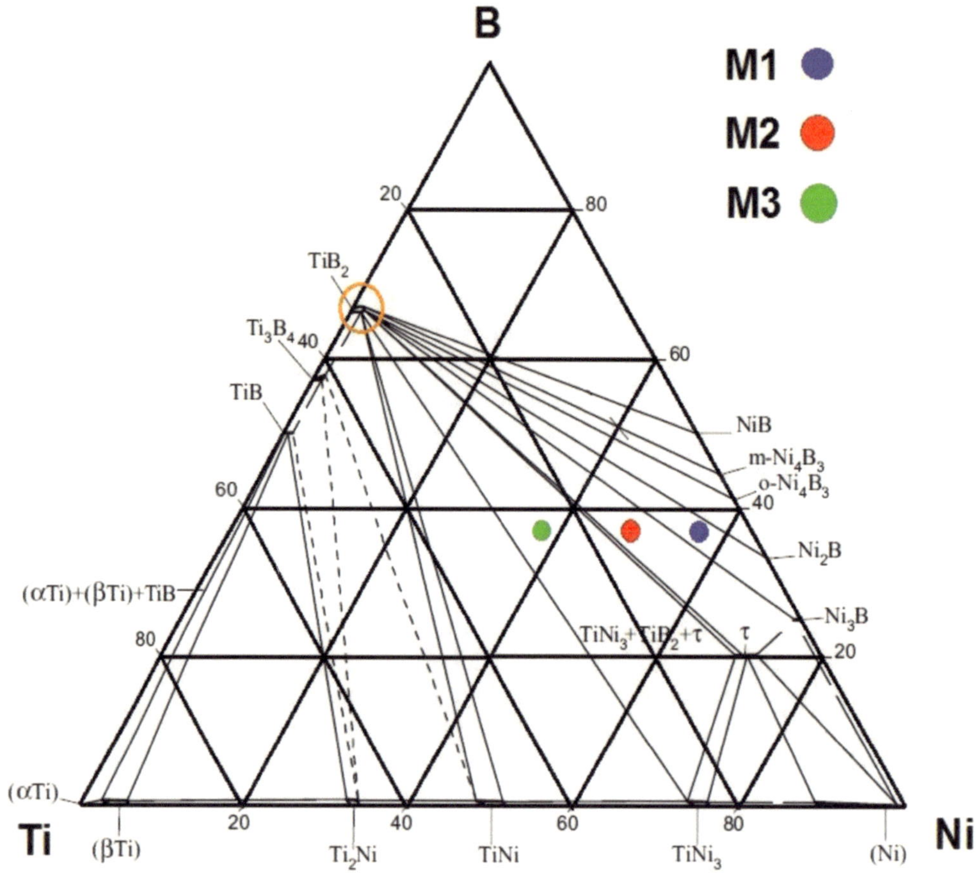

Figure 2. Isothermal section at 1073 K (800°C) of Ni-Ti-B in at.%, indicating the position of the M1, M2 and M3 mixtures respectively [2].

4. Results and discussion

4.1. Mechanical alloying

X-ray diffraction patterns taken each 21.6 ks (6 h) from the M1 mixture are shown in **Figure 3**. At 0 h of milling, the diffraction peaks of crystalline Ni, Ti and B were characterized. After 6 h of milling, the Ti and B peaks were not identified, observing the reduction in the relative intensity and widening of the nickel's peaks. Between 6 and 18 h of milling, the intermetallic compound Ti_3B_4 (unstable) and the (Ni) phase were identified. At the maximum milling time (24 h), only the diffraction peaks corresponding with a solid solution of nickel (Ni) were characterized, identifying an asymmetric widening and reduction of the relative intensity from the nickel's peaks, due to various factors such as the refinement of grain size and the increase of structure deformation, associated with the tendency to the formation of amorphous compound.

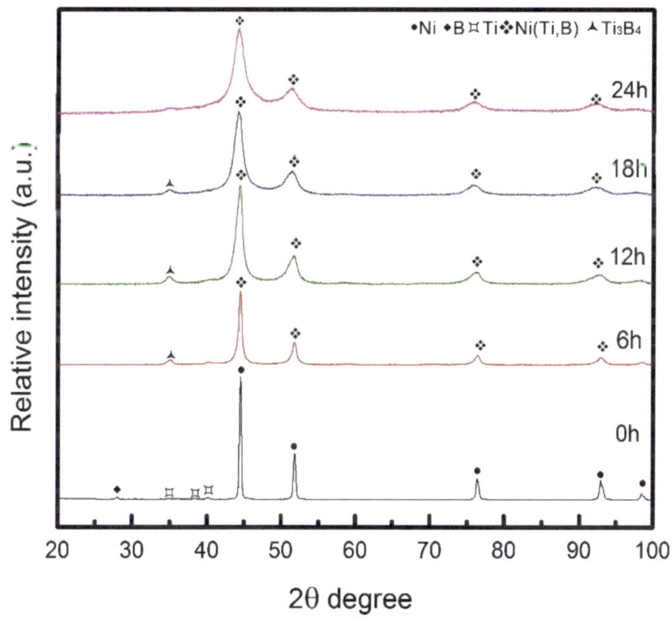

Figure 3. X-ray diffraction patterns for $Ni_{80}Ti_{10}B_{10}$ at different milling times (composition in wt.%).

Powder evolution during the milling of the mixtures M2 and M3 is shown in **Figures 4** and **5**, respectively. For the composition M2 after 12 h of milling, the TiB_2 phase was identified; with more milling time and until the 24 h of milling, the TiB_2 and (Ni) phases with an asymmetric widening and reduction of the relative intensity from the nickel's peaks were characterized.

Mixture M3 with more titanium concentration (24.36 at.%) reported the intermetallic compound Ti_3B_4 (unstable) and the (Ni) phase after 6 h of milling. With more milling time (12, 18 and 24 h), Ti_3B_4 was not detected and were characterized the crystalline phases of (Ni), TiB_2 and Ni_2B until the maximum milling time (24 h) with a less widening and more relative intensity from the nickel's peaks with respect to the M1 and M2 mixtures. This condition was because the deformation energy was taken advantage in the TiB_2 and Ni_2B phases transformation and not in the formation of the solid solution over saturated of nickel (Ni).

Thermodynamic data of this system based on experimental information show that the most negative enthalpy of formation for the three types of Ti borides corresponds with the TiB_2 phase [10]. The structure of the titanium diboride phase is isomorphous [11] with a large homogeneity ranged from 61 to 70 at.% B, characteristic of the M2 and M3 mixtures at the maximum milling time.

According with the **Figure 2** in the binary composition line of Ni-B, are identified the NiB, Ni_2B and Ni_3B phases with the increase of nickel content respectively; however, the formation of Ni_2B in the milled samples was more evident with the increase of titanium content. These means that the formation of the TiB_2 phase was more stable than the nickel borides and therefore, the largest amount of boron in the mixture reacted with titanium, resulting in a minimum relation of B/Ni in equilibrium; but enough to the formation of Ni_2B phase [12, 13].

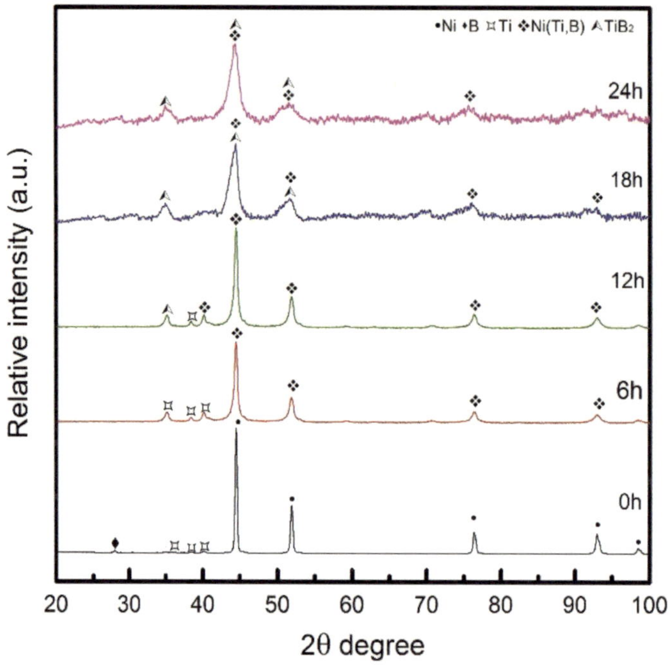

Figure 4. X-ray diffraction patterns for $Ni_{70}Ti_{20}B_{10}$ at different milling times (composition in wt.%).

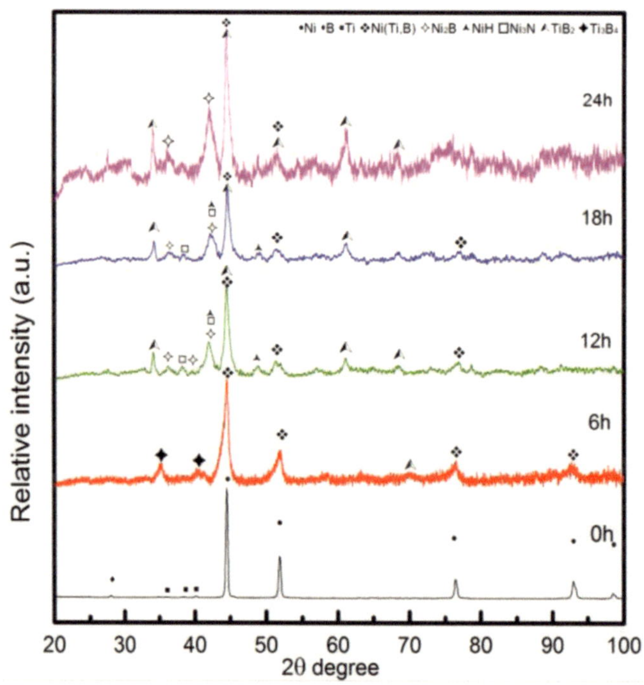

Figure 5. X-ray diffraction patterns for $Ni_{70}Ti_{30}B_{10}$ at different milling times (composition in wt.%).

Powder morphology after milling (24 h) for each composition is shown in **Figure 6**. M1 mixture has an acicular particles (**Figure 6a** and **b**) with a minimum particle size of 416.12 nm. Powder density determined by the Archimedes' principle was 15.89 g/c.c. With the increase of titanium content (M2 and M3 mixtures), the characterized morphology showed composite particles constituied by nodular powder as matrix, with the nucleation of the TiB_2 crystalline phases like inserted needles, indicated with the black arrows in **Figure 6d** and **f**. The minimum particle size was 823.07 and 741.4 nm for the M2 and M3 mixtures, respectively. The increase of particle size in these two last mixtures respecting to the M1 mixture was due to the formation of a composite particle conformed by nodules and needles. The average density reported for both mixtures (M2 and M3) was 15.5 g/c.c. similar with the M1 mixture.

Figure 6. Characteristic particle shapes after milling (24 h). (a) and (b) Acicular powder particles in the M1 mixture; (c)–(f) nodular powder particles with inserted needles of TiB_2 phase in the M2 and M3 mixtures.

4.2. Heat treatment after milling

XRD patterns from the M1 mixture after milling and heat treatment (**Figure 7**) show the reflections corresponding with the Ni_3B, TiB_2 and TiB phases where the deformation energy stored in the (Ni) phase at the maximum milling time was thermally activated to transform into the Ti and Ni borides. According to the number of reflections, the Ni_3B phase is more crystalline and stable with respect to the TiB_2 and TiB phases. According with the titanium content was expected more nickel boride phases (Ni_3B and Ni_2B) as is indicated in **Table 2**; however the reactions betweeen titanium and boron were more stable to transform in the two crystalline phases (TiB_2 and TiB) with a Gibbs free energy more negative with respect the reactions between Ni-B and Ni-Ti [14–16].

For the mixtures M2 and M3 (more titanium content respectivelly) after heat treatment (**Figures 8** and **9**), the phases identified in the milling change to Ni_3B and TiB_2, moving the equilibrio towards the formation of Ni_3B with less boron content, but with a higher mass relation because the relative intensity of the Ni_3B peaks was taller than the TiB_2 peaks, resulted of a more coalescence and growth of Ni_3B with respect the TiB_2 phase [17, 18]. Certain consistency was conserved with the equilibium phases indicated in **Table 2**, with the exception that was not identified the eutectic T phase. The representative reaction in the Ni-Ti-B system synthesized by MA and heat treated is as follows:

$$Ni + Ti + B \rightarrow (\gamma Ni) + TiB_2 \overset{\Delta}{\rightleftarrows} Ni_3B + TiB_2 \qquad (4)$$

Powder morphology after heat treatment for each composition is shown in **Figure 10**. Composite particles were characterized in all the mixtures and are formed by particles with a nodular morphology as matrix and particles with thin flakes morphology as second phase

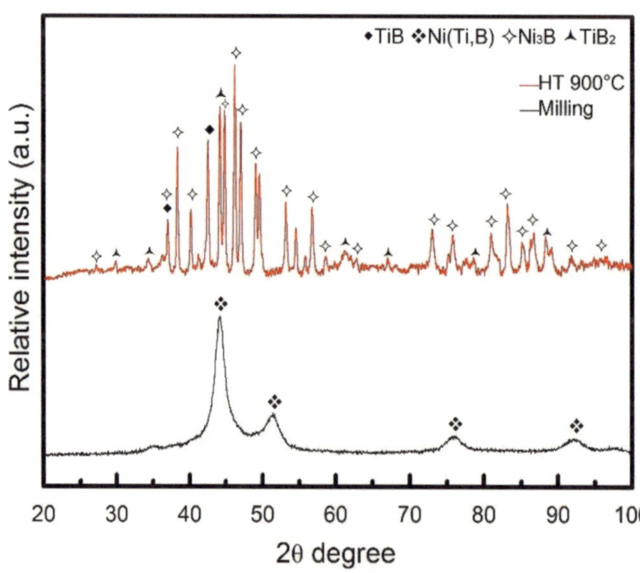

Figure 7. X-ray diffraction patterns for $Ni_{80}Ti_{10}B_{10}$ after milling and heat treatment (composition in wt.%).

http://dx.doi.org/10.5772/intechopen.73593

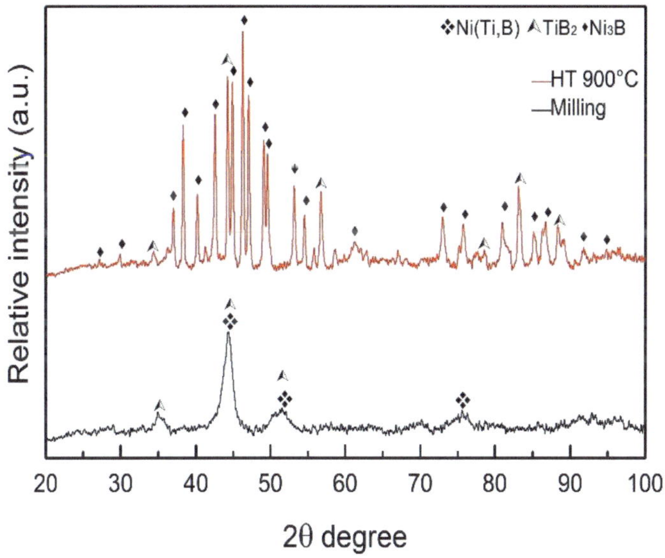

Figure 8. X-ray diffraction patterns for $Ni_{70}Ti_{20}B_{10}$ after milling and heat treatment (composition in wt.%).

inserted at the matrix. The average particle size like composite was of 639.45 nm with a density powders that increased after heat treatment of 15.63–16.30 g/c.c. This density change was correlated with the more density of the phases Ni3B (orthorhombic), combined with the hexagonal TiB2 phase in the mixtures after heat treatment and with as a result of the nucleation and growth mechanism of both phases.

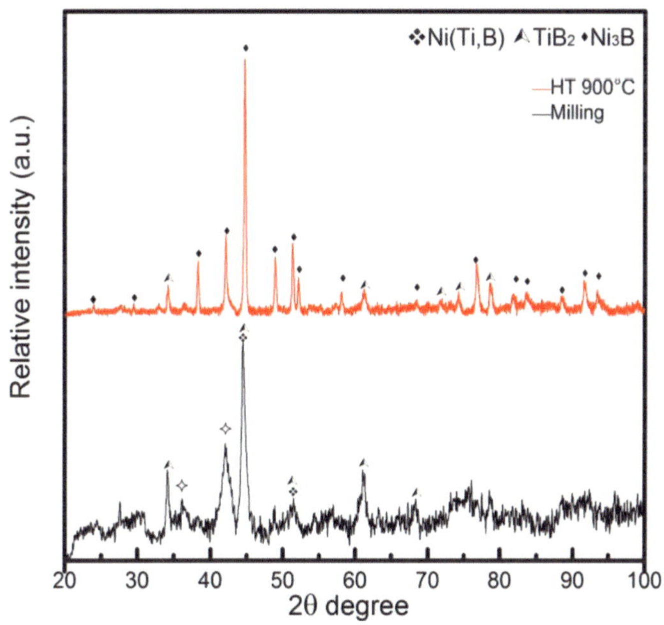

Figure 9. X-ray diffraction patterns for $Ni_{60}Ti_{30}B_{10}$ after milling and heat treatment (composition in wt.%).

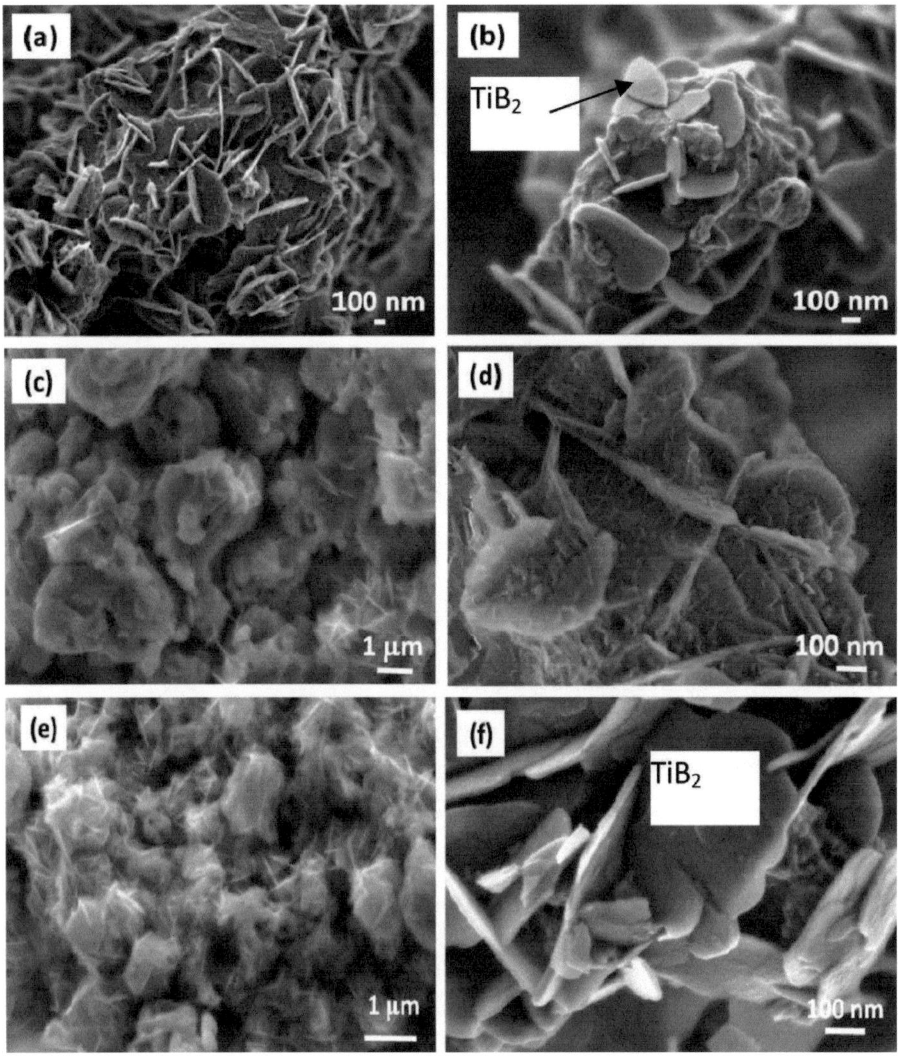

Figure 10. Nodular morphology particles with inserted thin flakes in all the mixtures after heat treatment at 1173 K (900°C). M1 (a and b), M2 (c and d) and M3 (e and f).

5. Conclusions

Ni-Ti-B system synthesized by mechanical alloying reported for the compositions with high content of nickel, a transformation sequence that consisted in the formation of (ɣ Ni) and TiB_2 phases after milling. With the increase in the titanium content, the formation of the TiB_2 phase was more evident, as was observed in the M2 and M3 mixtures.

Formation of the (ɣ Ni) phase by the accumulation of deformation energy during the milling was the first step for the subsequent transformation of the orthorhombic Ni_3B (O boride) and TiB_2 (Hex) phases thermally activated. The transformation sequence for the synthesis of Ni_3B and TiB_2 phases by MA was:

$$Ni + Ti + B \rightarrow (\gamma Ni) + TiB_2 \overset{\Delta}{\rightarrow} Ni_3B + TiB_2$$

Nanocomposite materials conformed by particles with nodular morphology and inserted needles were characterized after milling. High resolution microscopy showed the nucleation of thin flakes from the TiB_2 phase in Ni_3B matrix after heat treatment.

Titanium and nickel not showed the formation of a solid solution with the milling, so that was not enough the activation energy for the formation of the $TiNi_3$ intermetallic compound; indicating a Gibbs free energy ($\Delta G°$) more negative for the transformation of TiB_2 and Ni_3B phases.

TiB_2 and Ni_3B intermetallic compounds, have a great potential in the development of coatings for cutting tools [19]; specialty when is necessary high speed of machining with the minimum dimensional variation (tolerances) and associate with their thermal stability, high hardness and wear resistance.

Acknowledgements

This work was supported by the Sectorial Fund CONACYT-SENER (project number 259334) Energy Sustainability and was developed at the facilities of the Center of Research and Technological Development in Electrochemistry (CIDETEQ) with the INNOVA-COATINGS research group.

Author details

Jorge Morales Hernández[1]*, Verónica N. Martínez Escobedo[1], Héctor Herrera Hernández[2], José M. Juárez García[3] and Joel Moreno Palmerin[4]

*Address all correspondence to: jmorales@cideteq.mx

1 Centro de Investigación y Desarrollo Tecnológico en Electroquímica, Querétaro, México

2 Universidad Autónoma del Estado de México, Área de Electroquímica y Corrosión de Materiales, Estado de México, México

3 Centro Nacional de Metrología, El Márquez Querétaro, México

4 Universidad de Guanajuato, Guanajuato, México

References

[1] Merk N, Morris DG, Stadelmann P. Acta Metallurgica. 1987;**35**(9):2213-2225

[2] Semenova E, Rokhlin L, Dobatkina T, Kolchugina N. Chapter boron-nickel-titanium. In: Refractory Metal Systems. Vol. 11E2. 2010. pp. 153-162

[3] Huang L, Wang HY, Li Q, Yin SQ, Jiang QC. Journal of Alloys and Compounds. 2008;**457**:286-291

[4] Yi HC, Woodger TC, Guigné JY, Moore JJ. Metallurgical and Materials Transaction B. 1998;**29B**:867-875

[5] Eckert J. Materials Science and Engineering. 1997;**A226-228**:364-373

[6] Suryanarayana C. Progress in Materials Science. 2001;**46**:1-184

[7] Mohammad Sharifi E, Karimzadeh F, Enayati MH. Advanced Power Technology. 2011;**22**:526-531

[8] Mohammad Sharifi E, Karimzadeh F, Enayati MH. Journal of Alloys and Compounds. 2010:508-512

[9] Wen-Ming T, Zhi-Xiang Z, Yu-chen W, Jian-Min W, Jun L, Jun-Wu L. Transaction of Nonferrous Metals Society of China. 2006;**16**:613-617

[10] Nakama Y, Ohtani H, Hasebe M. Materials Transaction. 2009;**50**(5):984-993

[11] Battezati L, Antonione C, Baricco M. Journal of Alloy and Compounds. 1997;**247**:164-171

[12] Neikov OD, Naboychenko SS, Dowson G. Handbook of Non-ferrous Metal Powders, Technologies and Applications. Elsevier; 2009

[13] Xi L, Kaban I, Nowak R, Bruzda G, Sobczak N, Stoica M, Eckert J. Journal of Materials Engineering and Performance. 2016:3204-3210

[14] Selva Kumar M, Chandrasekar P, Chandramohan P, Mohanraj M. Materials Characterization. 2012;**73**:43-51

[15] Karolus M, Panek J. Journal of Alloy and Compounds. 2016;**658**:709-715

[16] Calka A, Radlinski AP. Journal of the Less-Common Metals. 1990;**161**:23-26

[17] Sklad PS, Yust CS. Science of Hard Materials. Plenum Press; 1983. pp. 911-912

[18] Calka A, Radlinski AP. Materials Science and Engineering. 1991;**A134**:1350-1353

[19] Bobzin K. CIRP Journal of Manufacturing Science and Technology. 2017;**18**:1-9

Properties and Catalytic Effects of Nanoparticles Synthesized by Levitational Gas Condensation

Young Rang Uhm

Additional information is available at the end of the chapter

http://dx.doi.org/10.5772/intechopen.72158

Abstract

One of the gas-phased methods of the levitational gas condensation (LGC) process was developed to obtain nanopowders with high purity. The instrument designed by unique concept using magnetically levitated melted droplet of metal is easily operated to synthesize nanopowder with highly defected surface. The complex compounds are also easily prepared using micron powder feeding (MPF) system in the instrument. The metals, ceramics, and carbon-coated metal nanoparticles prepared using the LGC are introduced in this chapter. Various applications such as magnetic and catalytic properties are also introduced. Nanoparticles prepared using LGC showed significantly enhanced catalytic activities during chemical reaction due to the high level of defects on their surface structure. The new heterogeneous catalysts of the solid nanoparticles were introduced in this chapter.

Keywords: levitational gas condensation (LGC), catalytic activities, metals and metal oxide nanoparticles, carbon encapsulated nanoparticles

1. Introduction

Among the various methods for preparing nanopowders, almost all of the processes face important challenges, such as poor control of size distribution, surface contamination, the agglomeration of the particles, and so on [1–5]. Many attempts have been made to develop processes and techniques that can synthesize nanoparticles with specific functional properties [4–6]. Dry methods such as the levitational gas condensation (LGC) process have been developed to obtain high-purity nanopowders while suppressing the agglomeration of the produced particles [7–15]. The catalytic effects of nanopowders are influenced not only by the reduced size of the particles but also by their increased surface area [16, 17]. The surface of metal oxides exhibits nonstoichiometry, resulting from oxygen defect structures [17]. Particles prepared by the LGC process show enhanced catalytic activities due to the high level of defects on their surface structure. In this chapter, the synthesis processes and resulting properties of various nanoparticles prepared by LGC are introduced. The sections are focused on three aspects:

1. Levitational gas condensation (LGC): The unique instrument used for the synthesis of the nanoparticles is introduced in this section [18–20].
2. Magnetic metal and carbon-encapsulated metal nanoparticles: The produced magnetic metal (Ni and Fe) and carbon-encapsulated metal (Ni@C and Fe@C) nanoparticles showed a noncollinear magnetic structure between the core and surface layer of the particles. The morphologies and the dispersion stability kinetics in the solvents are introduced. Also, the carbon-encapsulated metal nanoparticles were successfully applied as a catalyst for the multicomponent Biginelli reaction [7, 9, 12, 18–23].
3. Nonmagnetic metal and metal oxides: Cu oxides, Bi, and NiO nanopowders prepared by LGC are introduced. Cu oxide and NiO alloy nanopowders are widely applied as heterogeneous catalysts in oxidizing processes used in organic synthesis. Nanopowders of bismuth (Bi) with its low melting temperature were applied as a sensor electrode for detecting heavy metals in water [9, 14, 17, 24, 25].

2. Levitational gas condensation (LGC)

The LGC method is a kind of gas-phased method in which an electric current is flowed to two inductor coils, which are each wound opposite directions. The electric current following in different direction in each coil induces a magnetic field and creates a magnetic moment, which opposes gravity on the inside lower part of the coil [11, 18]. To synthesize a nanopowder, a melted metal is continuously evaporated and condensed in the levitated condition, suspended in the magnetic field. This method is shown in **Figure 1(a)**. In this study, we modified the inductor with a downward spiral type of coil, so that the levitation region produced by the magnetic field would be more stable for the melted droplet by generating an equivalent magnetic flux density, as shown in **Figure 1(b)**.

The total LGC system is illustrated in **Figure 2**. The nanoparticles formed at the surface of the liquid droplet are then flowed to a filter by the gas stream using a vacuum pump:

2.1. Levitational gas condensation (LGC) starting materials with a melting temperature of over 900°C

Metallic atoms were evaporated from an overheated surface and condensed by cold inert gas and then collected from the filter. To stabilize the powder surface and prohibit oxidation, the powders were passivated with thin oxide layers. The LGC apparatus consists of a high-frequency induction generator, levitation and evaporation chamber, and oxygen concentration control unit.

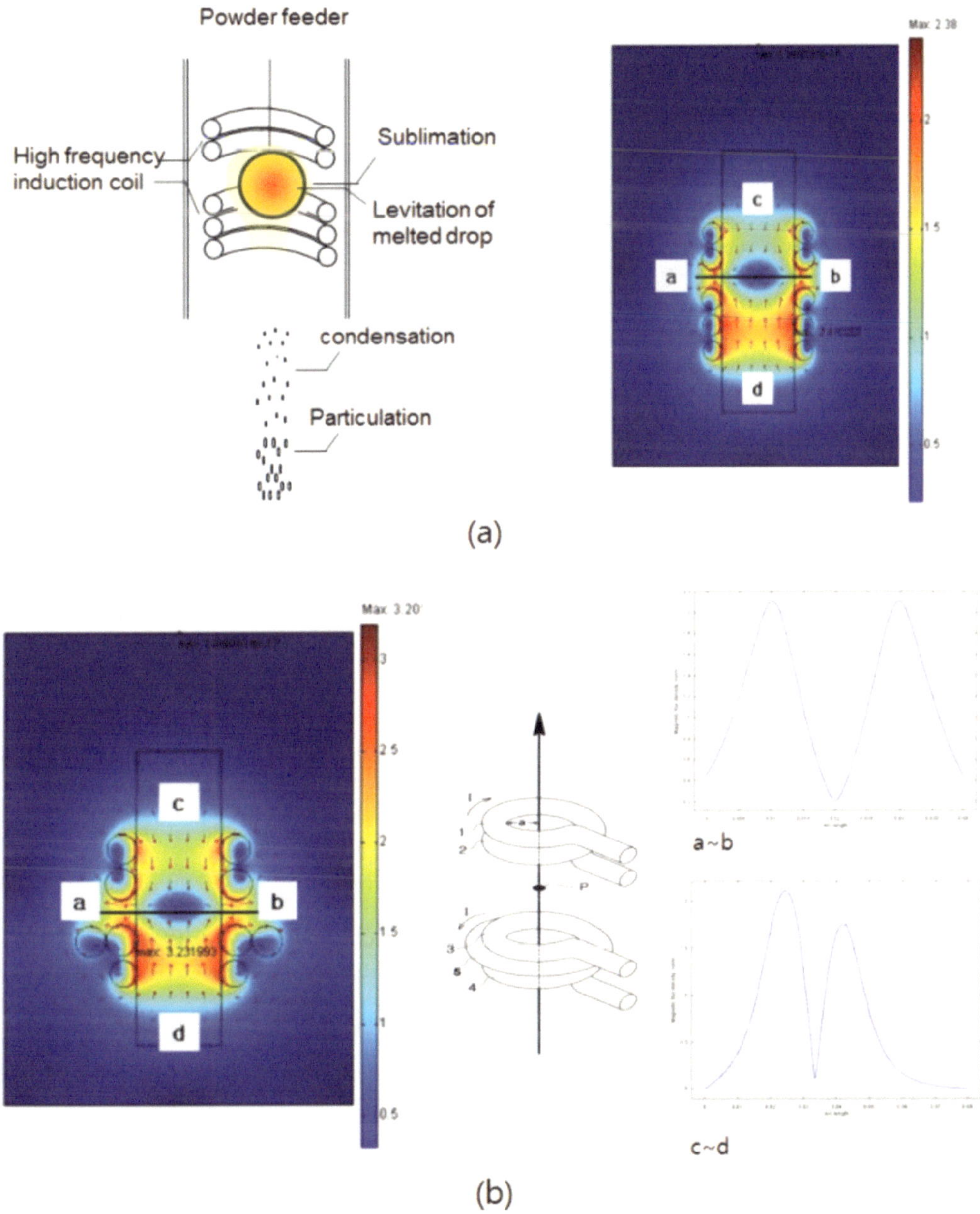

Figure 1. (a) The mechanism of forming nanoparticles and the contour maps of magnetic flux density depending on shape of inductors for a cylinder type and (b) a spiral type and its strength at in-plane (a~b) and vertical (c~d) direction [11].

The operating values used for the induction generator were 6, 5, 4.5, and 3 kW for the Ni, Fe, Cu, and Ag, respectively [7–15]. These values depended on the melting temperature as well as the magnetic permeability of the metals. The melting temperature of iron (1535°C) is higher than

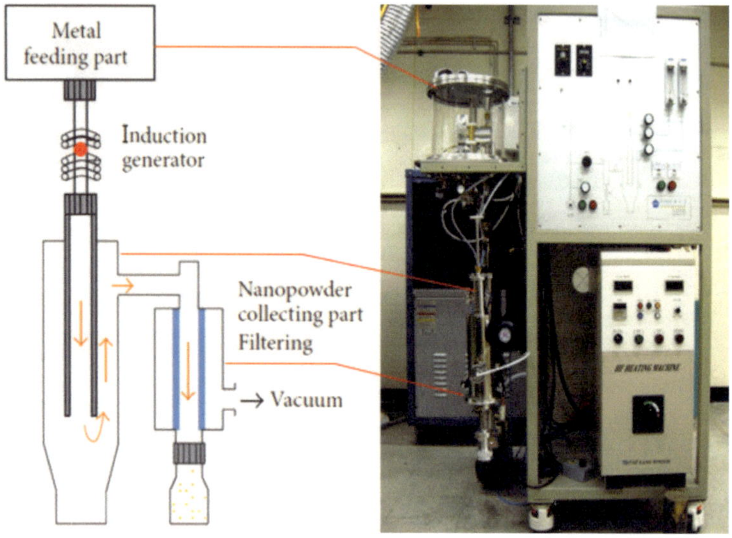

Figure 2. The concept of system for LGC.

that of Ni (1450°C). However, the high magnetic permeability of Ni affects the magnetic force and levitation of the melted droplets. Accordingly, the input power for Ni needs to be increased up to 6 kW, the maximum power for the inductor. Preparing pure Ti nanopowders using the LGC is impossible. To do so, the temperature of the inductor must be increased up to 2000°C; however, the starting materials have to be very thin and strongly passivated by a layer of titanium oxide. The seed materials need to be fully melted before levitation, because only liquid seeds can be suspended in the inductor, due to their lower density. However, it takes too long time to melt the Ti seed to produce liquid droplet for levitation.

We also supplied the starting materials for the melted liquid droplet during synthesis. We utilized a metal wire feed system, which is very convenient for fabricating nanopowder continuously. The amount of material fed over time can be controlled. The average size of the nanopowder is increased with increasing feed speed because of the increased amount of material introduced to the liquid droplet. The optimal feeding speed is between 10 and 30 mm/s during fabrication. If the feeding speed is very slow, below 10 mm/s, the size of the liquid droplets decreased, and they disappear. In contrast, if the feeing speed is increased to over 30 mm/s, the prepared powders have a wire shape because the particles are connected with each other. By adding oxygen to the inert gas, metal oxide particles or metal particles oxidized on the surface may be obtained. **Figure 3** shows metal and metal oxide particles prepared by LGC. The gas condensation (GC) method was used to obtain nanocrystalline powders of pure metal and nano-oxides with different compositions. Oxide nanopowders, such as Fe_3O_4, γ-Fe_2O_3, and Cu_2O, were produced by the LGC of metal wires at elevated pressure in an ($Ar + O_2$) atmosphere [8, 16].

High-purity Ni@C and Fe@C nanopowders were synthesized using the LGC method. The LGC apparatus consisted of a high-frequency induction generator, operating at 6 kW for Ni and 5 kW for Fe, a reaction chamber for the levitated liquid seed, and a unit to control the

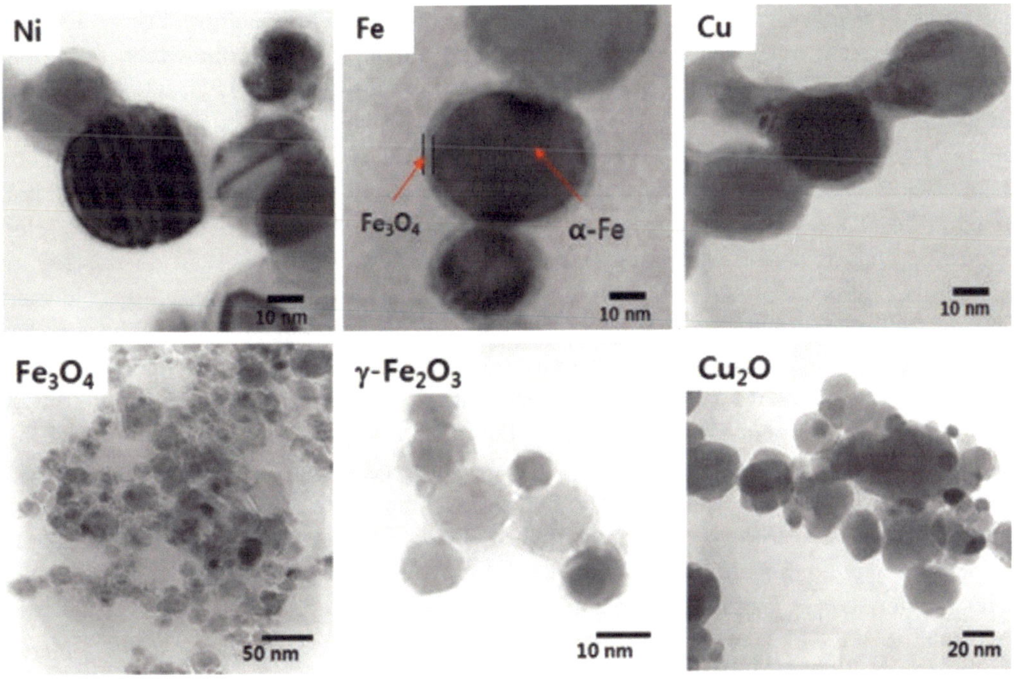

Figure 3. TEM images for metals and ceramic nanopowders.

methane (CH_4) concentration. The starting materials were Ni and Fe wires with a diameter of 4 mm. The Ni and Fe wires were fed into a melted droplet using the wire feeding system at a feeding rate of 2 mm/min. An ingot of 85 mg, which was used as the seed material for the levitation and the evaporation reactions, was melted by using electric induction. The pressure of the mixed Ar and CH_4 gas in the chamber was 100 Torr. CH_4 was 10% of the mixture gas. The inductor was heated up to a temperature of 2000°C, and the metallic atoms were evaporated from the overheated surface of the liquid droplet and condensed by cold inert gas and then collected into the filter. At the same time, the molecular CH introduced into the chamber was converted to atomic C and H with high activity under high temperature. The highly active C atoms react with the Ni and Fe atoms, and the H atoms are converted to H molecules. The newly created H gas is vented out of the reaction chamber by continuous vacuum operation. The results indicated that all of the as-made materials were composed of nanocapsules with uniform particle size at and below 10 nm. The nanocapsules consisted of outer multi-shells of carbon [26–28].

2.2. Starting materials with a melting temperature below 900°C: Zn, Sn, Bi, and ZnO

For most metals, the optimal design of the material heating method allows a metal drop to be heated and kept in a noncontact condition in the evaporation zone by high-frequency magnetic field. However, this method does not ensure the optimal heating of light-volatile metals such as Zn, Sn, and Bi. Therefore, another material evaporation method using a refractory crucible was applied, for heating and evaporation. This method is generally suitable for evaporating materials with high vapor pressures at moderate temperature. **Figure 4** shows a simple diagram of the device for obtaining light-volatile metal nanopowders. The apparatus consists of a high-frequency induction generator operating at 2.5 kW, a levitation and evaporation chamber, and an oxygen concentration control unit [11]. The wire feeding velocity (V_{Zn}) and mixed Ar and O_2 gas pressure in the chamber were 50 mm/min and 100 Torr, respectively. The mechanism of ZnO formation using the LGC method was analyzed. First, a liquid droplet, which is levitated against gravity by the magnetic force due to the coupled induction coils, is heated up to the temperature of 1560°C at 2.5 kW. Then Zn clusters are evaporated from the overheated surface of the liquid droplet and condensed by cold inert gas and collected into the filter. At the same time, molecular O_2 introduced into the chamber is converted to atomic O with high activity under high temperature. The highly active O atoms can diffuse into the Zn clusters and react with the Zn atoms. A large amount of the Zn phase was observed at and below an oxygen flow rate of V_{O2} = 0.05 ℓ/min, whereas mixtures of ZnO and small amounts of the Zn phase were observed under O_2 flow rates in the range from V_{O2} = 0.11ℓ/min to V_{O2} = 0.21 ℓ/min. However, at and above 0.21 ℓ/min of O_2 flow rate, levitation was impossible. Some metals, such as Bi and Sn, have insufficient tensile strength to prepare wire. In these cases, the micron-sized powders were used as the parent materials. The powder starting materials were supplied by a powder feeding (PF) system. A detailed explanation of the PF is provided in Section 2.3.

Bi powders were prepared using metal bismuth powder as the starting material [28]. The powder was supplied by the feeding system into a graphite crucible at a rate of 20 mg/min. The crucible was heated by induction currents up to T = 700–900°C. Bi particles entering the crucible were evaporated within 1–2 sec and carried by argon flow from the hot zone. The argon flow rate was varied in the range of 80–170 ℓ/h, at pressures in the range of 70–300 torr. The dependence of the mean sizes of the bismuth particles on gas pressure at a flow rate of 80 ℓ/h was 25, 70, and 120 nm for 70, 150, and 300 torr, respectively. Thus, the optimal conditions for obtaining Bi powder were realized at an argon pressure of 70 torr and a rate of 80 ℓ/h. A simple diagram of the process for forming nanoparticles from light-volatile seed in a crucible to produce volatile nanoparticles is represented in **Figure 5**.

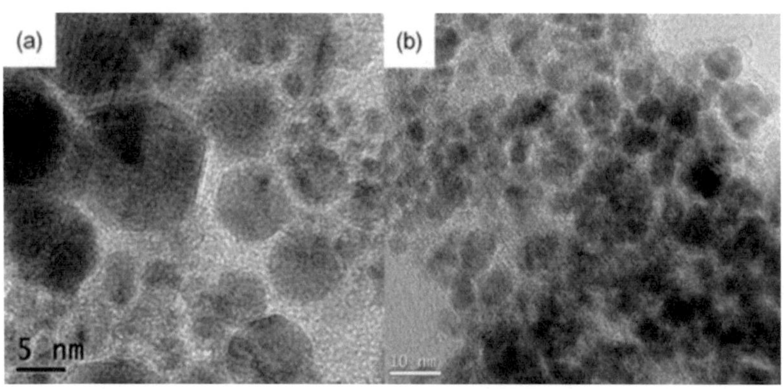

Figure 4. High-resolution TEM images of carbon encapsulated (a) Ni and (b) Fe.

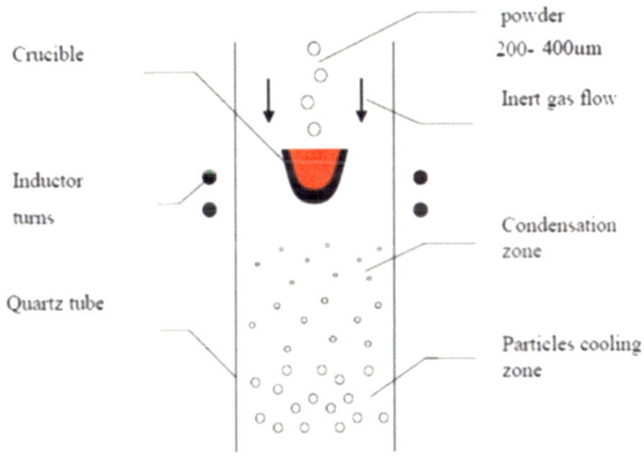

(a)

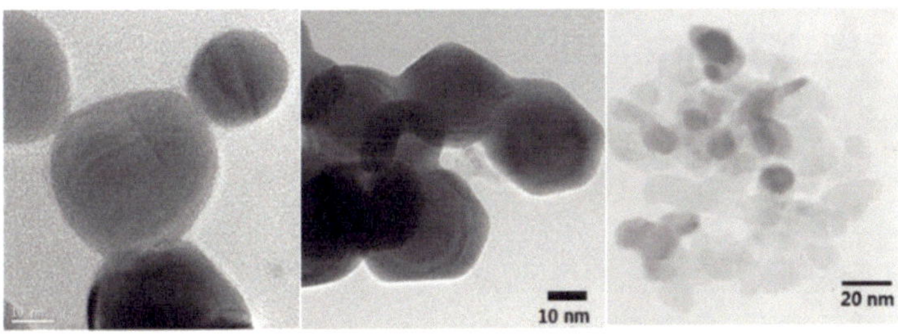

(b)

Figure 5. (a) Elementary diagram of the process for forming nanoparticles from light-volatile seed in a crucible and (b) TEM images for the Sn, Bi, and ZnO nanoparticles prepared by LGC using light-volatile seed.

2.3. Powder starting materials: NiFe2O4 and Ti-Ni

The wire feeding (WF) system was used for synthesizing metal, ceramic, and carbon-encapsulated materials. This system easily supplies seed parent materials continuously. However, it was impossible to synthesize several complicated metal-doped materials such as ferrites, perovskite, garnet, metal-doped ZnO, Ti-Ni, and Al-Ni-Co using the wire feeder in the LGC system, because the parent materials could not be prepared as wire. A newly modified micron powder feeding (MPF) system overcomes this problem of the LGC system [15, 20]. The MPF system can be used for synthesizing brittle metals, alloys, and complex doped materials. Commercial elemental powders of Ti (99.9 at.%, ~500 μm), Ni (99.9 at.%, ~500 μm), and Fe (99.9 at.%, ~ 500 μm) were used as the starting powders for the synthesis of Ti-Ni alloy and Ni-ferrite nanopowder, using the LGC. The Ti and Ni powders were mixed by pestle and mortar to achieve the desired equi-atomic composition and were then incorporated into the micron powder feeding system, which consisted of a rotating part to supply the Ti and Ni micron powders to the melted droplet and a vibrating part for mixing the powder. The Ti and Ni micron powders were fed into the powder feeding system at a feeding rate of 38 mg/min. An 83 mg Ti-Ni alloy ingot, which was used as the seed material for the levitation and

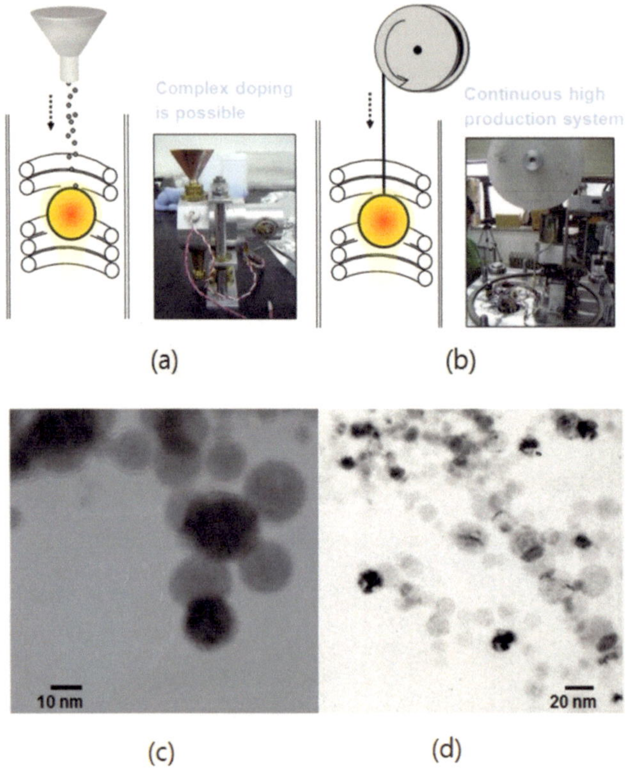

Figure 6. (a) Micron powder feeding (MPF) system and (b) wire feeding (WF) system in the LGC instrument. TEM images for (c) Ti-Ni alloys and (d) $NiFe_2O_4$.

evaporation reactions, was melted by an electric induction heating with an applied power of 6 kW at an argon gas pressure of 100 torr. The evaporated powders were filtered and finally passivated by partial oxidation. The starting materials were the mixed micron powders of Ni and Fe, which has a size ranging from 100 to 500 μm. The amount of micron powder fed into the liquid seed droplet was controlled at 80 mg/min. The mixed Ar and O_2 gas pressure in the chamber was 100 torr [15, 20] (**Figure 6**).

3. Magnetic metal and carbon-encapsulated metal nanoparticles

3.1. Magnetic properties of Ni, Fe, Ni@C, and Fe@C

Magnetic nanoparticles have attracted much attention because of their use in nano-fluids for biomedical application, thermally conductive fluids, various catalysts, etc. However, metallic nano-fluids end to be inherently vulnerable to oxidation, dissolution, and agglomeration during synthesis. In particular, agglomeration of the particles in a solvent is a serious problem when preparing nano-fluids. To overcome these problems, encapsulating the particles in a protective shell has been recommended to improve the chemical stability of the metal nanoparticles and their dispersion stability in the solvent. It is also worth noting that after encapsulation with a carbon coating layer, these materials are not prone to agglomeration because the coating reduces their magnetic interaction. In addition, the surface diffusion processes can preserve the chemical and structural properties of the nanopowder for a long time in many chemically aggressive conditions. A graphitic carbon shell in particular is regarded as an ideal coating since it is light and shows high stability in both chemical and physical environments [29–31].

The contents of the Ni and Ni@C nanoparticles synthesized by LGC using the micron powder feeding system were confirmed by XRD pattern. The XRD results for Ni and Ni@C showed the lattice parameters and the positions of the main peaks of the Ni powders. A small amount of NiO phase and amorphous graphitic layers was found in the XRD patterns and in the TEM images as mentioned in Section 1 [18]. The diffraction peaks at 44.4°, 51.8°, and 76.3° are due to the (1 1 1), (2 0 0), and (2 2 0) planes of fcc-Ni, respectively. The Ni powders synthesized by the LGC method showed low saturation magnetization. These results were attributed to the spin-canting effect and oxide phase on the surface [32]. The magnetic properties would be weak due to the antiferromagnetic NiO phase on the powder surface. The saturation magnetization was Ms. = 42 emu/g, as shown in **Figure 7(a)**. The slightly shifted hysteresis loop for the Ni sample can be explained by exchange bias between the ferromagnetic core of Ni and the antiferromagnetic surface of the NiO. The initial magnetization curve is not explained by the size effect. In previous studies, the virgin magnetization curve slightly spills over the limited hysteresis loop at 655 Oe. We assume that this effect is enhanced when the size of the particles is reduced, as suggested in a previous study. With decreasing particle size, the defects and the different magnetic structure on the surface of the particles are increased. The nature of this irreversibility in high magnetic fields follows a physical model and can be explained by a spin-glass or spin-canting behavior. The hysteresis loop of the as-made M@C materials in magnetic fields up to 2 T reveals their intrinsic magnetic behavior, indicated by the magnetization (M), the remanent magnetization (Mr), and the coercive force (Hc) of the M@C samples. The saturation magnetization demonstrates that

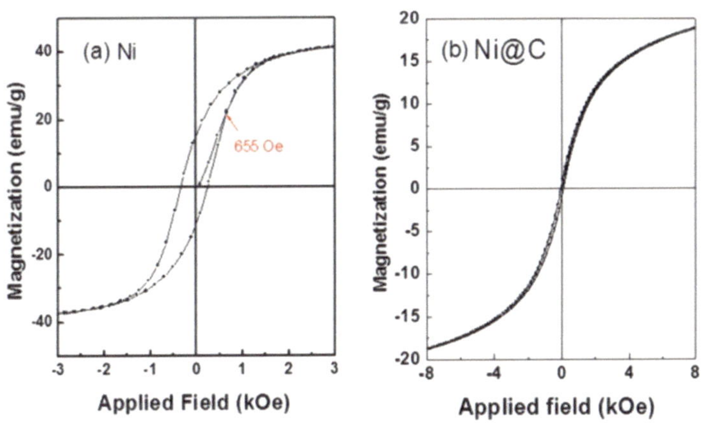

Figure 7. M-H loops for (a) Ni and (b) carbon-encapsulated Ni measured at 20°C.

the carbon-coated Ni nanocrystallites exhibited a superparamagnetic behavior at room temperature, which is related to the demagnetization effect arising from the additional energy of the magnetic fields outside the graphitic carbon encapsulation as shown in **Figure 7(b)**. The coercive force (Hc) and magnetization (M) were 76.6 Oe and 19.6 emu/g, respectively. The ratio of remanence to the saturation magnetization (Mr/M) was 0.04. The low magnetization compared with the Ni nanoparticles without the carbon shell is due to the coexistence of nonmagnetic carbon and the large percentage of surface spin due to the disordered magnetization orientation of the nanoparticles. The magnetic properties are influenced by both the particle size and the surface properties of the particle [33, 34].

A typical hysteresis loop of the Fe nanopowder at room temperature shows a saturation magnetization of Ms = 157 emu/g and coercivity of Hc = 836 Oe as shown in **Figure 8(a)**. An estimated single domain size of 14 nm for spherical iron particles with no shape anisotropy is reported. The size of the iron nanopowder is large enough to show very large value of coercivity. The hysteresis loops of the as-made Fe@Cs in magnetic fields up to 1 T reveal their intrinsic magnetic behavior, as shown in **Figure 8(b)**.

The hysteresis loops indicate that the carbon-coated Fe nanocrystallites exhibit superparamagnetic behavior at 50 and 300 K. The magnetization was not saturated in the applied fields up to 1 T, as shown in **Figure 6(b)**. In the nanoparticles, one can observe superparamagnetic behavior, which is related to the demagnetization effect arising from the additional energy of the magnetic fields outside the graphitic carbon encapsulation. The coercive force (Hc) and the magnetization (M) at 50 K were 130 Oe and 69.6 (emu/g), respectively. In a previous study, the Mössbauer spectrum for Fe@C nanopowder was measured at room temperature. The relative fraction of the α-Fe, Fe3C, and γ-FeC phases was determined to be about 27.6, 26.3, and 46.1%, respectively. The low magnetization compared with metal nanoparticles without a carbon shell was due to the coexistence of nonmagnetic carbon and the large percentage of surface spins due to the disordered magnetization orientation of the nanoparticles. The magnetic performance of the Ni@C and Fe@C samples was demonstrated in a liquid

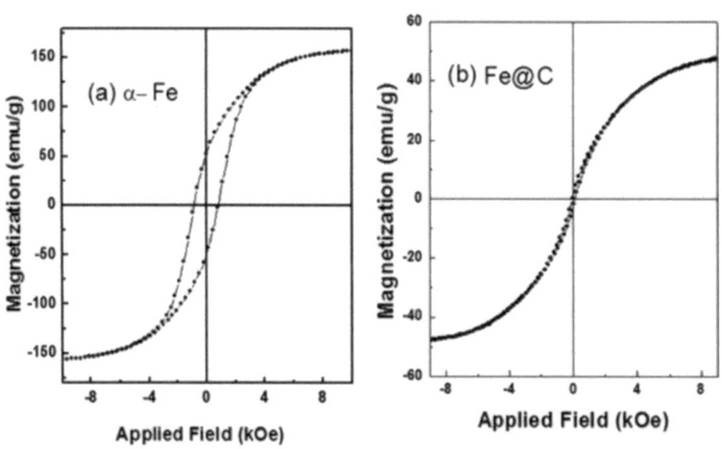

Figure 8. M-H loops for (a) α-Fe and (b) carbon-encapsulated Fe measured at 20°C.

phase (in ethanol and polyethylene glycol) by placing a magnet bar near the glass bottle. The carbon-encapsulated magnetic metals moved under the magnetic force. This suggests that Ni@C and Fe@C materials would be ideal adsorbents and catalyst supports because they are magnetically separable [35].

3.2. Dispersion stabilities of Ni, Fe, Ni@C, and Fe@C

To evaluate the dispersion stability and agglomeration phenomena of the carbon-encapsulated Ni and Fe nanoparticles in solvents of ethanol and ethylene glycol (EG), their time-dependent sedimentation behavior was investigated using transmission profile measurements obtained with a Turbiscan Lab [36–38]. The transmission profiles were taken every 1 h for 60 h when the suspending medium was ethanol. It was found that the transmission intensity decreased at the sample top owing to clarification and increased at the sample bottom due to sedimentation. A very stable Ni@C dispersion was observed without showing any clarification or sedimentation in EG. In contrast, a progressive fall signal was observed as a function of time in the middle region of Ni nanoparticles which had an average particle size of 20 nm. This can be explained by flocculation-induced particle growth. **Figure 9(a)** shows the Turbiscan screen data taken every 1 h for 3 days. The time-dependent transmission rates on the top, middle, and bottom show the same tendencies. The clarification in the top region and the progressive fall in the middle region of the ΔT signal were not observed in all suspensions. These imply that flocculation due to a coalescing reaction between the nanoparticles was insignificant. A very stable Fe@C dispersion, without any clarification on the top layer or sedimentation on the bottom layer, was observed in ethanol and EG. The viscosity of the solvent affected the dispersion stability kinetics. The dispersion stability of the solvents increased in the following order: water, ethanol, and ethylene glycol (or poly ethylene glycol). **Figure 9(b)** shows the effect of the solvent on the dispersion stability, as measured by using Turbiscan Lab, as well as the calculated mean value of the kinetics for each transmission (ΔT) profile as a function of time. The suspensions prepared in water displayed a rapid change in the mean ΔT values. As a result, sedimentation of the Fe@C nanoparticles in suspensions of water commenced as soon as the suspension was prepared. For the suspensions prepared in ethanol and EG, the variation in the mean ΔT was much less. However, this value increased continuously. Visual inspection confirmed that the suspension was stable, but sedimentation slowly occurred. However, coalescence between the Fe@C nanoparticles rarely occurred in the suspension because the carbon shell layer prevented agglomeration of the particles. The variation in the mean ΔT for the suspension prepared in EG was the smallest. The mean value of ΔBS increased when the particles were smaller than the wavelength of the incident light (880 nm). The tendency of ΔBS was similar to those of ΔT. From these results, for three kinds of solvent, we determined EG to be the most suitable solvent [38, 39].

3.3. Catalyst for the multicomponent Biginelli reaction

In this study, we introduce the catalytic effects of the Ni and Ni@C nanopowders observed during the synthesis of S-enantiomer from 3,4-dihydropyrimidine (DHPM). The synthesis

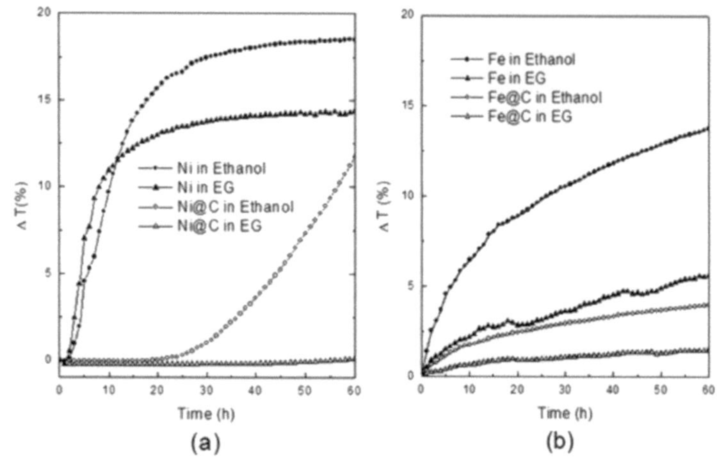

Figure 9. Variations in the mean ΔT for the (a) Ni and Ni@C suspensions and (b) Fe and Fe@C suspensions prepared in various solvents (ethanol and ethylene glycol).

of 4-Aryl-substituted DHPM compounds by the Biginelli reaction has attracted great attention in synthetic organic chemistry due to their pharmacological and therapeutic properties such as antibacterial and antihypertensive activity as well as their behavior as calcium channel blockers. Given the versatile biological activity of DHPM, development of an alternative synthetic methodology is of paramount importance [40–42]. This has led to the development of several new synthesis strategies involving combinations of Lewis acids and transition metal salts such as mainly homogeneous catalysts, which give high yields. However, in spite of their potential utility, many of these methods involve expensive reagents, long reaction times, high temperatures, and stoichiometric amounts of catalysts and result in unsatisfactory yields. Therefore, discovering a new, inexpensive catalyst for the Biginelli-type reaction under neutral and mild conditions is of prime importance. The starting materials used in this study were ethyl acetoacetate (I) (0.25 mmol), benzaldehyde (II) (0.25 mmol), and urea (III) (0.3 mmol). First, the benzaldehyde (II) (0.25 mmol), urea (III) (0.3 mmol), 0.1 g of catalyst (Ni or Ni@C), and chiral modifier of L-proline (0.025 mmol) were mixed and react in ethanol (50 ml) at 70°C for 2 hours. In the second step, ethyl acetoacetate (I) (0.25 mmol) was added and reacted under microwave for 3h. The ratio of the s-enantiomer in the as-prepared sample was characterized by high-performance liquid chromatography (HPLC) with a chiral column (Chiralcel OD-H) [13].

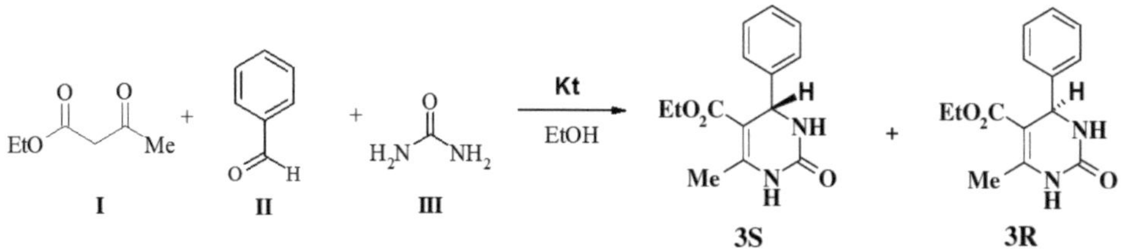

Catalysts	Yield (%) Racemic	HPLC		Δ (ee. %)
		S-enantiomer	R-enantiomer	
Ni	47	53.2	45.8	7.4
Ni@C	70	59.8	40.2	19.6

Table 1. Synthesis of 3,4-dihydropyrimidine based on the Biginelli reaction using nanosized catalysts of Ni and Ni@C.

Vigorous agitation appeared to be an extremely important factor influencing stereo selectivity. The results of stereo selectivity are represented in **Table 1**. The simultaneous use of a heterogeneous catalyst along with the chiral modifier allowed the ratio between stereoisomer in the Biginelli reaction to be changed in some experiments in favor of the S-enantiomer, with an excess of about 19.6%. The best results were obtained when using carbon-encapsulated Ni nanoparticles as the catalyst, L-proline as the chiral modifier, and methanol as the solvent. The catalytic reaction with Ni@C showed higher stereo selectivity than with Ni. The carbon shell influences the catalytic effect during synthesis [43].

4. Nonmagnetic metal and metal oxides: Cu oxide, Bi, and NiO

4.1. Catalytic activities of Cu oxide and ZnO

Cu oxides are widely applied in various organic syntheses such as reduction and oxidation processes, various condensation processes, for the syntheses of complex compounds, etc. The surface of the nanocrystalline Cu oxide includes a defect structure, resulting in non-stoichiometry. Such materials in themselves have the advantages of both homogeneous and heterogeneous catalysts. The aim of our investigation was the development of an effective catalytic and reaction systems based on nanocrystalline Cu oxides, with high reactivity at ambient temperature. To test the catalytic reaction, both the reaction of the liquid-phase oxidation of 2,3,5-trimethyl-1,4-hydroquinone (TMHQ) and the catalase activity were chosen. The oxidation of TMHQ is an intermediate stage of the hydroxylation of 2,3,6-trimethyl phenol in the synthesis of tocopherol. The process of TMHQ oxidation was carried out in a thermostatically controlled chamber, under agitation in a mixed water and methanol solution (1:1 in volume) at 50 ± 0.2°C. The rate of air supply was 6.2 ℓ/h. The reaction was carried out using the parent material (0.66 mmol) and the nanopowders (1 mmol). To compare the catalytic properties of the Cu oxides, powders with various sizes were synthesized. The size control of the powder was carried out by altering the feeding velocity of the Cu wire from 20 to 80 mm/min. Phase control of the Cu, Cu2O, and CuO was carried out by controlling the pressure of the inert gas. In **Table 2**, the crystallite conditions of the copper oxides are displayed. The dehydrogenation (oxidation) of the TMHQ in solution was to practically form 2,3,5-trimethyl-1,4-quinone (TMQ) (selectivity on TMQ > 99.5%). The kinetic curves of TMHQ oxidation in the presence of the nanocrystalline Cu oxide particles (Samples 1, 2, and 3) are given in **Figure 10**. Samples containing mainly pure Cu (Cu 78%, Cu_2O 22%, sample a)

	Average particle size (nm)	Conditions for synthesis: feeding speed, draft velocity (l/min), and pressure in chamber	Phase composition, wt. %			Oxidized yield of TMHQ	H_2O_2 conversion
			Cu	Cu_2O	CuO		
a	20	Slow feeding (20~30 mm/s) $0.0 \le V_{O2} \le 0.05$, 80 torr	78	22	—	No reaction	10.2
1	90	Fast feeding (60 mm/min) $V_{O2} = 0.2$, 120 torr	4	96	—	Initial	60.1
2	35	Slow feeding(20~30 mm/min) $V_{O2} = 0.2$, 100 torr	—	100	—	Active	99.8
3	35	Slow feeding (20~30 mm/min) $0.1 \le V_{O2} \le 0.15$, 100 torr	10	85	5	Active	82.3

Table 2. The concentration of Cu oxide nanopowders, the reaction yields for dehydrogenation of TMHQ, and hydrogen peroxide (catalase activity).

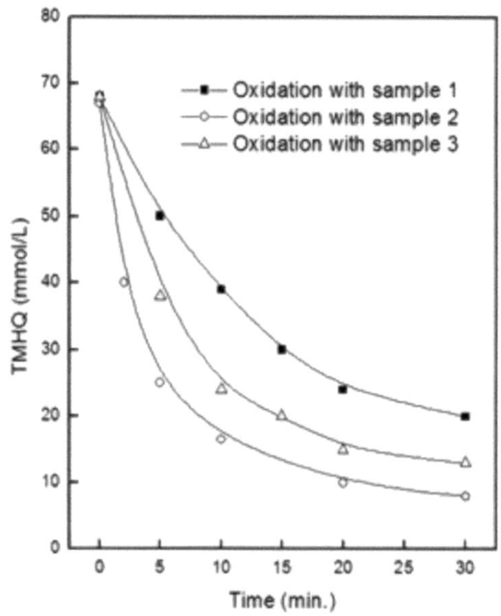

Figure 10. Kinetic curves for the dehydrogenation of TMHQ using Cu oxide catalysts with average particle size of 35 and 90 nm.

in the structure showed rare catalytic reaction. Catalytic activity depended on both the average particle size and the oxide phase such as the concentration of Cu_2O in the nanopowders. Sample 1 with mainly Cu_2O phase and an average size of 90 nm performed as catalyst with theoretical yield. Samples 2 and 3 with an average size of 35 nm showed significantly active effect. These samples contained a small quantity of CuO, not exceeding 0.15 mol fraction. The results of the oxidation of TMHQ are represented in **Table 2**. The catalytic yield of Sample 1 compared with Samples 2 and 3 was relatively very low. The Samples 2 and 3 with the same range of particle sizes included different ratios of Cu and Cu oxide phase. The yield of

TMHQ oxidation depended on the particle size and amount of the Cu_2O phase. Obviously, oxidation of the parent material substantially occurred under the action of the fixed oxygen, which was activated in the matrix of nanopowders [44–46].

The catalase activity is an informative parameter about the catalytic properties of the materials in the redox process. It was simulated by measuring the ability of the catalase in the decom position of hydrogen peroxide to isolate molecular oxygen. Decomposition of the hydrogen peroxide was carried out in a thermostatically isolated chemical reactor (10 ml). A mixture of water and methanol (1:1 in volume) was agitated with a stirring rod at 50 ± 0.2°C. The reaction was carried out with hydrogen peroxide (1.7 mmol) and nanopowder (2 mg). The catalytic activity of the Co, Mn, Fe, and Cu hydroxides was also estimated by catalase activity. The aim of this work was to solve a scientific problem related to the chemical intoxication mechanisms of water phenol solutions and its derivatives during their cleaning. The data in **Table 2** show the results PF the catalase activities for the same nano-Cu oxide samples. The reaction of Sample 2 with a size of 35 nm, containing mainly Cu_2O, showed much higher activity than Sample 1 with a size of 90 nm and the same oxide phase. The size of the particles was the most significant factor. The particle size affects the surface state. The specific surface areas of Samples 1, 2, and 3 were 17, 35, and 38 m^2/g, respectively. The specific surface area is related to the particular manufacturing method. The LGC method produces nano-scaled powders with high specific surface area [47].

The photocatalytic activity of the ZnO was evaluated based on the photodegradation of phenol aqueous solutions under different irradiation conditions. For experiments under UV-visible light, 100 mL of 50 ppm phenol in aqueous solution with 0.5 g catalytic powders was loaded in a glass container and stirred with a magnetic stirrer a under irradiation form, a Hg-Xe lamp. Total organic carbon (TOC) values as a function of time were measured after filtration under reduced pressure. **Figure 11** shows the photo mineralization of phenol with UV-visible light (solar simulator) in the presence of ZnO. Obviously, when ZnO is added to the phenol, the total organic carbon (TOC) value was reduced to 60% [48–51].

4.2. Electrochemical analyses using Bi nanopowder

The anodic stripping voltammetry (ASV) method is a powerful electrochemical technique for trace metal analysis. The traditional electrodes for ASV measurements are mercury-drop electrode and a mercury-film electrode, due to high sensitivity of the mercury [52–56]. However, mercury is very toxic. The toxicity of the mercury has led its usage to be completely banned in some countries. In this study, we focused on searching for alternative environment-friendly electrode materials. The Bi-film electrode has been considered replacing the mercury-film electrode due to its nontoxicity. The properties of Bi materials show not only excellent resolution of neighboring peaks but also insensitivity to the dissolved oxygen in the solution. However, there are still some problems to use the electrode such as a low detection limit comparing to mercury electrode and complication of preparing electrode including processes of additional washing or polishing of the carbon surface and dissolving Bi ions into the solution for the pre-deposition of the Bi on the film electrode. In order to overcome the above weaknesses of the Bi-film electrode, a Bi nanopowder-labeled electrode with

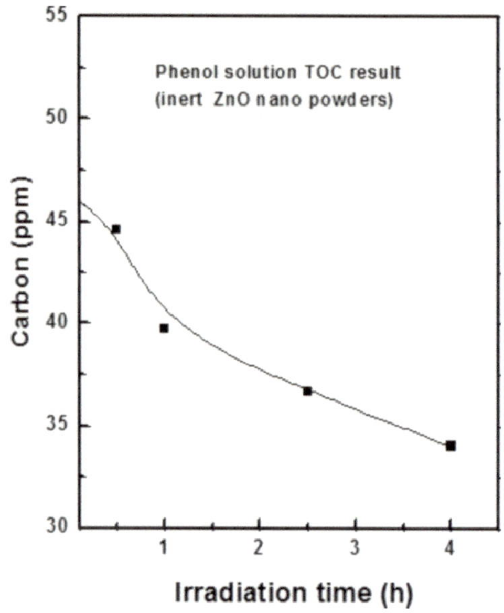

Figure 11. Photo mineralization of phenol with sunlight (TOC: total organic carbon content at times) in the presence of ZnO (Hg-Xe lamp with a wavelength of 200 ~ 2500 nm and 1 kW of power).

a larger electrochemical active surface area was fabricated [57–59]. In this study, the nano-Bi-fixed electrode sensor and a nanosized Bi-binding technology were developed to improve the electrochemical characteristics of Bi for detecting heavy metals. For this purpose, the Bi nanopowder was synthesized using the LGC method and was then coated on a conductive carbon layer using a Nafion solution. **Figure 12** illustrates the attached Bi working electrode and the analysis system setup for measuring Zn, Cd, Pb, and Ta [14, 24, 58–62].

The working electrode was prepared using conductive carbon ink (DongYoung Chemical Co., LTD, in South Korea) painted flexible polyester film by a semiautomatic screen printing instrument. Then the prepared carbon ink with a thickness of 80 μm on painted thick film was

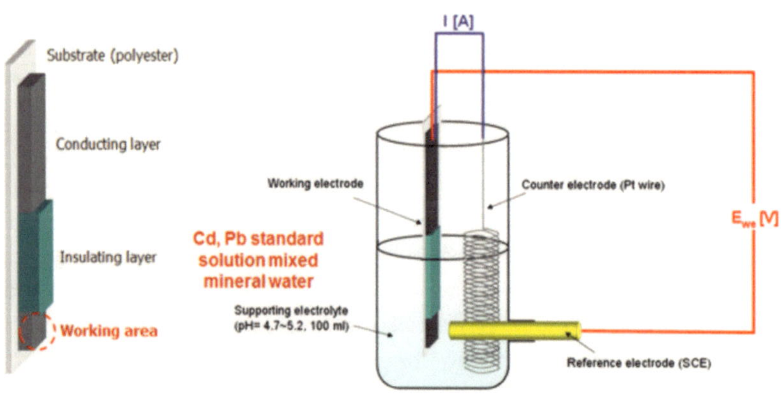

Figure 12. Illustration for working electrode and total system for electrochemical analyses.

partially covered by an insulating layer. Bi nanopowders were well dispersed into 20 ml of distilled water using an ultrasonic treatment. A Nafion solution (Fluka) was added in to the Bi-dispersed suspension for strong chemical bonding between nanopowder and the carbon paste. Finally, the Bi nanopowder-dispersed suspension was dropped onto the working area and dried in the air at room temperature. As the concentration of Nafion in suspension was increased, the value of pH was decreased due to the strong acidity of the Nafion.

When the Bi nanoparticles were dispersed in distilled water without Nafion, the zeta potential showed a positive value [14]. However, as Nafion was added in the suspension, the zeta potential changed to a negative value. The amount of Nafion should be optimized to be 200 μℓ for dispersion stability and the phase stability of Bi nanoparticles. The sensor electrodes were prepared using the screen-printed carbon surface with the Bi nanoparticles strongly attached by Nafion. A platinum wire and a saturated calomel electrode (SCE) were used as a counter electrode and a reference electrode, respectively. The supporting electrolyte was a 0.1 M NaAc and 0.025 M HCl solution of pH 5.0. The prepared nanoparticles are confirmed by XRD as shown in **Figure 13(a)**. Also, the screen-printed Bi nanoparticles dispersed in Nafion on the electrode could be observed by TEM, as shown in **Figure 13(b)**.

Figure 14 shows results of the anodic stripping voltammograms (ASV) using the Bi nanopowder-attached electrode for measuring various concentrations of Cd and Pb ions in solution. The ASV showed well-defined peaks at −0.85 V and −0.65 V corresponding to the oxidation of Cd and Pb, respectively. **Figure 15** demonstrates the dependence of the stripping peak current density Ip on the Cd and Pb concentrations over a range of 3~30 ppb (deposition potential = −1.35 V and deposition time = 3 min). From the linearity between the metal concentration and the peak current, the values of the sensitivity of the nano-Bi-fixed electrodes were determined to be 9.01 ± 0.012 and 7.15 ± 0.007 $\mu A/ppb \cdot cm^2$ for Cd and Pb, respectively. The estimated detection limits of the nano-Bi-fixed electrode were 0.31 and 0.42 ppb for Cd and Pb, respectively, on the basis of the signal-to-noise characteristics (S/N = 3) under a 10 min accumulation. These values are much lower than the domestic and the international content limits of Cd and Pb ions in drinking water, which are listed in **Table 3**, indicating the excellent

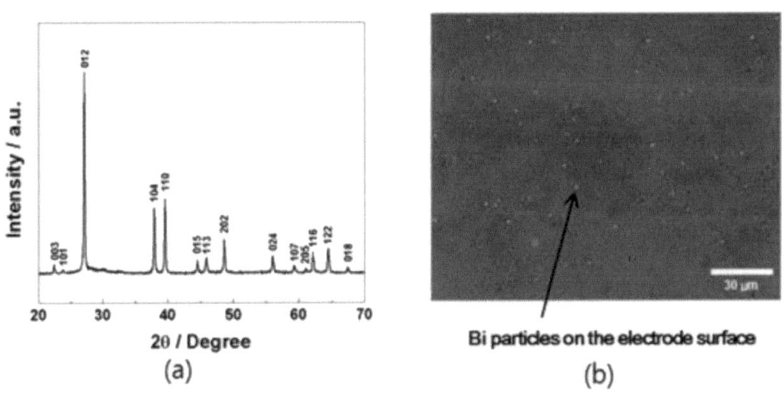

Figure 13. XRD pattern for (a) Bi nanopowders and (b) SEM image for screen-printed Bi.

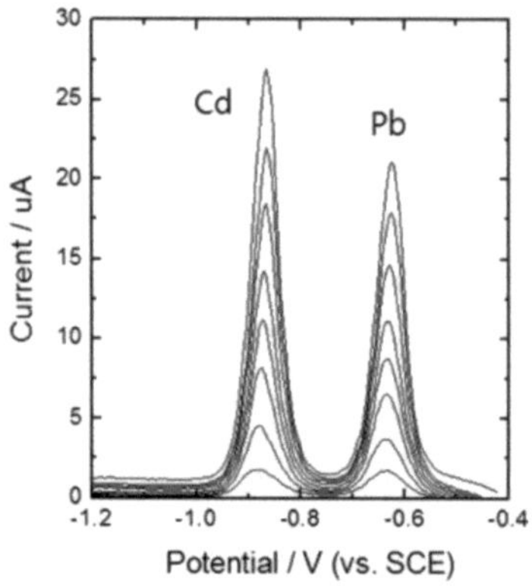

Figure 14. Square wave anodic stripping voltammograms experimentally measured on the nano-Bi-fixed electrode for various concentrations of Zn, Cd, and Pb ions.

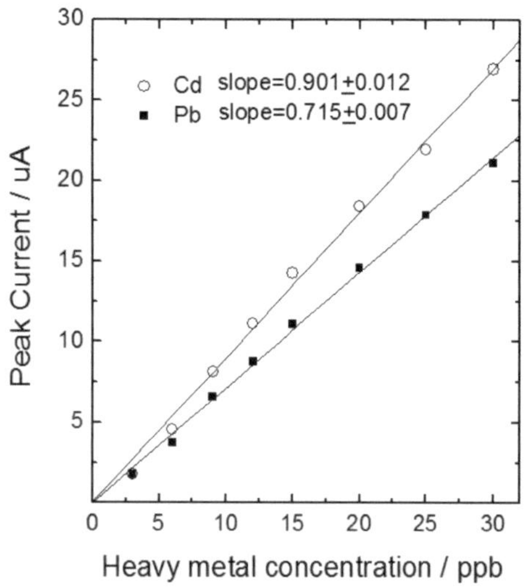

Figure 15. Dependence of the stripping peak current density Ip on the Cd and Pb concentrations over the range of 3 ~ 30 ppb (deposition potential = −1.35 V; deposition time = 3 min).

detection of the Bi nanopowder-fixed electrode. Consequently, the low toxicity of the Bi nanopowder-fixed electrode with high sensitivity about heavy metals promises the development of an attractive sensor for monitoring toxic chemical species in environmental matrices with a clean methodology [62].

4.3. Catalytic effect of NiO in the Biginelli and Hantzsch reaction

Nickel oxide powders were obtained by the gas condensation method in an argon-oxygen mixture flow. The argon flow rate was 130 ℓ/h, oxygen concentration 8.5 vol %, pressure equal to 90 torr, and Ni feed rate 1.8 g/h. XRD analysis showed the following phase composition: NiO 90.7%, Ni 9.3%, and mean size of nickel oxide particles 13 nm. According to the electron microscopy, the particles proved to have a nearly uniaxial shape. The nickel phase detected by the XRD method appears to result from the incomplete oxidation of some particles, and it is located in the center of the particles, while the outer layers must certainly be in an oxidized state (**Figures 16** and **17**).

The NiO nanoparticles were applied for the synthesis of both dihydropyridine (DHP) and dihydropyrimidine (DHPM) as mentioned in Section 3.3. The most plausible pathway for DHP prepared by the Hantzsch reaction has been shown to involve the interaction of benzaldehyde with one molecule of β-dicarbonyl compound 1 to give chalcone 3, while another molecule of β-dicarbonyl compound 1 is transformed into enamine 2. In route A, enamine 2 is condensed with an aldehyde and ethyl acetoacetate 1 in the reflux in a suitable solvent (methanol or ethanol) [63–65]. Route B involves the reaction of chalcone 3 with enamine 2, and it seems to give better yields of products and easier purification. In the presence of aqueous ammonia, compound 3 undergoes a partial decomposition into benzaldehyde and diketone 1, thus giving a rise to the formation of symmetrical analogues nitrendipine 16a, b. When the Hantzsch reaction is carried out at 22–25°C in the presence of L-proline and nanosized NiO, the ratio of enantiomers of nitrendipine is changed in favor of the S-enantiomer by 3.4%.

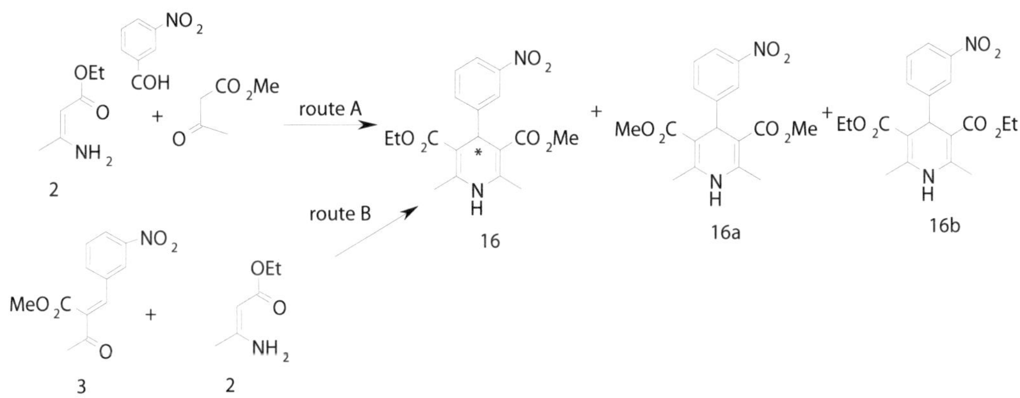

The Biginelli reaction for synthesizing DHPM was carried out in the presence of L-proline and nanosized NiO (obtained by the Institute of Metal Physics) to change the ratio of enantiomers of 3S in Section 2.3 is changed in favor of S-enantiomer by 15.4%. Our future plans involve studying the factors that affect enantiofacial discrimination for the Hantzsch and Biginelli reaction, such as the nature of nanosized metal oxides or chiral modifiers, reaction time, temperature, and solvent. We also plan to synthesize new nanosized metals and their oxides, as well as chiral modifiers.

Heavy metal	unit	Korea	IBWA	FDA	WHO/FAO
Cd	ppb	5	5	5	3
Pb	ppb	50	5	5	10

Table 3. Domestic and international content limits of cd and Pb ions in drinking water.

Figure 16. TEM image of NiO nanopowder.

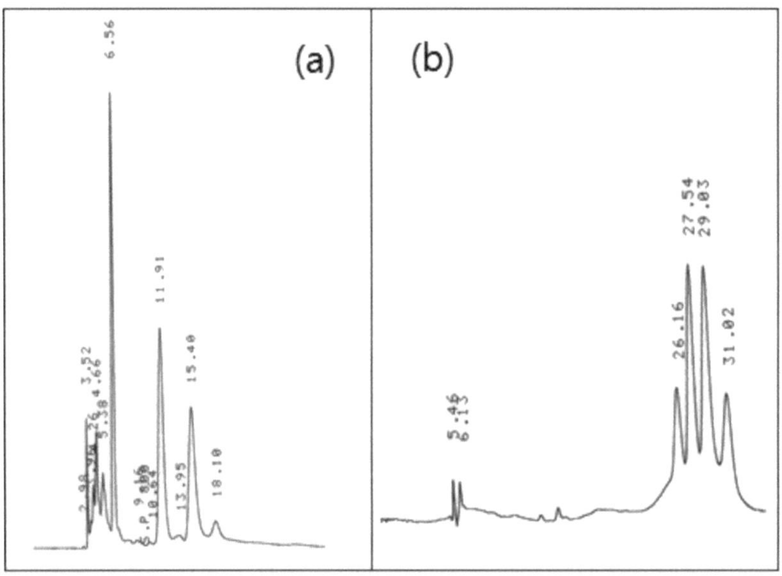

Figure 17. Chromatogram of a racemic mixture of (a) DHPM, enriched with S-enantiomer by 15.4%. The product was obtained in the presence of L-proline and NiO nanopowder mic mixture of (b) DHP, enriched with S-enantiomer by 3.4%.

5. Conclusion

Ceramics, such as NiO, ZnO, and Cu_2O, magnetic nanoparticles, including γ-Fe_2O_3, Fe_3O_4, and $NiFe_2O_4$, and metals, such as Cu, Ni, Zn, Sn, Ag, Au, Bi, and carbon-encapsulated metals (Ni and Fe), were synthesized by levitational gas condensation (LGC) method using wire feeding (WF) and micron powder feeding (MPF) systems. The magnetic properties have been characterized using a vibrating sample magnetometer (VSM). The size and shape of the nanopowders were investigated by transmission electron microscopy (TEM). The surface effect influenced the magnetic behaviors of nanopowders. Bi metals were dispersed in Nafion. The Bi particles could be applied as sensor electrode instead of mercury-based electrolyte. The particle size of carbon-coated metal with diameters in the range of up to 10 m was smaller than those of metals without a carbon shell. The dispersion stability kinetics of carbon-coated nanopowders showed good dispersion. The best results were obtained when using carbon-encapsulated Ni nanoparticles as a catalyst, L-proline as a chiral modifier, and methanol as a solvent. The catalytic reaction of Ni@C showed enhanced stereoselectivity. Also, the simultaneous use of the heterogeneous catalyst and chiral modifier may lead to an increase in the selectivity of the Biginelli reaction. Nanoparticles prepared using LGC showed significantly enhanced catalytic activities during chemical reaction due to the high level of defects on their surface structure.

Author details

Young Rang Uhm

Address all correspondence to: uyrang@kaeri.re.kr

Radioisotope Research Division, Korea Atomic Energy Research Institute (KAERI), Daejeon, Republic of Korea

References

[1] Ruoff RS, Lorents DC, Chan B, Malhotra R, Subramoney S. Single crystal metals encapsulated in carbon nanoparticles. Science. 1993;**259**(5093):346-348

[2] Scott JHJ, Majetich SA. Morphology, structure, and growth of nanoparticles produced in a carbon arc. Physical Review B. 1995;**52**(17):12564-12571

[3] Lu C-H, Yeh C-H. Influence of hydrothermal conditions on the morphology and particle size of zinc oxide powder. Ceramics International. 2000;**26**(4):351-357

[4] Park JB, Jeong SH, Jeong MS, Kim JY, Cho BK. Synthesis of carbon encapsulated magnetic nanoparticles by pulsed laser irradiation of solution. Carbon. 2008;**46**:1369-1377

[5] Junichi N, Chie O, Osamu O, Nobuyuki N. Synthesis structures and magnetic properties of carbon encapsulated nanoparticles via thermal deposition of metal acetylide. Carbon. 2006;**44**(14):2943-2949

[6] Gonsalves KE, Rangarajan SP, Wang J. In: Nalwa HS, editor. Handbook of Nanostructured Materials and Nanotechnology. San Diego: Academic Press Ed; 2000

[7] Han BS, Rhee CK, Lee MK, Uhm YR. Synthesis of nano crystalline Ni and Fe by levitational gas condensation method. IEEE Transaction Magnetics. 2006;**42**(11):3779-3781

[8] Uhm YR, Kim WW, Rhee CK. Phase control and Mössbauer spectra of nano γ-Fe2O3 particles synthesized by the levitational gas condensation (LGC) method. Physica Status Solidi A. 2004;**201**(8):1934-1937

[9] Yermakov AY, Uimin MA, Mysik AA. Magnetism and structure of nanoparticles and mesoscopic systems. Materials Science Forum. 2002;**386-388**:455-464

[10] Uhm YR, Lee HM, Rhee CK. Structure and magnetic properties of carbon-encapsulated Ni nanoparticles. IEEE Transactions on Magnetics. 2009;**45**(6):2453-2455

[11] Uhm YR, Han BS, Lee MK, Hong SJ, Rhee CK. Synthesis and characterization of nanoparticles of ZnO by levitational gas condensation. Materials Science and Engineering A. 2007;**449-451**:813-816

[12] Uhm YR, Lee HM, Rhee CK, Kim CS. Magnetic properties and dispersion stability of carbon encapsulated Fe nanoparticles. Journal of the Korean Physical Society. 2010; **57**(6):1609-1613

[13] Uhm YR, Lee HM, Fedorova O, Ovchinnikova I, Valova M, Rusinov G, et al. Synthesis of carbon encapsulated metal (Ni and Cu) nano particles and applications for chiral catalysts. Research on Chemical Intermediates. 2010;**36**:867-873

[14] Lee G-J, Lee HM, Uhm YR, Lee MK, Rhee CK. Square-wave voltammetric determination of thallium using surface modified thick-film graphite electrode with bi nanopowder. Electrochemistry Communications. 2008;**10**:1920-1923

[15] Han BS, Uhm YR, Lee MK, Park JJ, Rhee CK, Kim GM. Synthesis of Ti50Ni50 alloy nanopowders synthesized by modified levitational gas condensation method. Solid State Phenomena. 2007;**119**:263-266

[16] Uhm YR, Han BS, Lee MK, Rhee CK. The heat treatment of nanocrystalline Cu2O to enhance the catalytic effect at the surface layer. Solid State Phenomena. 2006;**118**:635-638

[17] Uhm YR, Lee MK, Rhee CK. Study on the catalytic properties of copper oxide nanoparticles synthesized by levitational gas condensation (LGC)method. Journal of the Korean Magnetic Society. 2007;**17**(2):99-102

[18] Uhm YR, Rhee CK. Synthesis and magnetic properties of Ni and carbon coated Ni by Levitational gas condensation (LGC). Journal of Nanomaterials. 2013;**2013**:1-6

[19] Uhm YR, Lee HM, Lee GJ, Rhee CK. Synthesis of metal and ceramic magnetic nanoparticles by Levitational gas condensation (LGC). Journal of Magnetics. 2009;**14**(2):75-79

[20] Uhm YR, Han BS, Lee HM, Rhee CK. Synthesis and characterization of NiFe2O4 nanoparticles synthesized by levitational gas condensation (LGC). Solidstate Phenomena. 2008; **135**:123-126

[21] Uhm YR, Park JH, Kim WW, Cho C-H, Rhee CK. Magnetic properties of nano-size Ni synthesized by the pulsed wire evaporation (PWE) method. Materials Science and Engineering B. 2004;**106**(3):224-227

[22] Caoa H, Huangb G, Xuana S, Wu Q, Gua F, Li C. Synthesis and characterization of carbon-coated iron core/shell nanostructures. Journal of Alloys and Compounds. 2008;**448**:272-276

[23] Liu BH, Ding J, Zhong ZY, Dong ZL, White T, Lin JY. Large-scale preparation of carbon-encapsulated cobalt nanoparticles by the catalytic method. Chemical Physics Letters 2002;**358**(1-2):96-102

[24] Lee G-J, Lee H-M, Rhee C-K. Bismuth nano-powder electrode for trace analysis of heavy metals using anodic stripping voltammetry. Electrochemistry Communications. 2007;**9**(10):2514-2518

[25] Uhm YR, Park JH, Kim WW, Lee MK, Rhee CK. Novel synthesis of Cu2O nano cubes derived from hydrolysis of cu nano powder and its high catalytic effects. Materials Science and Engineering A. 2007;**449-451**:817-820

[26] Uhm YR, Rhee CK, Park J-J, Jun S-H. Morphologies and dispersion stabilities of carbon encapsulated metal (Ag and Cu) nanoparticles synthesized by pulsed-wire evaporation (PWE). Research on Chemical Intermediates. 2013;**39**:3387-3398

[27] Jun S-H, Uhm YR, Rhee CK, Song R-H, Park S-J, Kim H-W. Preparation of carbon-encapsulated Ag nanoparticles for dispersion in La0.6Sr0.4Co0.3Fe0.7O3−δ. Journal of the Korean Physical Society. 2011;**59**(6):3648-3651

[28] Wang ZH, Choi CJ, Kim BK, Kim JC, Zhang ZD. Characterization and magnetic properties of carbon-coated cobalt nanocapsules synthesized by chemical vapor condensation process. Carbon. 2003;**41**(9):1751-1758

[29] Savino V, Yann L, Rocio C, Oscar B, Brigitte B, Morales M. Continuous production of inorganic magnetic nanocomposite for biomedical applications by laser pyrolysis. Journal of Magnetism and Magnetic Materials. 2006;**5**(12):971-975

[30] Xu B, Guo J, Wang X, Liu X, Ichinose H. Synthesis of carbon nanocapsules containing Fe, Ni, or co, by arc discharge in aqueous solution. Carbon. 2006;**4**(13):2631-2634

[31] Jacob DS, Genish I, Klein L, Gedanken A. Carbon-coated core-shell structured copper and nickel nanoparticles synthesized in an ionic liquid. Journal of Physical Chemistry B. 2006;**110**(36):17711-17714

[32] Kinemuchi Y, Ishizaka K, Suematsu H, Jiang W, Yatsui K. Magnetic properties of nanosize NiFe2O4 particles synthesized. Thin Solid Films. 2002;**407**(1-2):109-113

[33] Yu C, Qiu JS. Preparation and magnetic behaviour of carbon-encapsulated cobalt and nickel nanoparticles from starch. Chemical Engineering Research and Design. 2008; **86**(8):904-908

[34] Hayashi T, Hirono S, Tomita M, Umemura S. Magnetic thin films of cobalt nanocrystals encapsulated in graphite-like carbon. Nature. 1996;**381**(6585):772-774

[35] Scott J, Majetich S. Morphology, structure, and growth of nanoparticles produced in a carbon arc. Physical Review B. 1995;**52**(17):12564-12571

[36] Lee MK, Uhm YR, Rhee CK, Lee YB. Organic suspension behavior of rutile TiO2 nanoparticles with high specific surface area. Materials Transactions. 2010;**51**(12):2157-2161

[37] Park K-Y, Uhm YR, Son KJ, Park D-Y. Electrophoretic deposition of Ni nanoparticles from suspension on to a Ni plate. Research on Chemical Intermediates. 2015;**41**(9):6721-6730

[38] Woo SH, Lee MK, Rhee CK. Sedimentation properties of TiO2 nanoparticles in organic solvents. Solid State Phenomena. 2007;**119**:267-270

[39] Shaw DJ. Introduction to Colloid and Surface Chemistry. 4th ed. London: Butterworth; 1992 306 p

[40] Rodriguez H, Suarez M, Perez R, Petit A, Loupy A. Solvent-free synthesis of 4-aryl substituted 5-alkoxycarbonyl-6-methyl-3,4-dihydropyridones under microwave irradiation. Tetrahedron Letters. 2003;**44**:3709-3712

[41] Sharma SD, Gogoi P, Konwar D. A highly efficient and green method for the synthesis of 3,4-dihydropyrimidine-2-ones and 1,5-benzodiazepines catalysed by dodecyl sulfonic acid in water. Green Chemistry. 2009;**9**:153-157

[42] Lewandowski K, Murer P, Svec F, Frechet JM. A combinational approach to recognition of chirality: Preparation of highly enantioselective aryl-dihydropyrimidine selectors for chiral HPLC. Journal of combinational Chemistry. 1999;**1**(1):105-112

[43] Uhm YR. Magnetic properties and application of caltalysts in Biginelli reaction for the Ni and Ni@C synthesized by Levitational gas condensation (LGC). Journal of the Korean magnetics Society. 2017;**27**(3):1-5

[44] Metelitza DI, Naumchik IV, Karasyova EI, Plozov GI, Shadyro OI. Inhibition of peroxidase-catalyzed oxidation of aromatic amines by substituted phenols. Applied Biochemistry and Microbiology. 2003;**39**(4):352-362

[45] Mercier C, Chabardes P. Organometallic chemistry in industrial vitamin a and E synthesis. Pure and Applied Chemistry. 1994;**66**(7):1508-1518

[46] Chen L, Tang R, Li Z, Liang S. An efficient and mild oxidation of α-isophorone catalysed ny N-hydroxyphthalimide and copper chloride. Bulletine Chemical Society. 2012; **33**(2):459-463

[47] Yermakov AY, Feduschak TA, Uimin MA, Mysik AA, Gaviko VS, Chupakin ON, Shishmakov AB, Kharchuk VG, Petrov LA, kotov YA, Vosmerikov AV, Korolyov AV. Reactivity of nanocrystalline copper oxide and its modification under magnetic field. Solid State Ionics. 2004;**172**:317-323

[48] Muthaukumaran S, Gopalakrishanan R. Structural, FTIR and photoluminescence studies of cu doped ZnO nanopowders by co-precipitation method. Optical Materials. 2012; **34**:1946-1953

[49] Mohan R, Krishnamoorthy K, Kim S-J. Enhanced photocatalytic activity of cu-doped ZnO nanorods. Solid State Communications. 2012;**152**(5):375-380

[50] Han BS, Uhm YR, Kim GM, Rhee CK. Novel synthesis of nanorod ZnO and Fe-doped ZnO by the hydrolysis of metal powders. Journal of Nanoscience and Nanotechnology. 2007;**7**(11):4158-4160

[51] Han BS, Uhm YR, Rhee CK. Synthesis for nanoflower and rod of ZnO by a surfactant free and low temperature method. Surface Review and Letters. 2010;**17**(2):173-176

[52] Akcay H, Oguz A, Karapire C. Study of heavy metal pollution and speciation in Buyak Menderes and Gediz river sediments. Water Research. 2003;**37**:813-822

[53] Jadwiga O. The determination of Zn, cd, Pb and cu in soil extracts by anodic stripping voltammetry. Water, Air, and Soil Pollution. 1989;**45**:43-48

[54] Selehattin Y, Sultan Y, Gulsen S, Murat S. Direct determination of Zn heavy metal in tap water of Canakkale (TURKEY) by anodic stripping voltammetry technique. International Journal of Electrochemistry and Science. 2009;**4**:288-294

[55] Raj J, Raina A, Mohineesh, Dogr TD. Direct Determination of Zinc, Cadmium, Lead, Copper Metal in Tap Water of Delhi (India) by Anodic Stripping Voltammetry Technique. In: E3S Web of Conference. Rome, India: EDP Sciences; April 2013. 09009. pp. 1-4

[56] Novotry K, Turzikova A, Komarek J. Speciation of copper, lead and cadmium in aquatic systems by circulating dialysis combined with flame AAS. Fresenius Journal of Analytical Chemistry. 2000;**366**:209-212

[57] Morales MG, Marin MRP, Blazquez LC, Pinilla-Gil E. Applicability of the bismuth bulk rotating disk electrode for heavy metal monitoring in undisturbed environmental and biological samples: Determination of Zn(II) in rainwater, tap water, and urine. Analytical Methods. 2014;**6**:8668-8674

[58] Granado Rico MA, Olivares Martin M, Pinilla-Gil E. Modification of carbon screen-printed electrodes by adsorption of chemically synthesized bi nanoparticles for the voltammetric stripping detection of Zn(II), cd(II), and Pb (II). Talanta. 2009;**80**(2):631-635

[59] Lee G-J, Kim CK, Lee MK, Rhee CK. Reliability evaluation of nano bi/silver paste sensor electrode for detecting trace metals. Journal of Nanoscience and Nanotechnology. 2012;**12**(7):5673-5677

[60] Lee G-J, Kim C-K, Lee MK, Rhee CK. A study on optimization of Nano-sized bismuth binding technology using a Nafion solution. Journal of the Korean Physical Society. 2010;**57**(6):1667-1671

[61] Lee HM, Lee G-J, Uhm YR, Lee HJ, Lee MK, Rhee CK. Synthesis of nanocrystalline bismuth and its application to the detection of trace metals. Journal of Nanoscience and Nanotechnology. 2010;**10**(1):309-313

[62] Cesarino I, Gouveia-Caridade C, Pauliukaite P, Cavalheio ETG. Characterization and application of bismuth film modified graphite-polyurethane composite electrodes. Electroanalysis. 2010;**22**(13):1437-1445

[63] Marinkovic V, Agbaba D, Vladimirov S, Stankovic S. Simultaneous HPLC determination of nitrendipine and impurities of the process of synthesis. Journal of Pharmaceutical and Biomedical Analysis. 2001;**24**:993-998

[64] Oliver Kappe C. Recent advances in the Biginelli Dihydropyrimidine synthesis new tricks from an old dog. Accounts of Chemical Research. 2000;**33**(12):879-888

[65] Atwal KS, Rovnyak, O'Reilly GC, Schwartz. Synthesis of selectivity functionalized 2-Hetero-1,4-dihydropyrimidines. Journal of Organic Chemistry. 1989;**54**:5898-5907

Chapter 5

Vibrational Behavior of Single-Walled Carbon Nanotubes Based on Donnell Shell Theory Using Wave Propagation Approach

Muzamal Hussain and Muhammad Nawaz Naeem

Additional information is available at the end of the chapter

http://dx.doi.org/10.5772/intechopen.73503

Abstract

This chapter is concerned with the vibration analysis of single-walled carbon nanotubes (SWCNTs). This analysis is based on the Donnell thin shell theory. The wave propagation approach in standard eigenvalue form has been employed in order to derive the characteristic frequency equation describing the natural frequencies of vibration in SWCNTs. The axial modal dependence is measured by the complex exponential functions implicating the axial modal numbers. Vibration frequency spectra are gained and evaluated for physical parameter like length-to-diameter ratios. The dimensionless frequency is also investigated in armchair and zigzag SWCNTs with in-plane rigidity and mass density per unit lateral area for armchair and zigzag SWCNTs. These frequencies of the SWCNTs are computed with the aid of the computer software MATLAB. These results are compared with those obtained using molecular dynamics (MD) simulation and the results are somewhat in agreement.

Keywords: vibration analysis, wave propagation approach, Donnell thin shell theory, single-walled carbon nanotube

1. Introduction

Iijima [1] discovered the carbon nanotubes (CNTs) in 1991 and the uses of carbon nanotubes (CNTs) have been originate in various areas such as electronics, optical, medicine, charge detectors, sensors, field emission devices, aerospace, defense, construction and even fashion. To study their remarkable properties, a bulk of research work was performed for their high springiness and characteristic ratio [2], a very effective Young modulus and tensile potency [3], well-bonding strength and superconductivity between carbon atoms [4]. Study of free vibrations of these tube have has been examined with regard to their properties and material behavior. For their useful applications, it needs more explorations to examine vibration characteristics of SWCNTs.

Poncharal et al. [5] and Treacy et al. [6] conducted the experiments, the resonance frequency of multi-walled carbon nanotubes for clamped-free excited by electrical loads or thermal were detected in a transmission electron microscope (TEM). Thermal vibrations of SWCNTs have been performed for natural frequencies by Molecular dynamics (MD) to predict the Young's modulus by Zhao et al. [7].

The behaviors and material properties of CNTs using either or continuum mechanics modeling or atomistic modeling have been conducted in Wang et al. [8]. For the analysis of CNTs, when compared to continuum mechanics modeling, atomistic modeling is an easy approach and relatively inexpensive. Consequently, the development of continuum mechanics model has attracted much attention of researchers; especially after Yakobson et al. [9] showed that the results obtained using continuum mechanics modeling and molecular dynamics (MD) simulations are in good agreement.

A comprehensive molecular dynamics (MD) study for the contraction and thermal expansion behaviors on different mode of vibration analysis carried by Cao et al. [10]. The fundamental frequency for deformed clamped-clamped shift of SWCNTs under torsion, bending and axial loadings investigated by them. Lordi and Yao [11] performed molecular dynamic (MD) simulations to determine the Young's modulus and thermal vibration frequencies of SWCNTs using the universal force field with various clamped-free conditions based on the Euler beam theory. Carbon nanotubes model of chiral SWCNTs for analyzing their resonant frequency are developed by Hsu et al. [12] and these tubes were observed under a thermal vibration. The model used for implicating the shear deformation and rotatory inertia was Timoshenko beam model. Chawis et al. [13] and Bocko et al. [14] used nonlocal theory of elasticity for the vibration analysis of SWCNTs. An analysis of vibration characteristics of SWCNTs was examined by Yang et al. [15] and initiated this analysis is based on Timoshenko beam model for nonlocal theory. A number of end conditions have interpreted by Azrar et al. [16–17] for the vibrations of these tubes. Recently, vibration behaviors of SWCNTs have been investigated by some researchers [18, 19, 20].

To examine the feasibility of SWCNTs as a nano-resonator, the molecular structural mechanics method was employed by Li and Chou [21] .The predicted fundamental frequencies were perceptive to dimensions such as diameter, length along with boundary conditions clamped-free or clamped SWNTs, but the frequencies are correlatively imperceptive to chirality of the tubes. The vibration and buckling aspects of carbon nanotubes using nonlocal Donnell shell theory was examined by Ansari et al. [22, 23].

Vibration analysis of SWCNTs is examined by using the present approach with clamped-clamped and clamped-free vibration. Single-walled carbon nanotubes (SWCNTs) have three distinctive structures as: (i). armchair (ii). zigzag (iii) chiral. These structures have different properties but their vibrational behavior is less clear according to the regarding situation. Vibration analysis of armchair and zigzag type of carbon nanotubes is executed for following boundary conditions: clamped-clamped (C-C), and clamped-free (C-F). Variations of dimensionless frequencies are attained for length-to-diameter ratio.

Here an analytical investigation of single-walled carbon nanotubes is conducted for extracting their vibration characteristics. The study of free vibration of SWCNTs is done based on cylindrical shell model. This analysis based on the Donnell thin shell theory. These shell dynamics equations are solved by wave propagation approach. The Donnell shell theory based on WPA is, therefore, another choice of powerful research technique of CNTs whose results are applicable in the limit of acceptable statistical errors than the earlier used BM and other approaches [3–6, 22, 23]. The shell frequency equation is formulated in the eigenvalue form. To provide the complete characteristic of vibrational behavior of SWCNTs by using wave propagation approach is studied in the present chapter. Results are obtained for various material parameters. The dimensionless frequency is also investigated in armchair and zigzag SWCNTs with in-plane rigidity. Now the gape is that there is no research to find directly the dimensionless frequencies of SWCNTs based on cylindrical shell model by using wave propagation approach. However, to the best of authors' knowledge, to find the frequency of SWCNTs, there is no research works on the vibration analyses of zigzag, armchair SWCNTs based on cylindrical shell using wave propagation approach. These frequencies of the SWCNTs are computed with the aid of the computer software MATLAB and these results are compared against MD simulation results in order to assess the accuracy and validity of the cylindrical shell model for predicting the vibration frequencies of SWCNTs.

2. Theoretical formation

2.1. Cylindrical shell model for the vibration of SWCNT

Carbon nanotubes have two kinds, which are single-walled carbon nanotubes and multi-walled carbon nanotubes. Actually multi-walled carbon nanotubes are singled walled carbon nanotubes that are coaxially interposed with different radii. When a graphene sheet rolled up into one time, then it becomes a SWCNTs to produce a hollow cylinder but with end caps. A schema of graphene sheet and single-walled carbon nanotube are shown in **Figure 1**.

Armchair and zigzag nanotubes are made when chiral angle is equal to 0 and 30 respectively and both are the limiting cases with (m, m) and (m, 0). The structure of single-walled carbon nanotubes is similar to the circular cylinders with regard to geometrical shapes as shown in **Figure 2**. So, the motion equations for cylindrical shells are utilized for studying the free vibrations of SWCNTs. According to the Donnell thin shell theory (He et al. [24]), the governing equation of motion for free vibration of a CNTs is used. Where v_1, v_2, and v_3 are the longitudinal, circumferential, and radial displacements of the shell, R is the radius of the shell, Eh is the in-plane rigidity, ρh is the mass density per unit lateral area, t is the time and ν is the Poisson ratio.

It is assumed that for the representation of the modal deformation displacement functions in the axial, circumferential and radial directions are $v_1(x, \theta, t)$, $v_2(x, \theta, t)$ and $v_3(x, \theta, t)$ correspondingly. The three unknown displacement functions for SWCNTs executing vibration, a system of PDE is given as:

$$\frac{\partial^2 v_1}{\partial x^2} + \frac{1-\nu}{2R^2}\frac{\partial^2 v_1}{\partial \theta^2} + \frac{1+\nu}{2R}\frac{\partial^2 v_2}{\partial x \partial \theta} - \frac{\nu}{R}\frac{\partial v_3}{\partial x} = \frac{(1-\nu^2)\rho h}{Eh}\frac{\partial^2 v_1}{\partial t^2} \tag{1}$$

$$\frac{1+\nu}{2R}\frac{\partial^2 v_1}{\partial x \partial \theta} + \frac{1-\nu}{2}\frac{\partial^2 v_2}{\partial x^2} + \frac{1}{R^2}\frac{\partial^2 v_2}{\partial \theta^2} - \frac{1}{R^2}\frac{\partial v_3}{\partial \theta} = \frac{(1-\nu^2)\rho h}{Eh}\frac{\partial^2 v_2}{\partial t^2} \tag{2}$$

$$\frac{\nu}{R}\frac{\partial v_1}{\partial x} + \frac{1}{R^2}\frac{\partial v_2}{\partial \theta} - \left(\frac{1}{R^2} + \frac{(1-\nu^2)}{Eh}.D\left(\frac{\partial^4 v_3}{\partial x^4} + 2.\frac{1}{R^2}\frac{\partial^4 v_3}{\partial x^2 \partial \theta^2} + \frac{1}{R^4}\frac{\partial^4 v_3}{\partial \theta^4}\right)\right) = \frac{(1-\nu^2)\rho h}{Eh}\frac{\partial^2 v_3}{\partial t^2} \tag{3}$$

where $D = \frac{Eh^3}{12(1-v^2)}$ denotes the effective bending stiffness.

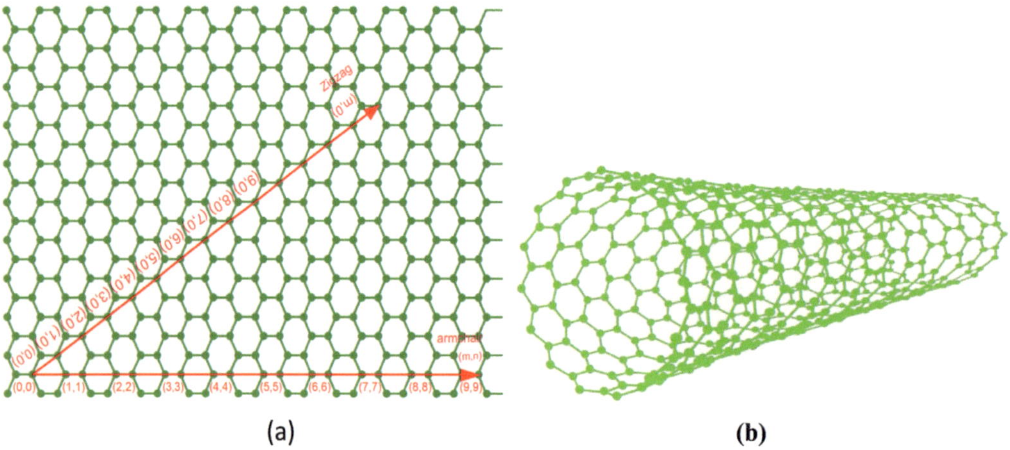

Figure 1. Hexagonal lattice (a) graphene sheet (b) single-walled carbon nanotube.

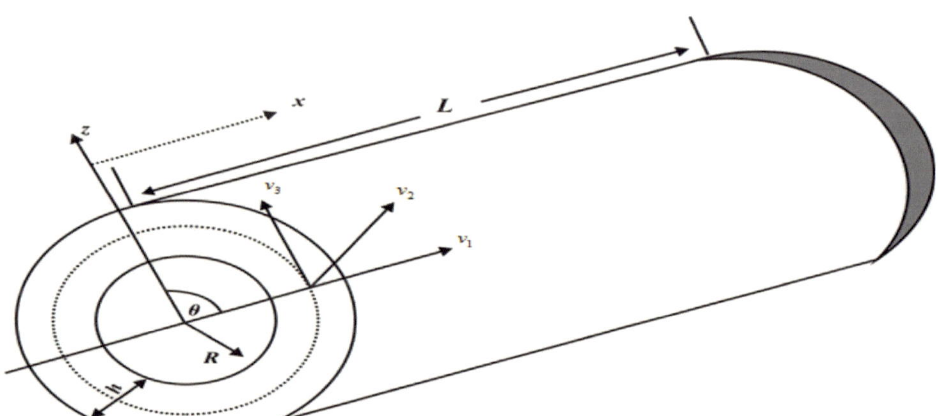

Figure 2. Geometry of SWCNTs.

2.2. Applications of the wave propagation approach

An efficient and a simple technique which corporate as wave propagation approach is employed for the solution of CNT problem in the form of differential equation. Before this, present method has been successively used for the study of shell vibrations [25–27]. The axial coordinate and time variable are denoted by x, t correspondingly and the circumferential coordinate signifies by θ. The functions $v_1(x,\theta,t)$, $v_2(x,\theta,t)$ and $v_3(x,\theta,t)$ are used to designate their respective displacement deformation function. So for modal deformation displacements are written in the assumed expression as:

$$v_1(x,\theta,t) = p_m e^{-ik_m x} \cos(n\theta) e^{\omega t} \tag{4}$$

$$v_2(x,\theta,t) = q_m e^{-ik_m x} \sin(n\theta) e^{\omega t} \tag{5}$$

$$v_3(x,\theta,t) = r_m e^{-ik_m x} \cos(n\theta) e^{\omega t} \tag{6}$$

where p_m, q_m and r_m stand for three vibration amplitude coefficients in the axial, circumferential and radial directions. The axial half and the circumferential wave numbers are denoted by m and n respectively and angular frequency is designated by ω. The formula for fundamental frequency f which is written as: $f = \omega/2\pi$. Where k_m is the axial wave number related with an end conditions. Using the expressions for $v_1(x,\theta,t)$, $v_2(x,\theta,t)$, $v_3(x,\theta,t)$ and their partial derivatives in applying the product method by substituting the modal displacement functions for partial differential equations, the space and time variable are split.

After putting Eqs. (4)–(6), into Eqs. (1)–(3), the above equations is transmuted in matrix representation after the arrangement of terms, and to designate the vibration frequency equation for SWCNTs, an eigenvalue problem is formed:

$$\begin{bmatrix} L_{11} & L_{12} & L_{13} \\ L_{21} & L_{22} & L_{23} \\ L_{31} & L_{32} & L_{33} \end{bmatrix} \begin{pmatrix} p_m \\ q_m \\ r_m \end{pmatrix} = \frac{(1-\nu^2)\rho h}{Eh} \omega^2 \begin{bmatrix} 1 & 0 & 0 \\ 0 & 1 & 0 \\ 0 & 0 & 1 \end{bmatrix} \begin{pmatrix} p_m \\ q_m \\ r_m \end{pmatrix} \tag{7}$$

The form of non-zero solution of (p_m, q_m, r_m) yields the vibration frequency and associated modes for SWCNTs. The expressions for the terms L_{ij} 's are given in the Appendix-I. Where the roots of the equation furnish the frequencies. The lowest root corresponds to the frequency of vibration. It is clear that the frequency should be minimized with respect to the wave numbers m, n in order to obtain the frequency of vibration.

3. Result and discussion

The vibration frequency spectra for SWCNTs are evaluated by Eq. (3) based on Donnell thin cylindrical shell theory. Variations of the frequencies are obtained with regard to the material properties and tube thickness. Keeping in view of this aspect, the natural frequencies of the longitudinal clamped-free vibration of SWCNTs with a length 6.92 nm are first determined by the MD simulation. The E/ρ ratio as measured from molecular dynamic simulation is $3.6481 \times 10^8 m^2/s^2$. By applying this ratio on the longer tube having length 14.4 nm is simulated by MD simulation. The results were found to be same, demonstrating the unconventionality of the ratio on the length. We shall adopt the material properties and tube thickness as suggested by Zhang et al. [18], i.e. the in-plane rigidity Eh =278.25 Gpa·nm, E/ρ=$3.6481 \times 10^8 m^2/s^2$, Poisson's ratio ν = 0.2. The in-plane stiffness or rigidity is computed as Eh = 278.25 GPa.nm which is based on the E/ρ ratio, to calculate the natural and dimensional frequencies of vibration using wave propagation approach, the ratio E/ρ= 3.6481×10^8 m^2/s^2 is used throughout this study. For example, the range of reported thickness is from 0.0612 to 0.69 nm in [9, 28–29] and v varies from 0.14 to 0.34. Considering a diameter $d = 6.86645 \times 10^{-10}$ m, the vibration frequencies for Single-walled carbon nanotubes of various length-to-diameter ratios are calculated using Eq. (3). In present model, the effects of different length-to-diameter ratio for clamped-clamped and clamped-free boundary condition have been considered and matched quantitatively with MD results as well as for the validity and to assure the accuracy. In this study, all frequency results are presented in THz unless otherwise stated. In present study, the frequencies of SWCNTs are obtained by using the some parameters which are compared with MD simulation and continuum shell. However the MD results were obtained for a clamped-clamped SWCNT. Two sets of result are compared as shown in the **Tables 1** and **2**.

It can be observed that the results which are obtained from present model, the values are nearer to the molecular dynamics results when the length-to-diameter ratio is greater than 10.26. From **Table 1**, one can notice that the average percentage error between MD results is approximately 3.6%. This fact shows that the results obtained by present method and earlier MD simulation model are in good agreement.

3.1. Vibration of clamped-clamped and clamped-free SWCNTs

In the application of micro-oscillators and micro or nano-strain sensors, the carbon nanotube sensor is generally clamped with both ends [30]. The clamped-clamped and clamped-free single-walled carbon nanotubes have been performed by atomistic simulations [20–21, 31]. In this section, the distinctive first and third mode frequencies for the set of clamped-clamped single-walled carbon nanotubes is given by present models for their vibration frequencies and

L/d	Frequencies (THz)		
	Present	MD	Percentage error
6.67	0.67832	0.64697	4.85
8.47	0.44146	0.43335	1.87
10.26	0.30922	0.30518	1.32
13.89	0.17360	0.18311	−5.19

Table 1. Comparison of frequencies of C-C SWCNT with MD simulation for the first vibration mode.

L/d	Frequencies (THz)		
	Present	MD	Percentage error
4.67	0.17074	0.23193	−26.38
6.47	0.09048	0.12872	−29.70
7.55	0.06678	0.1000	−31.61
8.28	0.05566	0.07935	−29.85
10.07	0.03777	0.05493	−31.23

Table 2. Comparison of frequencies of clamed-free SWCNT for the first vibration mode.

compared with molecular dynamic simulations and Timoshenko beam model with thicknesses h = 0.34 nm. The results are in good agreement with the MD and Timoshenko beam model results showing same trend in the open literature.

The first and third mode natural frequencies accessed by present model and compared with MD simulations are depicted graphically in **Figure 3** respectively. It can be observed from **Tables 1** and **2** that the length-to-diameter ratio of the set of C-C SWCNTs is somewhat dissimilar from that of C-F SWCNTs. For the prediction of mechanical characters [7, 19–21] from atomistic studies and the experimental studies [5, 6, 33–38] are often used for the clamped-free carbon nanotubes. The frequencies for the first and third modes obtained from present model which is compared with molecular dynamic simulation and Timoshenko beam model are shown in **Figure 3**. It can be readily seen that higher frequencies are produced by higher modes and when length-to-diameter ratio rises at each mode then frequency falls down smoothly as shown in **Figure 3**. The relationship between the length-to-diameter ratios and

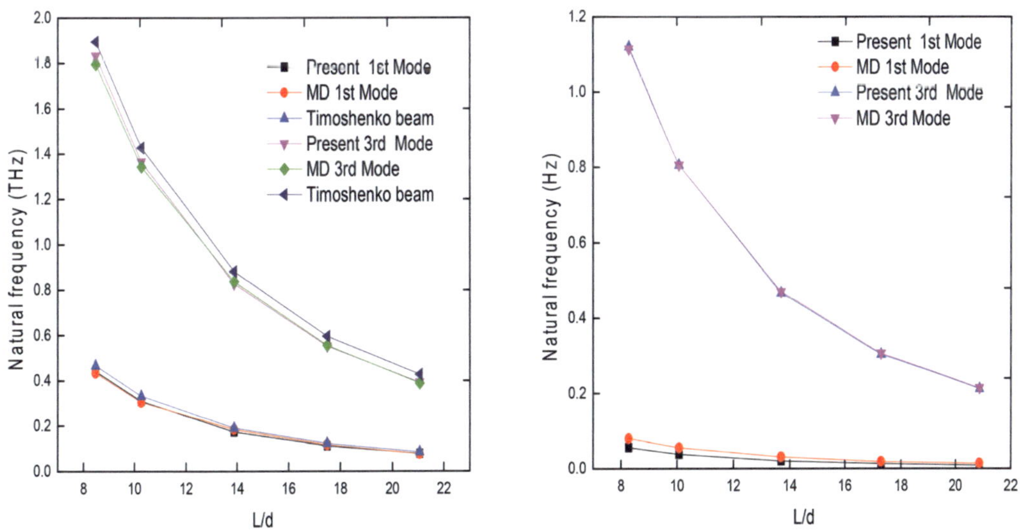

Figure 3. Comparison of numerically obtained results for clamped-clamped and clamped-free frequencies of SWCNTs for first and third mode versus length-to-diameter ratio *L/d* with MD by Cao et al. [32] and Timoshenko beam model [18].

natural frequencies is inversely proportional indicates that the vibrations are very sensitive due to long tube and since the SWCNTs are of almost the same diameter. The results for SWCNTs given by MD are little bit higher than the frequencies investigated by the present model. In MD simulation, the frequencies of length-to-radius ratio are 8.28 is 0.0793 and at 20.89 is 0.0138. But in the present model, the frequencies at 8.28 are 0.05566 and at 20.89 is 0.00883, when compared to the MD results.

3.2. Vibration of SWCNTs with dimensionless frequency

Furthermore, the parametric study for the vibrational behavior of SWCNTs with dimensionless is carried out and presented in **Figure 4**. Alibeigloo et al. [39, 40] and Soldatos et al. [41] used the dimensionless frequency for multi-walled carbon nanotubes and for thin cylindrical shell with respect to length-to-radius ratio respectively. This frequency is associated with frequency Ω through the following formula: $\Omega = \omega R\sqrt{\frac{\rho}{E}}$. A variation of non-dimensional frequency versus length-to-diameter ratio is presented in **Figure 4**. This figure shows that, increasing the value of length-to-diameter ratio as well as there is a decrease in dimensionless frequency. From the physical point of view, it is noted that when the length of SWCNTs becomes small, the effect of the atomic interactions among a reference point and all other atoms becomes significant.

Figure 4 shows an armchair and zigzag CNT, vary the length-to-diameter ratio from 8.3 to 20.9 will change the dimensionless frequency from 0.0385 to 0.0063 THz in case of clamped-clamped boundary condition. Likewise, in clamped-free condition it changes from 1.2827 to 0.5094 THz in armchair case. It may be seen from the above **Figure 4** that the resulting value of dimensionless frequency decreases with the increase in length-to-diameter ratio. Now in

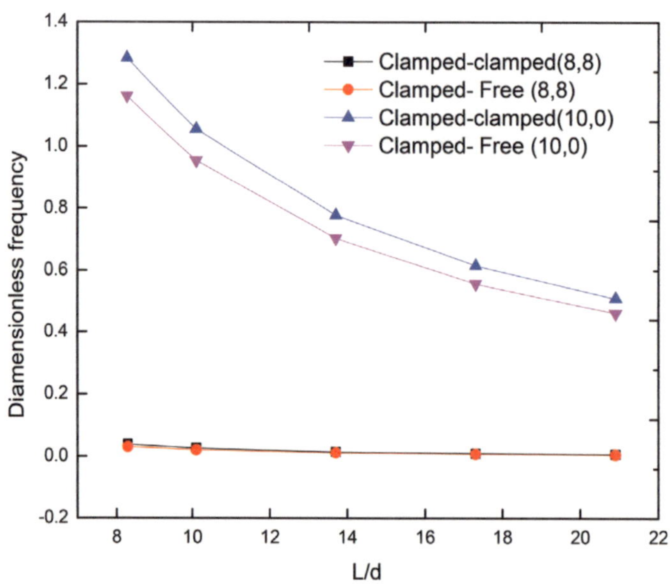

Figure 4. Variations of dimensionless frequencies of CC and CF armchair and zigzag SWCNTs.

zigzag CNT, changing the length-to-diameter ratio from 4.86 to 35.53, the dimensionless frequency changes from 0.0303 to 0.0049 THz in case of clamped-clamped boundary condition. Likewise, in clamped-free condition it varies from 1.16605 to 0.4609 THz.

3.3. Vibration of SWCNTs with in-plane rigidity

For the results generated so far, the nanotube in-plane rigidity has been taken to be *Eh* =278.25 Gpa·nm. However, there exist some inconsistencies concerning this quantity in the literature. The reported CNT in-plane stiffness is largely scattered, ranging from *Eh* =300 Gpa·nm to *Eh* = 400 Gpa·nm [42]. Both set of **Figure 5** is presented to investigate the influence of the in-plane rigidity *Eh* variation on the dimensionless frequency of a (7, 7) armchair and (9, 0) Zigzag SWCNT with different boundary conditions likewise as clamped-clamped and clamped-free boundary conditions. These figures shows that for all the selected boundary conditions, dimensionless frequency calculated via shell model are sensitive to the nanotube in-plane rigidity *Eh* and also the larger the in-plane rigidity in-plane rigidity *Eh*, the higher the dimensionless frequency. The difference is more considerable for shorter length CNTs.

Previous study reveals that the bending rigidity of SWCNTs should be considered as an independent material parameter not linked to the representative thickness by the classic bending rigidity formula, i.e., $D = Eh^3/12\left(1-\nu^2\right)$ and the actual bending rigidity of SWCNTs is lesser than its classical counterpart [43, 44]. For shorter length-to-diameter ratio, the value of dimensionless frequencies for clamped-clamped at *Eh* = 300 Gpa·nm, *Eh* = 400 Gpa·nm is 0.04863, 0.05788, respectively which shows that a slight increase in frequency due to increase of in-plane rigidity *Eh*. Same trend is observed for dimensionless frequency. For the present shell model with in-plane rigidity *Eh*, the values of the C-F single-walled carbon nanotubes respectively, which are a little lower than those of corresponding CC SWCNTs with bending rigidity are plotted in **Figure 5**.

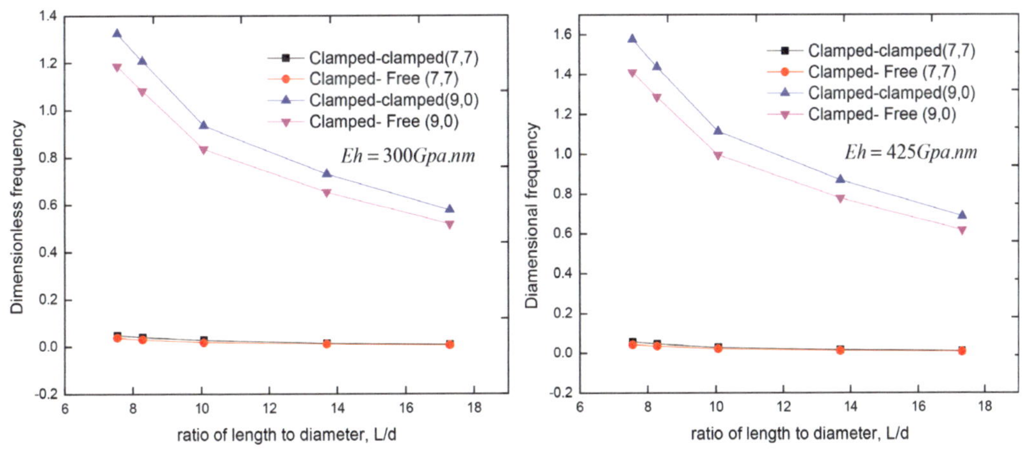

Figure 5. Variations of dimensionless frequencies Ω of CC and CF armchair and zigzag SWCNTs when *Eh* = 300 GPa.Nm and *Eh* = 425Gpa.Nm.

3.4. Vibration with mass density per unit lateral area

Figure 6 is presented to investigate the influence of the mass density per unit lateral area variation on the dimensionless frequency of a (12, 12) armchair and (14, 0) zigzag SWCNT with boundary conditions: clamped-clamped and clamped-free. These figures shows that for all the selected boundary conditions, the frequency calculated via shell model are sensitive to the nanotube mass density and also the larger the mass density per unit lateral area ρh, lower the frequency. It is observed that applying the mass density per unit lateral area ρh to the present shell model, yields the slight decrease of the frequency. For shorter length-to-diameter ratio, the value of dimensionless frequencies at ρh =740.52 nm, 800.64 nm and 820.80 nm is 0.3667, 0.03342, 0.03128 respectively for clamped-clamped and 1.74285, 1.71298, 1.6001 respectively for clamped-free, which shows that decreases in frequency. For the present shell model, the values of length-to-diameter ratio for C-C SWCNTs, which are a little higher than those corresponding C-F SWCNTs values with mass density per unit lateral area are plotted in **Figure 6**.

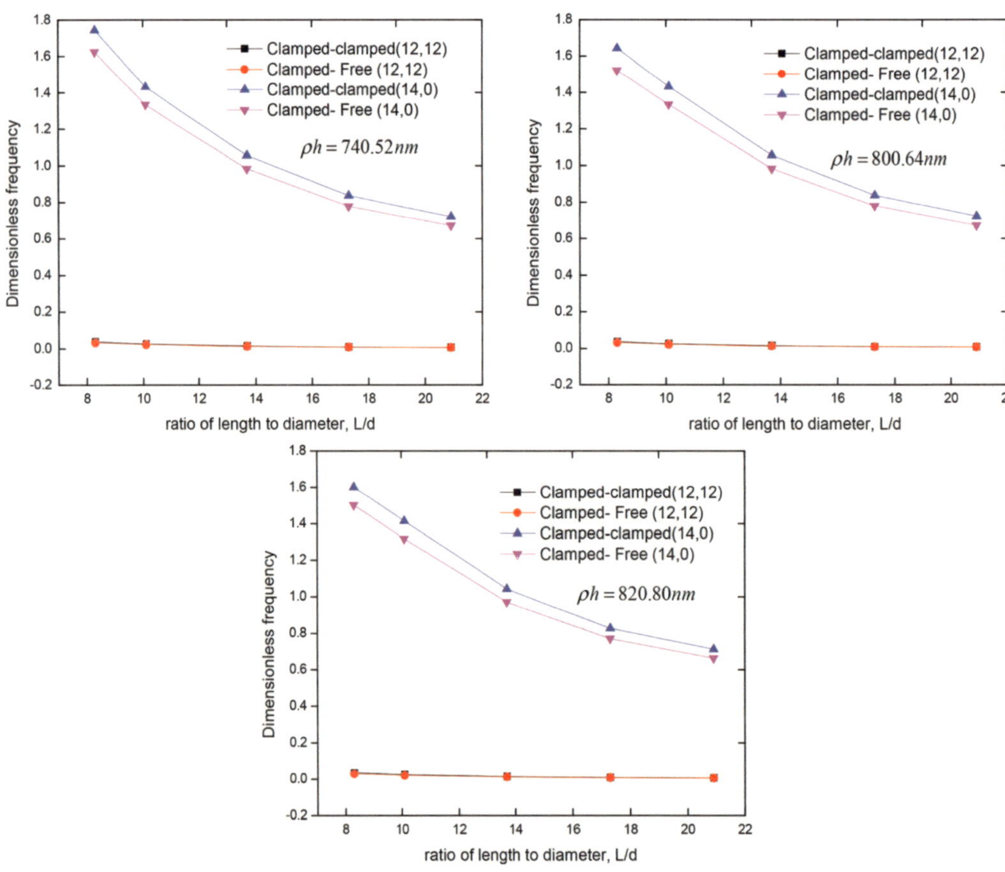

Figure 6. Variations of dimensionless frequencies of CC and CF armchair and zigzag SWCNTs when $\rho h = 740.52nm$, $\rho h = 800.64nm$ and $\rho h = 820.80nm$.

http://dx.doi.org/10.5772/intechopen.73503

4. Conclusions

The vibration behavior of CF and CC SWCNTs are extensively investigated by present model compared with MD simulation. With properly chosen parameters, the present models can reproduce satisfactory frequencies that are in reasonable agreement with those results obtained by MD simulations and Timoshenko beam model. The effects of the length-to-diameter ratio for armchair and zigzag CNTs with in-plane rigidity, mass density per unit lateral area on the dimensionless frequencies are also examined with present models. It is found that the frequencies decreases smoothly when length-to-diameter ratio would increases and higher mode of vibration occurred when the frequencies are higher. For a clamped-free SWCNT, their exist an inverse proportionality which is observed between the resulting frequency and length-to-diameter ratio. For clamped-clamped SWCNTs, the results took a similar trend but in this case frequency values are much higher. The results are obtained numerically for different boundary conditions and plotted in graphical forms. In the field of CNTs vibrations, wave propagation approach presents a good application. A better cylindrical shell model is needed to furnish more accurate prediction of the vibration frequencies of SWCNTs, such as the nonlocal shell theory that incorporates the effect of small length scale effect.

Appendix 1

$$L_{11} = k^2{}_m + \tfrac{1-v}{2R^2} n^2,\ L_{12} = ik_m \tfrac{1+v}{2R} n,\ L_{13} = ik_m \tfrac{v}{R},\ L_{21} = -n \tfrac{1+v}{2R} ik_m,\ L_{22} = \tfrac{1-v}{2} k^2{}_m + \tfrac{n^2}{R^2}.$$

$$L_{23} = \tfrac{n}{R^2},\ L_{31} = -\tfrac{v}{R} ik_m,\ L_{32} = \tfrac{n}{R^2},\ L_{33} = \tfrac{1}{R^2} + \tfrac{(1-\nu^2)}{Eh} D\left(k^4{}_m + 2\tfrac{1}{R^2} k_m{}^2 n^2 + \tfrac{n^4}{R^4}\right).$$

Author details

Muzamal Hussain* and Muhammad Nawaz Naeem

*Address all correspondence to: muzamal45@gmail.com

Department of Mathematics, Government College University Faisalabad (GCUF), Pakistan

References

[1] Iijima S. Helical microtubules of graphitic carbon. Nature. 1991;**345**:56-58

[2] Falvo MR, Clary GJ, Taylor RM II, Chi V, Brooks FP Jr, Washburn S, Superfine R. Bending and buckling of carbon nanotubes under large strain. Nature. 1997;**389**:532-534

[3] Li CY, Chou T. A structural mechanics approach for the analysis of carbon nanotubes. International Journal of Solids and Structures. 2003;**40**:2487-2499

[4] Sakhaee - Pour A, Ahmadian MT, Vafai A. Vibrational analysis of single-walled carbon nanotubes using beam element. Thin-Walled Structures. 2009;**47**:646-652

[5] Poncharal P, Wang ZL, Ugarte D, De Heer WA. Electrostatic deflection and electromechanical resonators. Science. 1999;**283**:1513-1516

[6] Treacy MMJ, Ebbesen TW, Gibson JM. Exceptionally high Young's modulus observed for individual carbon nanotubes. Nature. 1996;**381**:678-680

[7] Zhao Q, Gan ZH, Zhuang OK. Electrochemical sensors based on carbon nan-otubes. Electroanalysis. 2002;**14**:1609-1613

[8] Wang CM, Ma YQ, Zhang YY, Ang KK. Buckling of doublewalled carbon nanotubes modelled by solid shell elements. Journal of Applied Physics. 2006;**99**:114317

[9] Yakobson BI, Brabee CJ, Bernhole J. Nanomechanics of carbon tubes: Instability response. Physical Review Letters. 1996;**76**:2511-2514

[10] Cao GX, Chen X, Kysar JW. Strain sensing of carbon nanotubes: Numerical analysis of the vibrational frequency of deformed sing-wall carbon nanotubes. Physical Review B. 2005; **72**:195412

[11] Lordi V, Yao N. Young's modulus of single-walled carbon nanotubes. Journal of Applied Physics. 1998;**84**:1939-1943

[12] Hsu JC, Chang RP, Chang WJ. Resonance frequency of chiral single –walled carbon nanotubes using Timoshenko beam theory. Physics Letters A. 2008;**372**:2757-2759

[13] Chawis T, Somchai C, Li T. Non local theory for free vibration of single walled carbon nanotubes. Advanced Materials Research. 2013;**747**:257-260

[14] Bocko J, Lengvarský P. Vibration of single-walled carbon nanotubes by using nonlocal theory. American Journal of Mechanical Engineering. 2014;**2**(7):195-198

[15] Yang J, Ke LL, Kitipornchai S. Nonlinear free vibration of single-walled carbon nanotubes using nonlocal Timoshenko beam theory. Physica E: Low-dimensional Systems and Nanostructures. 2010;**42**:1727-1735

[16] Azrar A, Azrar L, Aljinaidi AA. Length scale effect analysis on vibration behavior of single walled carbon nanotubes with arbitrary boundary conditions. Revue de Mécanique Appliqué et Théorique. 2011;**2**(5):475-484

[17] Azrar A, Azrar L, Aljinaidi AA, Hamadiche M. Dynamics instability analysis of multi-walled carbon nanotubes conveying fluid. Advanced Materials Research. 2013;**682**:153-160

[18] Zhang YY, Wang CM, VBC T. Assessment of Timoshenko beam models for vibrational behavior of single–walled carbon nanotubes using molecular dynamics. Advances in Applied Mathematics and Mechanics. 2009;**1**(1):89-106

[19] Li CY, Chou TW. Vibrational behaviors of multiwalled-carbon-nanotube-based nanomechanical resonators. Applied Physics Letters. 2003;**84**:121-123

[20] Strain and pressure sensing using single-walled carbon nanotubes. Nanotechnology. 2004; **15**:1493-1496

[21] Li CY, Chou TW. Single-walled carbon nanotubes as ultrahigh frequency nano-mechanical resonators. Physical Review B. 2003;**68**:073405

[22] Ansari R, Rouhi H, Sahmani S. Calibration of the analytical nonlocal shell model for vibrations of double-walled carbon nanotubes with arbitrary boundary conditions using molecular dynamics. International Journal of Mechanical Sciences. 2011;**53**:786-792

[23] Ansari R, Sahmani S, Rouhi H. Rayleigh – Ritz axial buckling analysis of single-walled carbon nanotubes with different boundary conditions. Physics Letters A. 2011;**375**:1255-1263

[24] He XQ, Eisenberger M, Liew KM. The effect of van der waals interaction modelling on the vibration characteristics of multiwalled carbon nanotubes. Journal of Applied Physics. 2006;**100**:124317

[25] Sun S, Cao D, Chu S. Free vibration analysis of thin rotating cylindrical shells using wave propagation approach. Archive of Applied Mechanics. 2013;**83**:521-531

[26] Hussain M, Naeem MN. Vibration analysis of single-walled carbon nanotubes using wave propagation approach. Mechanical Sciences. 2017;**8**(1):155-164

[27] Li X. Study on free vibration analysis of circular cylindrical shells using wave propagation. Journal of Sound and Vibration. 2008;**311**:667-682

[28] Vodenitcharova T, Zhang LC. Length scale effect analysis on vibration behavior of single walled carbon nanotubes with arbitrary boundary conditions. Physical Review B. 2003; **68**:165401

[29] Hernandez E, Goze C, Bernier P, Rubio A. Elastic properties of single-wall nanotubes. Applied Physics A Materials Science & Processing. 1999;**68**:287

[30] Hauptmann P. Resonant sensors and applications. Sensors and Actuators, A: Physical. 1991;**26**:371-377

[31] Cao GX, Chen X, Kysar JW. Strain sensing of carbon nanotubes: Numerical analysis of the vibrational frequency of deformed sing-wall carbon nanotubes. Physical Review B. 2005; **72**:195412

[32] Cao GX, Chen X, Kysar JW. Thermal vibration and apparent thermal contraction of single-walled carbon nanotubes. Journal of the Mechanics and Physics of Solids. 2006;**54**: 1206-1236

[33] Purcell ST, Vincent P, Journet C, Binh VT. Tuning of nanotube mechanical resonances by electric field pulling. Physical Review Letters. 2002;**89**:276103

[34] Duan WH, Wang CM, Zhang YY. Calibration of nonlocal scaling effect parameter for free vibration of carbon nanotubes by molecular dynamics. Journal of Applied Physics. 2007; **101**:024305

[35] Li C, Chou T-W. Modeling of carbon nanotubes and their composites. In: Nanomechanics of Materials and Structures. Dordrecht: Springer; 2006. pp. 55-65

[36] Gibson RF, Ayorinde EO, Wen YF. Vibrations of carbon nanotubes and their composites: A review. Composites Science and Technology. 2007;**67**:1-28

[37] Swain A, Roy T, Nanda BK. Vibration behavior of single walled carbon nanotube using finite element. International Journal on Theoretical and Applied Research in Mechanical Engineering. 2013;**2**(4):129-133

[38] Huang Y, Wu J, Hwang KC. Thickness of graphene and single-wall carbon nanotubes. Physical Review B. 2006;**74**:245413

[39] Alibeigloo A. Three-dimensional free vibration analysis of multi-layered graphene sheets embedded in elastic matrix. Journal of Vibration and Control. 2012

[40] Alibeigloo A, Shaban M. Free vibration analysis of carbon nanotubes by using three dimensional theory of elasticity. Acta Mech. 2013;**224**:1415-1427

[41] Soldatos KP, Hadjigeorgiou. Three-dimensional solution of the free vibration problem of homogeneous isotropic cylindrical shells and panels. Journal of Sound and Vibration. 2009;**137**:369-384

[42] Wang CY, Zhang LC. A critical assessment of the elastic properties and effective wall thickness of single-walled carbon nanotubes. Nanotechnology. 2007;**19**:075705

[43] Ru CQ. Effective bending stiffness of carbon nanotubes. Physical Review B. 2008. (2000); **62**:9973-9976

[44] Ansari R, Rouhi H. Nonlocal Flugge shell model for the axial buckling of single walled carbon nanotubes: An analytical approach. International Journal of Nano Dimension. 2015;**6**(5):453-462

Chapter 6

Nanosynthesis Techniques of Silica-Coated Nanostructures

Kwok Wei Shah

Additional information is available at the end of the chapter

http://dx.doi.org/10.5772/intechopen.74097

Abstract

Core-shell nanomaterials are fast-emerging hybrid nanocomposites in area of nanotechnology, materials science and biochemistry, which are fast attracting research attention. Nanostructured nanomaterials are utilized in wide industry fields such as electronics, biopharmaceutical, biomedicine, optics and biocatalysis. Owing to the additional exterior shell-coating material, the primary core material's functionality, biocompatibility, chemical stability and colloidal dispersibility can be greatly enhanced. Silica, in particular, is found to be an excellent exterior shell-coating material, has been widely researched for the synthesis of core-shell nanocomposite materials. So far, there have been numerous publications devoted to silica-coating techniques using hydrophobic silanes, such as tetraethylorthosilicate (TEOS) or tetramethyl orthosilicate (TMOS), via the classic Stober method. Recently, there has been strong interest in the use of water-soluble silanes such as MPTMS (3-(mercaptopropyl)-trimethoxysilane), MPTES (3-(mercaptopropyl-triethoxysilane), MTMS (3-(methyltrimethoxysilane)) and sodium silicate for water-based silica-coating techniques have gained much attention, due to the fast-growing need to focus on process simplicity, large-scale fabrication and environmental-friendly synthesis techniques of silica-based core-shell nanomaterials. Hence, this chapter focuses on the recent development on silica-coating techniques for colloidal nanoparticles, particularly on water-based techniques and morphologies. In summary, we emphasize the importance of advanced nanomaterials in today's world and envisage there will be more breakthrough research on aqueous silica-coating techniques for silica-encapsulated core-shell nanomaterials.

Keywords: silica-shell coatings, core-shell nanomaterials, aqueous one-pot synthesis

1. Introduction

Core-shell nanomaterials are considered as highly functional hybrid nanomaterials with dual properties, originating from either core or shell materials. Core-shell nanomaterials have been widely researched due to its extraordinary ability to exhibit distinctive properties of both core and shell materials combined to deliver a wide spectrum of industry applications and requirements. Core-shell nanomaterials are commonly applied in different industries such as biomedical, bio-electronics, pharmaceutical applications, bio-catalysis, photoluminescence imaging, creating photonic crystals, etc. Especially for bioscience and medical research, the core-shell particles are mainly used for cell detection, bioimaging, targeted drug delivery, controlled drug release and bioengineering applications [1].

The benefits of an exterior shell-coating on a core particle are that surface reactivity and thermal stability can be greatly enhanced, while colloidal stability and dispersibility are vastly improved. Core-shell nanostructure itself can allow controlled release of center material, reduced usage of precious materials, hollow-core templating, etc. Hollow particles can be prepared using as a template by utilizing core-shell nanoparticles and after removing the core either by dissolution or calcination. These hollow-core nanoparticles are useful for various applications such as biocatalytic agent, biosupports, bioadsorbents, super light-weight structure, miniature vessels and thermal/electric insulative materials [1].

Silica is one of the most widely researched as exterior shell-coating material for encapsulating all sorts of core nanomaterials. Silica coating of nano-sized metals, semiconductors, magnetic and ceramic nanomaterials can enhance its large surface area, quantum confinement and photocatalytic, optical and magnetic properties. Silica is often chosen as the ideal shell material to encapsulate core nanomaterials in order to modify its exterior surface properties. Silica is also a favorite choice for core-shell nanostructures to enhance colloidal stability. The main advantages of silica as an excellent candidate for shell-encapsulation material predominantly lies in its exceptionally good colloidal stability, especially in water-based media, easy controllability of the synthesis, chemical inertness, controlled porosity, high processability and optical transparency. Silica possesses other advantages such as (i) lower Van der Waals interactions compared to bare core particles (Hamaker constant is smaller for core-shell) and (ii) charged molecules that can be grafted onto silica shell at silica/water interfaces under alkaline conditions. Thus, silica shell can provide effective steric-hindrance and electrostatic protection on its core material, as well as function as excellent dispersant to provide electrostatically stable colloids. Furthermore, silica shell allows core nanoparticle to be biocompatible and easily bioconjugated with functional groups, which is necessary for biomedical, diagnostic and therapeutic applications (magnetic resonance imaging, MRI) [2].

However, the simple large-scale production and environmental-friendly fabrication of silica-coated core-shell nanomaterials currently remain a great challenge/bottleneck for practical commercialization [3]. The effective growth of homogenous silica shells on core nanoparticles faces problems of low chemical compatibility between both components [4]. Most existing silica-coating methods are dependent upon prior priming the surface of core material with coupling agents, surfactants or polymers. The purpose of surface priming is to increase the compatibility of the core surface with silica, while providing particles with sufficient colloidal stability, so that colloids can be transferred into alcohol-based medium and classical Stober method [5] (using hydrophobic TEOS or TMOS in ethanol-water medium under alkaline conditions) is applicable for growing uniform silica shells on metal cores. A seminal publication on silica-coating gold nanoparticles was published by Liz-Marzán et al. [6]. Liz-Marzán et al. proposed silica-coating gold colloids in ethanol medium using a modified Stöber process. To achieve this, the authors surface-primed the nanoparticles with ATPMS before stabilizing the gold colloids by silica coating with a thin silica layer using sodium silicate in water medium, so that they can be stably transferred into ethanol medium. Since Liz-Marzán's pioneering work in 1996, many research studies have been carried out to simplify and shorten the tedious multiple steps procedure [7].

Recently, there has been an increasing number of studies on aqueous synthesis techniques of silica-coating nanoparticles due to its process simplicity, large scalability and environmental friendliness. The key advantage of synthesizing silica in aqueous solution will be that the problem of poor colloidal stability of core nanoparticles in alcohol mediums, due to reduced surface charges, can be resolved and, thus, obviates any need to perform tedious surface stabilization procedures [7]. Essentially, any aqueous synthetic route will allow users to overcome a key challenge of colloidal instability faced by existing silica-coating techniques, which occurs when metal colloids are transferred into organic solvents such as ethanol or isopropanol for silica coating via classic Stober synthesis techniques.

Water-based techniques enable scale-up production of silica-coated nanomaterials by harnessing the excellent colloidal stability of water-soluble nanoparticles in aqueous medium. The water solubility of silane precursors presents a highly attractive property to directly coat nanoparticles with silica in few simpler steps. Another advantage of an aqueous synthesis is that it promotes green chemistry in material synthesis through the use of inexpensive, non-toxic and ecologically friendly solvents such as water, in place of toxic solvents such as ethanol or isopropanol. By adopting a water based silica coating strategy, users can stably produce large quantities of core-shell nanostructures via simpler one-pot synthetic route. Essentially, by using water-soluble silane precursors, a water-based technique will allow fast and extensive silica coating of nanoparticle cores within a shorter time; without any need for transfer into alcoholic medium, dialysis or a change in reaction mixture, which are necessary for existing techniques such as Stöber process or reverse microemulsion techniques. A general synthetic route promotes green chemistry through sustainable material synthesis through the use of an inexpensive, non-toxic and ecologically friendly water medium, in place of organic toxic solvents such as ethanol or isopropanol. Furthermore, another important advantage of aqueous silica coating is that it is relatively neutral pH and lack of alcohol, thus being more suitable for encapsulating sensitive biomaterials such as enzymes and cells [8]. Aqueous silica-coating techniques are relatively straightforward processes that are less toxic, more cost-efficient and more environment-friendly. Hence, aqueous silica-coating techniques are not only less complex but also highly upscalable for commercial/industrial productions. These are the key reasons that motivated this chapter into summarizing the latest developments of new water-based techniques for direct silica coating of core nanoparticles.

	Pros	Cons
"Aqueous-based" silica-coating techniques	1. Facile one-step synthesis 2. Precisely controllable shell thickness 3. Shell growth is generally faster than non-aqueous routes 4. Functional groups may be intrinsically incorporated 5. Mild and less toxic, environmentally friendly 6. Most cost-effective and upscalable technique	1. May still require a small amount of acidic and/or basic catalysts 2. May require new surface functional groups. 3. Relatively new field, hence limited literature available. Generally applicable to hydrophilic core-materials only
"Solvent-based" silica-coating techniques	1. Shell thickness adjustable through reactant conditions 2. Extensively researched and mature technology 3. Mild, ambient temperature, low-pressure conditions 4. Nanometric resolution control of core/shell morphology 5. Structures and morphologies easily reproducible and tunable 6. Generally fast hydrolysis and condensation of silica	1. Complex reactions, multiple precursors, solvents, controls 2. Multiple reactants increase costs, time and environmental pollution, limited scale 3. Alcoholic and extreme pH conditions are cytotoxic 4. Processes to remove alcohol require evaporation and buffer solutions, increasing time, cost and complexity 5. Need surface functionalisation

Table 1. Pros & cons of aqueous and non-aqueous silica-coating routes.

Since water-based silica-coating of core-shell nanoparticles is a fast broadening field that is presently seeing remarkable developments, we summarized a collection of up-to-date work related to a variety of water-based silica core-shell nanomaterials and their synthesis techniques. In this chapter, four main types of aqueous synthesis routes related to water-soluble silane precursors have been widely used in recent publications, namely sodium silicate, MPTMS (3-(mercaptopropyl)-trimethoxysilane), MPTES (3-(mercaptopropyl-triethoxysilane) and MTMS (3-(methyltrimethoxysilane)). These silane precursors possess high water solubility. For example, the alkoxy group of MPTMS has been replaced with a sufficiently polar group. The substitution of the methoxy group in water-insoluble tetramethyl orthosilicate (TMOS) with a mercaptopropyl group ('MP') results in MPTMS that can be hydrolyzed in water prior to condensation to form core-shell structures at the nanoscale.

Table 1 summarizes the pros and cons of aqueous silica coating (using water-soluble silanes, namely sodium silicate, MPTMS, MPTES and MTMS) and solvent-based silica coating (using water-insoluble silanes, such as TEOS), which will be discussed in subsequent sections.

2. Sodium silicate-assisted silica coating

Silanization of diverse metal-core and semiconductor-core nanomaterial into core-shell systems has gathered much research attention due to enhancement of surface properties. Silica-shell encapsulation was useful in inhibiting particle growth, agglomeration and photo-induced decomposition. For example, synthesizing silica coatings encapsulating gold nanoparticles with Raman-active dye molecules was proven to be useful as a surface-enhanced Raman scattering (SERS) nanoprobes. Inert and biocompatible silica-shell coatings are useful for bioconjugation of quantum dots (QDs). Silica core-shell inhibits agglomeration of nanoparticles, hinders foreign biospecies from contaminating the nanoparticles' surface and aids to enhance the high photoluminescence for biosensing. Surface functionalisation of the silica outer shell with any various types of functional groups, such as carboxylate COOH, amine NH_2, phosphate, thiol SH and poly(ethyleneglycol) (PEG) groups, enables enhanced control in bioconjugation techniques [9].

A primary work on silica-coating gold nanoparticles was published by Liz-Marzán et al. [6] in 1996. The authors proposed silica-coating gold colloids in ethanol-based medium using a modified Stober process [5]. First, the gold colloidal surface must be activated with a surface-coupling agent, 3-aminopropyltrimethoxysilane (APTMS), to render the gold nanoparticle surface vitreophillic (that is receptive toward silica monomers or oligomers) before proceeding to coat them with water-soluble sodium silicate due to the low affinity between gold nanoparticles and sodium silicate. After APTMS treatment, the vitreophilic Au nanoparticles are coated with a thin inhomogenous silica shell (thickness < 2 nm) using sodium silicate in water at a low pH to stabilize the gold colloids, so that they can be stably transferred into alcoholic medium (**Figure 1**). However, the critical step of silanization with sodium silicate is poorly reproducible, which requires an extended time duration, typically from a few days to weeks, before a sufficient layer silica shell can be formed to stabilize the nanoparticles in the alcohol solution [10]. Using Stober process as the last step, Liz-Marzán coated a thicker homogenous silica layer (thickness > 5 nm) in ethanol-water mixture using silane precursor TEOS and the silanol groups provided by sodium silicate as anchor points.

Zhou et al. in 2005 [11] synthesized, based on a simple chemical precipitation, nano-sized CdSe quantum dots at normal temperature in slightly alkaline conditions. Also, silica encapsulation onto these QDs via a slow polymerization of SiO_2 using sodium silicate is done by adding ethanol. Cadmium selenide (CdSe) quantum dots were thinly encapsulated with a 1–2 nm thin silica in an aqueous sodium silicate at pH 10.5 over 5 days of vigorous stirring (**Figure 2**). However, despite the absence of aggregates proving improvement to dispersity, the coated particles were hardly monodispersed and without a distinctive core-shell

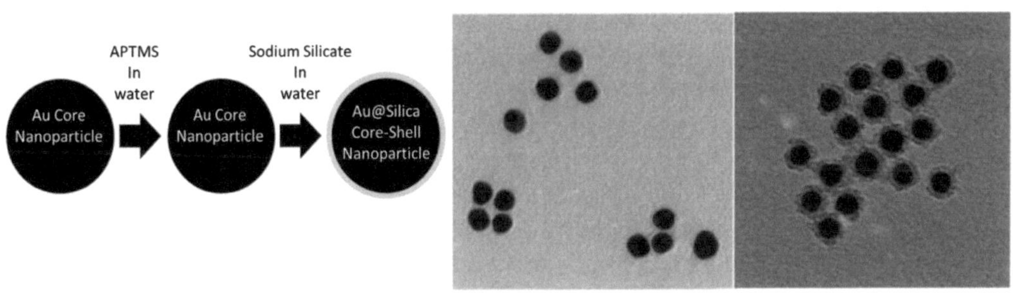

Figure 1. (Left) Schematic of aqueous synthesis procedure used by Liz-Marzán to coat thin silica shell on Au nanoparticles. (Right) TEM images of 15 nm-sized gold nanoparticles coated with 2 nm silica shells after addition of sodium silicate [6].

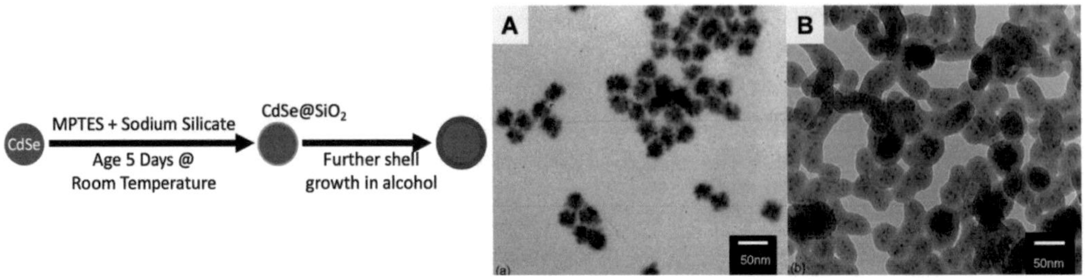

Figure 2. (Left) Schematic of aqueous synthesis procedure used by Zhou et al. to coat thin silica shell on CdSe nanoparticles using sodium silicate. (Right) TEM of CdSe NPs before (A) and after (B) encapsulation [11].

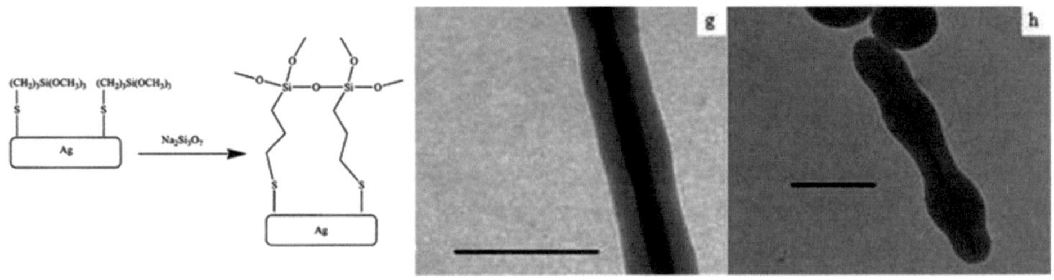

Figure 3. (Left) Schematic of 3-mercaptopropyltrimethoxysilane is binded via the sulfur to the nanowire surface with Ag, leaving three free $–OCH_2$ methoxy groups facing the water medium. Addition of sodium silicate produces silica, SiO_2. (Right) TEM images of Ag nanowires and silica-encapsulted Ag nanowires. Scale bar is 500 nm (g) and 100 nm (h). For thick layers of silica shells, the coating becomes inconsistent (h) [12].

structure. The authors attributed the success of silica-coating to the mercapto-groups present in the MPTMS, which are used to bridge the low affinity between CdSe core and silica shell. This principle would be repeated by Wolcott et al. [9] in the following year in their reported attempt to coat cadmium telluride (CdTe) quantum dots as they primed the CdTe nanoparticles with MPTMS prior to gradual silica growth under stirring over 72 h in aqueous sodium silicate solution.

Hunyadi et al. in 2006 [12] synthesized silver nanowires coated with a silica shell. Simona uses an indirect coating method that involved the use of a bifunctional linker, $(HS(CH_2)3Si(OCH_3)3$, 3-mercaptopropyltrimethoxysilane), so that the thiol SH groups bind Ag to the core surface [12]. Simona's approach is essentially seedless, surfactantless approach. Different volumes of sodium silicate $Na_2O(SiO_2)_3$ were added in alkaline pH and left overnight to allow formation of well-controlled layers of silica. The silica thickness was precisely controlled from $\sim$10 to $\sim$150 nm. The uniform silica coating over the whole surface of the nanowire is subjected to initial sodium silicate concentrations. Simona also found that MPTMS is critical to make a uniform layer of silver. When the MPTMS dosage exceeded 8.08 mM, the silica shell became undulating and inhomogenous (**Figure 3**). The reference experiment to synthesize SiO_2 on silver nanowires without the silane primer proved that the maximum shell thickness achieved was 5–10 nm only, regardless of the starting dosage of the silicate precursor. This outcome proves that silane performs a critical "surface-primer" function at interface of core-shell materials in the encapsulation process as a molecular binder.

http://dx.doi.org/10.5772/intechopen.74097

Kobayashi et al. in 2007 [13] carried out silica encapsulation of nano-sized copper prepared from water-based solutions of copper salts. The silica shells chemically stabilized Cu colloids. Hydrazine and citric acid were used to reduce Cu salts and CTAB surfactants to stabilize 50 nm Cu metallic core particles. Then, silane-coupling agent, ATPMS, is added for surface functionalisation, and then Cu nanoparticles were coated with sodium silicate. $Cu@SiO_2$ core-shell nanoparticles are synthesized with ~10 nm silica layer. The mixed solution was left for 15 min for APTMS to attach on the surface of the core particles. Then, sodium silicate was added to the colloid, followed by tuning to pH 10 using CTAB resin for cation exchange. The synthesis time for the silica shell was 24 h. The key problem for using Cu-based particles is their fast oxidation in air environment. This arises from their instability toward oxygen to form CuO. To improve stability and reduce oxidation of copper nanoparticles, surfactant capping surrounding the copper particles inhibits oxygen molecules found inside the water-based medium from reacting with the copper. Previously, Kobayashi performed chemical stabilization by silica protection of cobalt nanoparticles prepared through reducing of Co salts in water-based medium. The silica shell forms a homogenous protection layer habiting O_2 molecules from reacting with the metallic cobalt nanoparticles. The nanoparticles were stable and chemically unchanged under high temperature in O_2 environment. This method has benefits such as (i) extraordinary colloidal stable dispersions in water, (ii) facile surface functionalisation to synthesize colloids for non-aqueous mediums and (iii) facile control of particle-to-particle interactions.

Hu et al. in 2009 [14] synthesized $Fe_3O_4@SiO_2$ magnetic composite nanospheres for fast and recyclable removal of lead and mercury from water (**Figure 4**). The $Fe_3O_4@SiO_2$ nanospheres were synthesized with the dissolution of sodium silicate in deionized water, into which Fe_3O_4 was added. HCL was added as a catalyst to adjust pH to 6.0 and the mixture stirred for 3 h at 80°C. A thin layer of silica coating visibly encapsulates the iron oxide spheres; the deposition of silica was also confirmed through XRD.

Wang et al. in 2010 [15] demonstrated a slightly different route starting with a suspension of Fe_3O_4 in deionized water at 80°C under a nitrogen flow, the pH was adjusted to 6.0 with

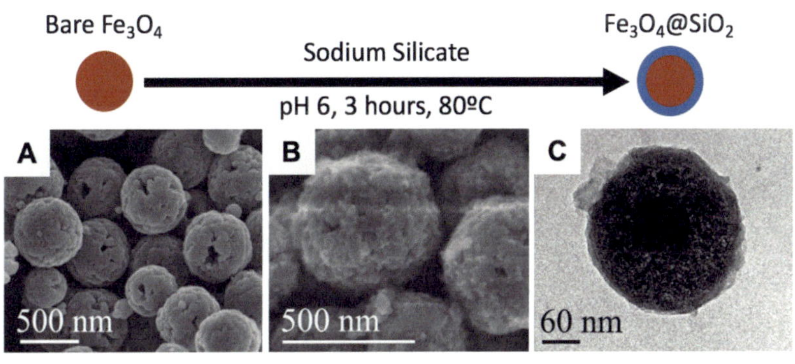

Figure 4. (Top row) Schematic of aqueous synthesis procedure used by Hu et al. to coat thin silica shell on Fe_3O_4 nanoparticles using sodium silicate. (Bottom row) SEM images of bare Fe_3O_4 microspheres (A, B) and (C) TEM images of $Fe_3O_4@SiO_2$ composite microparticles [14].

hydrochloric acid. Sodium silicate precursor was then added dropwise under strong stirring. The coating process was carried out for 5 h and the $Fe_3O_4@SiO_2$ nano particles collected via magnetic separation (**Figure 5**). In both Hu's and Wang's experiments, the silica shell was thin (<2 nm) and only just barely discernible under high-magnification TEM.

Li et al. in 2013 [16] proposed a shell-isolated nanoparticle-enhanced Raman spectroscopy (SHINERS) technique using core-shell structured gold@silica nanoparticles in aqueous solution with sodium silicate. The authors synthesized shell-isolated nanoparticles (SHINs) of sphere (55 and 120 nm diameter) finding that a 90°C reaction temperature accelerates the formation of silica shell on the gold nanoparticles. Gold nanoparticles were surface-primed using APTMS and coated with a 4 and 1 nm layer of silica using sodium silicate. Additionally,

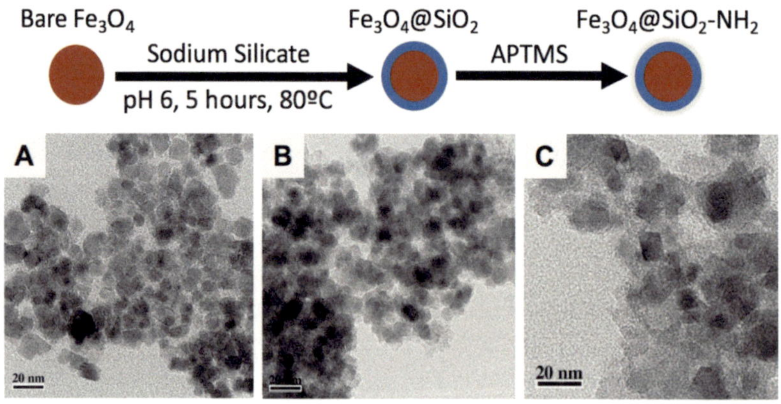

Figure 5. (Top row) Schematic of aqueous synthesis procedure used by Wang et al. to coat thin silica shell on Fe_3O_4 nanoparticles using sodium silicate. (Bottom row) TEM image showing (A) Fe_3O_4, (B) $Fe_3O_4@SiO_2$ particles and (C) $Fe_3O_4@SiO_2–NH_2$ [15].

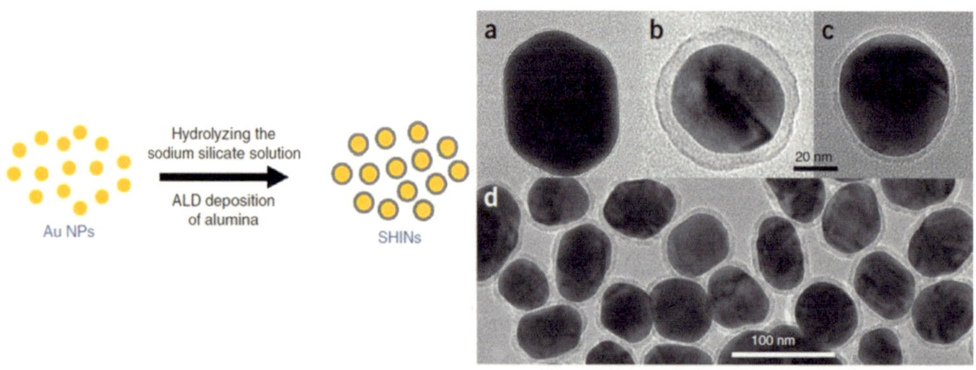

Figure 6. (Left) Schematic of aqueous synthesis procedure used by Li et al. to coat thin silica shell on Au nanoparticles using sodium silicate. (Right) TEM photos of gold@SiO_2 nanoparticles. (a–d) Various pH levels are used to grow silica shells and using APTMS as surface primer. High pH of 11 and low pH of 8 conditions with 1 h. heating using sodium silicate solution were used to prepare the SHINs (a and b). (c and d) gold@SiO_2 NPs were synthesized of ~10.2 for 0.5 and 1 h, respectively, at pH 10 [16].

the authors noted that excessively high pH conditions (>11) resulted in a thin silica layer with pinholes and low conditions (pH < 8) resulted in thicker silica shells. The shell will be dissolved by sodium hydroxide interfering with the silica growth if the pH is too high, e.g. pH greater than 11. This results in holes of nano to micron sizes within the thin silica shells. The silica shell grows too fast and thickens quickly for a low pH of 8 for the sodium silicate medium (**Figure 6**). Therefore, plasmonic field enhancement from the metal surface cannot extend beyond the shell, and no signal is acquired.

Shin et al. 2015 [17] demonstrated that a single layer of self-assembled core metal particles with silica shells can produce 1~2 nm empty nano-gaps by assembling silica-gold nanoparticles ($Au@SiO_2$) and chemical etching on different substrates. Colloidal silica shell gold-core nanoparticles were prepared by adding 5% v/v (3-aminopropyl)trimethoxysilane to gold nanoparticle seed solution. The gold colloid reacted with aqueous APTMS solution under vigorous stirring for 30 min at room temperature. After mixing, sodium silicate aqueous solution was added. A 5 nm silica layer is grown by placing in 90°C hot bath for 1 h under fast stirring and allowing the solution to cool to normal temperature. By reacting for 0.5 h in hot bath, the silica layer is grown to thickness of 2.5 nm only.

3. MPTMS-assisted silica coating

Silanization methods have also been developed for surface stabilization of core nanoparticles using sodium silicate, attaching artificial resin polymers (e.g., polyvinylpyrrolidone) or substituting the depleted sodium citrate on the surface of core nanoparticle adding extra fresh sodium citrate to metal cores. After surface stabilization steps, traditional Stober method is used for thick silica coating of the core. This requires conventional silane precursors (TMOS, TEOS) to be dissolved in ethanol-water mixture, which facilitates its hydrolysis and condensation. As these hydrophobic precursors are insoluble in aqueous medium and cannot undergo hydrolysis, large amount of alchohol is used to dissolve hydrophobic silanes prior to silica condensation process. Besides using such hydrophobic silane precursors, there arises strong interest to discover water-soluble silane precursors for silica encapsulation in aqueous-based medium, which allows silica-coating process to be much simpler, cost-efficient and environmentally friendly for various applications [18].

Recently, several research groups have reported the use of 3-mercaptopropyl-trimethoxysilane (MPTMS) as a water-soluble silane precursor to synthesize thick silica shells in fully aqueous medium. Previously, Niu [19], Nakamura [20], Lee [21], Shah [22] and Shang [23] successfully silica-coated metal nanoparticles, quantum dots and iron-oxide nanoparticles using water-based MPTMS in aqueous solution under mild conditions. Using MPTMS as a water-soluble silane precursor, direct alcohol-free silica coating in water-based medium becomes possible. This straightforward, one-pot method uses MPTMS as (i) surface primer, which renders particle-surface vitreophilic, (ii) a silane-precursor, which provides a thick silica shell layer >10 nm for particle stabilzation, (iii) surface modifier, which provides Thiol SH groups and (iv) a negative surface charge for dispersibility. Furthermore, speed, thickness and uniformity can be easily controlled through the concentration of precursor, reaction times and reaction temperatures, making aqueous routes far less complex than classic Stober method.

Niu et al. in 2010 [19] reported a facile route to fabricate homogenous, monodisperse, accurately size-tunable, SH-functionalized and magnetic silica-shell nanocomposite nanospheres (SHSSCNs) with sizes less than 100 nm. These nanocomposite nanospheres are synthesized based on the self-assembly of magnetic nanoparticles and a surface-priming copolymer polystyrene-block-poly(acrylic acid) (PS-b-PAA) in aqueous medium, followed by an easy silanization reaction with a silane precursor MPTMS to grow a silica shell on the interface of magnetite micelles at ambient temperature. (Step 1) Hydrophobic magnetite (Fe_3O_4) nanocores with a dimension of 5 nm were added to form magnetite-based micelles through their self-assembly in water with an amphiphilic block copolymer, PS-b-PAA. After filtration with deionized water, (Step 2) MPTMS was used to grow a hybrid silica layer by hydrolyzing and condensing the silane onto surface of the magnetic micelles under alkaline conditions. At the same time, SH groups were functionalised onto the surface of the nanocomposite particle in situ by using MPTMS silane as precursor, which is obviously distinct from traditional silica-encapsulation techniques reported to date. This is possible due to the fact that the core particles are encapsulated by a layer of polymeric shell, which allows the MPTMS silane to attach easily and directly onto the polymer shell's abundant surface hydroxyl groups. During the silica-encapsulation procedure, MPTMS performs as the silane precursor, thiol-functional linker and surface-capping stabilizers for the magnetite micelles and also for SH-modification of the nanoparticles in the silanization step.

Nakamura et al. in 2011 [20] used MPTMS for surface modification of quantum dots and synthesis of the silica shell to grow a layer of organosilica dense shell. MPTMS, quantum dot QD605 and NH_4OH were added and then heated for 3 h to grow the silica layer at 100°C. Extra steps like filtration or exchanging the reacting medium or adding alcohol were required. The growth of the silica shell was easily done in 3 h (**Figure 7**). The procedure is fast and easy, and the thiol groups inherent in silica shell enable easy surface bioconjugations to produce functionalised multiluminescent bioimaging markers. MPTMS and ammonia NH_4OH were mixed and then incubated to grow the silica layer. Fluorescent markers such as QD 605 and

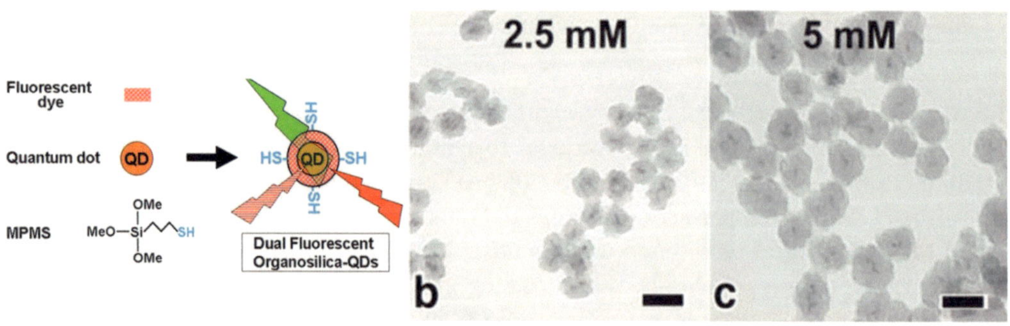

Figure 7. (Left) (a) Synthesis of dual fluorescent thiol-OS-QDs by Nakamura et al. to coat thin silica shell on Au nanorods using sodium silicate. Synthesis scheme for a single-step fabrication of double fluorescent thiol-OS-QDs. (Right) TEM images of thiol-OS-QDs with different silica shell thickness regulated by MPTMS dosage (b and c) [20].

rhodamine B were added. After reaction, the solution was centrifuged to discard unreacted contents. The aqueous silica encapsulation using Nakamura's ultra-fast technique took only 3 h, as opposed to using conventional reverse emulsion technique, which requires 1 day or using water-based sodium silicate, which requires 3 days. In addition, the presence of intrinsic thiol SH groups allows conjugation with thiol-reactive dyes or biomolecules, without the need for additional surface functionalization steps.

Lee et al. in 2011 [21] reported that for in vivo applications, such as biological imaging and medicine, nanoparticles must be safe in cell toxicity. Gold nanorods are often surface-capped and stabilized with CTAB cetyltrimethylammonium bromide, which will disrupt cell biology and cell growth as a cytotoxic surfactant. Silica coating and surface encapsulation of single nanoparticles can provide a safer substitute to surface CTAB for both decreasing cytotoxicity and improving colloidal stability. Lee synthesized silica-coated gold nanorods in aqueous medium. In his synthesis technique, Au nanorods were mixed to a solution of (3-aminopropyldimethylethoxysilane) and MPTMS for 12 h under stirring after which NH_4OH was added to form silica shell thickness < 2 nm. The authors found that the amount of amino-silane affected the growth of silica layer surrounding gold-NRs (**Figure 8**).

Shah et al. in 2014 [22] reported a novel direct silica coating in water-based silica-coating routes, which is alcohol-free one-step method using "pre-hydrolysed" MPTMS making silica coating far less complex than classic Stober method. This technique presents a facile, scalable, environmentally friendly process for thick coating of metal particles with a thick (5–30 nm) silica shell. The success of the process is using "pre-hydrolysed" MPTMS as silane precursor. Silica coating could be directly attributed to the strong and direct metal-thiol bond. Briefly, metal nanoparticles were washed and re-dispersed in deionized water before being combined into a pre-hydrolysed MPTMS in DI water. Ammonia was added as catalyst. While growth was easily scalable, the growth rate of the silica shell is controllable to 25, 11.7 and 10 nm/h. during condensation. The process slows down owing to eventual consumption of hydrolysed MPTMS. Different metal NPs (i.e., gold, silver, platinum) are successfully encapsulated with a silica layer under a completely alcohol-free environment, whose shell thickness can be easily and precisely synthesized by changing the growth duration (**Figure 9**). This novel water-based procedure is used for the fabrication of SERS-enhanced silver@SiO_2 nanoparticles without the

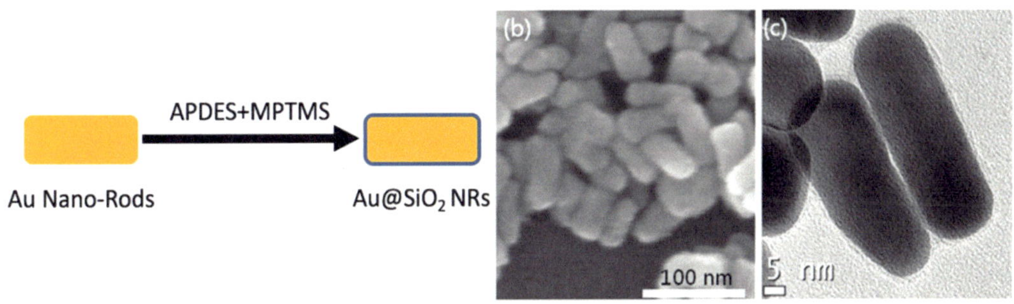

Figure 8. (Left) (a) Schematic of aqueous synthesis procedure used by Lee et al. to coat thin silica shell on Au nanorods using sodium silicate. (Right) Images of AuNRs sized 50 nm, (b) SEM image of core-shell Au nanorods and (c) TEM image of silica shell (<2 nm) AuNRs [21].

fluorescence background, which is used for bioimaging as SERS markers. The facile silica-encapsulation procedure developed here presents a highly potential encapsulating method and protection for the metallic core nanoparticles. The resulting highly negatively charged and SERS-enhanced metal@SiO_2 nanoparticles with thiol-functionalised surfaces hold great potential for biomedical applications. Such as fluorescence-free SERS-enhanced nanoparticles are useful for ultrasensitive bioimaging and biodetection applications.

Shang et al. in 2016 [23] synthesized iron oxide Fe_3O_4 nanoparticles coated with 10–20 nm thick silica shells using the same method reported by Shah et al. [22] after reacting for 2 h at room temperature with constant stirring (**Figure 10**). Hydrolysed MPTMS is formed by adding MPTM into deionized water and stirred [22]. Thereafter, Fe_3O_4 spheres were added into pre-hydrolysed MPTMS solution followed by adding ammonium hydroxide. Fe_3O_4@SiO_2-SH spheres (0.1 g) were dispersed in H_2O_2, in order to oxidize SH groups to SO_3H groups to form Fe_3O_4@SiO_2-SO_3H sulfonic acid-functionalised spheres. When Fe_3O_4@SiO_2 spheres and $PdCl_2$ solution were added, Pd^{2+} ions could be surface immobilized onto the surface of silica shell via coordination interaction between SO_3H groups and Pd^{2+} ions. Pd^{2+} ions immobilized on the

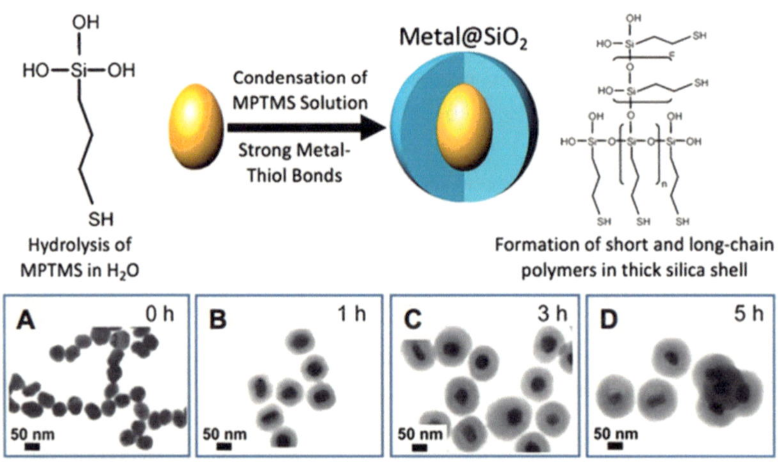

Figure 9. (Top row) Schematic of aqueous synthesis procedure used by Shah et al. to coat thin silica shell on Ag nanospheres using MPTMS. (Bottom row) Ag@SiO_2 (A) and Ag@SiO_2 after coating for (B) 1 h, (C) 3 h and (D) 5 h [22].

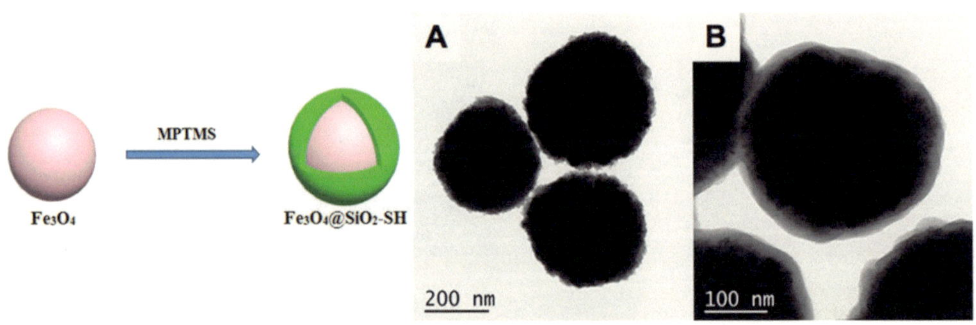

Figure 10. (Left) Schematic of aqueous synthesis procedure used by Shang et al. to coat thin silica shell on Ag nanospheres using MPTMS. (Right) (A, B) TEM micrographs of resulting Fe_3O_4@SiO_2-SH spheres [23].

surface of Fe_3O_4@SiO_2 particles oxidized the pyrrole monomer to give PPy-Pd layer surrounding Fe_3O_4@SiO_2-SO_3H spheres. The nanocomposite Fe_3O_4@SiO_2-SO_3H@PPy-Pd nanoparticles reduced 4-nitrophenol and undergo Suzuki coupling reactions as a nanocatalyst.

4. MPTES-assisted silica coating

Cui et al. in 2011 [24] proposed that highly stable Au@SiO_2 nanoparticles with functional groups were easily fabricated by pre-hydrolysis of MPTES in a water together with gold nanoparticles. In the preparation of organosilica shell, heterofunctional poly(ethyleneglycol) (SH-PEG-COOH) solution was added into solution containing Au nanoparticles and then MPTES is added. The solubility of MPTES in the aqueous solution is very low and tends to form an emulsion. NH_4OH was then added to adjust the solution pH to 8.5–9. The synthesis performed under static conditions for 48 h without stirring or shaking is the unique advantage of simplicity of the present method. The size of organosilica outer layer can be easily tuned by controlling the MPTES added. Steric hindrance caused by SH-PEG-COOH polymers, which are formed on gold surface, can inhibit gold particles aggregation, and therefore, polymer brushes perform a critical role in core-shell nanoparticles. Without any SH-PEG-COOH steric hindrance, gold particles will agglomerate and nucleate from the mixture even in MPTES solution. Au@SiO_2 particles appear to be rounded in shape and singular in nature. Every particle is sized about 55 nm and has a single gold core. The silica layer is about 15–75 nm in thickness (**Figure 11**). The silica thickness could be simply tuned by the concentration of MPTES. The surface charge of these Au@SiO_2 core-shell particles was measured to have 51.7 mV in zeta potential. Conventional Au@SiO_2 core-shell particles are measured to have 38 mV in zeta potential. These zeta measurements confirm the improved stability of the Au@SiO_2 nanoparticles upon organosilica shell coating.

Gao et al. in 2015 [25] synthesized multifunctional gold nanostars through the direct silica coating in aqueous solution using "pre-hydrolyzed" MPTES. Organosilica shells are synthesized by first hydrolyzing 3-mercaptopropyltriethoxysilane (MPTES) as silane precursor in water. Due to the large amount of SH groups provided by MPTES, the tedious step of surface modification is avoided, and the synthesis technique is easily controlled and simple. Gao reported this simple synthesis of highly functional anisotropic Au nanostars via directly coating silica on the surface

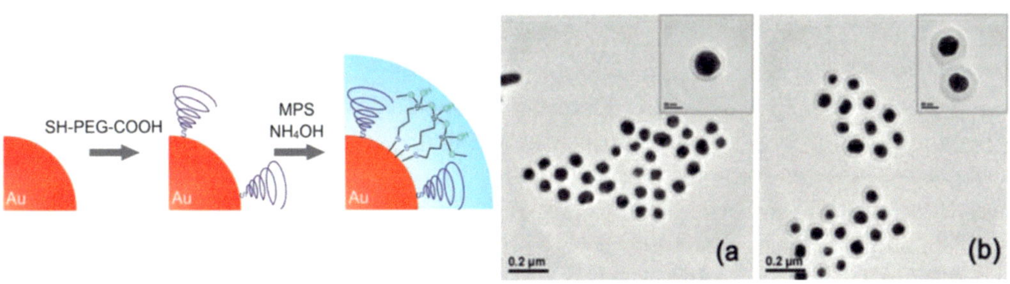

Figure 11. (Left) Proposed growth mechanism of Au@organosilica nanoparticles by Cui et al. using MPTES in water medium. (Right) TEM images of gold@silica core-shells with the 55 nm gold particles and varying silica thicknesses: (a = 15 and b = 20 nm) [24].

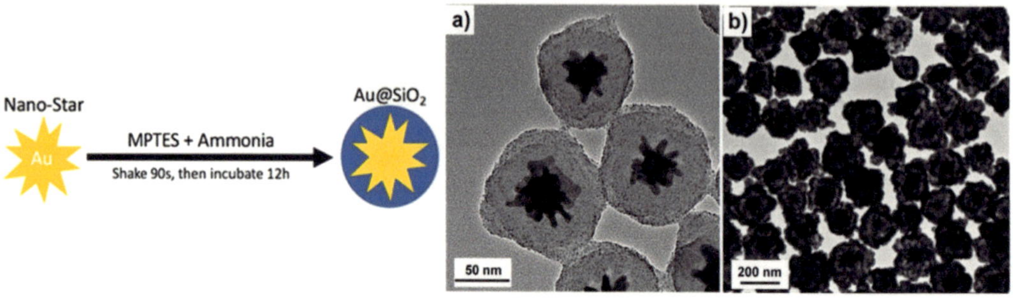

Figure 12. (Left) Illustration of the design of gold nanostars@SiO_2 coated using MPTES in water medium by Gao et al. (Right) TEM images of synthesized gold nanostars@SiO_2 (a) Scalebar 50 nm (b) Scalebar 200 nm [25].

of nano-sized star-shaped gold cores and subsequently immobilization of gadolinium chelates. Highly dispersed and uniformly sized Au-Gd nanostars@SiO_2 absorb strongly within the NIR spectrum and give out strong SERS waves which can strengthen magnetic resonance imaging. More silica coating was grown by adding MPTES and NH_4OH to the water-nanostars mixture and stirring for 1–2 min. The suspension was left for half a day to produce star-shaped Au@Silica. Obviously, 60 nm Au@SiO_2 nanostars and 20 nm silica shell, which are highly homogenous, are seen under TEM imaging (**Figure 12**). In addition, using this one-step procedure, in situ surface immobilization of SH groups onto Au@SiO_2 nanostars on the outer surface can be easily done for subsequent bioconjugation. This nanocomposite with distinct core-shell structure preserves the star shape and as a result gives efficient conversion from photo to thermal effect. The silica outer coating could prevent Raman markers from detachment loss. This ensures that Au nanostars@SiO_2 has greater Raman signals strength and sensing ability. A facile way via aqueous-coating silica directly on Au star-shaped cores to fabricate functional nanoparticles for applications in bioimaging and detection of cancer supports future research in silica-coated nanostructures for cancer therapy and diagnosis applications.

5. MTMS-assisted silica coating

Core-shell nano/microcapsule is a fast expanding area of research because of their wide applications, such as effective storage, controlled release and strong adsorption. Batch production of oil-loaded nano/microcapsules favors facile one-step fabrication methods.

Fei et al. in 2006 [26] synthesized monodisperse functionalised silica microcapsules using a one-pot procedure at room temperature. Silica encapsulation of oil microdroplet is reported. This is done without using any mediating agent or surfactant. Fei synthesized silica nano/microcapsules using self-silanization reaction at water-oil interface. N,N-diethyl-m-toluamide (DEET, Aldrich) was chosen as PCM to be silica-encapsulated. A water-oil solution of methyltrimethoxysilane (MTMS) and DEET was mixed into water and 3-aminopropyltrimethoxysilane (APTMS) and sonicated to get a white dispersion. When MTMS and APTMS are mixed, sol-gel reaction takes place and the solution reacts to give white-colored suspension. Adding MTMS or APTMS alone into water, there is no whitening or sol-gel reactions to be observed. Ottenbrite et al. [27] reported similar sol-gel effect. Monodisperse spheres are about 900 nm in diameter. The spheres had sizes

http://dx.doi.org/10.5772/intechopen.74097

of 900 nm and shell layer of 70 nm thick (**Figure 13**). During the hydrolysis and sol-gel reactions between MTMS and APTMS, MTMS diffused from micro-droplet to the oil surface. Microparticle size and shell dimensions can be easily adjusted by tuning the precursors dosage and concentrations, subjected to oil loading and sphere hardness requirements. In summary, silica-coated nanospheres were synthesized in a facile way [26]. In this one-step sol-gel technique, no use of intermediate agents or surfactant were required.

Jin et al. in 2010 [28] presented a one-step "sol-gel" method for the formation of silica-coated core-shell microcapsules of phase change materials (PCM) in aqueous medium without using surfactants or dispersants. The main issues associated with the use of polymeric shell materials for encapsulating core PCM materials are their low thermal conductivity and undesirable residues such as formaldehyde. Silica shell material has a much higher thermal conductivity and chemical stability than the polymeric materials. There have been many reports on the use of silica as the shell material of microcapsules or nanocapsules for controlled release and targeted drug delivery, and protecting active agents. Various methods have been proposed for making silica-shelled PCM microcapsules, including Pickering stabilization, water/oil/water emulsion templating using sodium silicate as the precursor, water/oil or oil/water emulsion templating using tetraethylorthosilicate (TEOS) as the precursor under acidic or basic conditions and hydrolysis-condensation reactions using tetraethylorthosilicate (TEOS) as the silica precursor with surfactants. Jin silica-coated nonadecane using methyltrimethoxysilane (MTMS) and 3-aminopropyl trimethoxysilane (APTMS) as silica precursors without any surfactants in an aqueous solution of distilled water. The synthesis was done under magnetic stirring to produce clear spherical, core-shell microparticles. A clear core-shell microstructure of few microns and a shell thickness about 500 nm (**Figure 14**). Positively charged amine groups of the APTMS suggest that MTMS in the oil phase diffuses to the oil-water interface where it reacts with the APTMS in the water phase to form a silica shell around the oil droplet. This effect was also observed by Ottenbrite [27] in their synthesis of organosilica nanoparticles, using aminopropyltriethoxysilane as the silica precursor. Jin found that increasing the volume of APTMS in water above 2% results in gelling, and no microcapsules were formed.

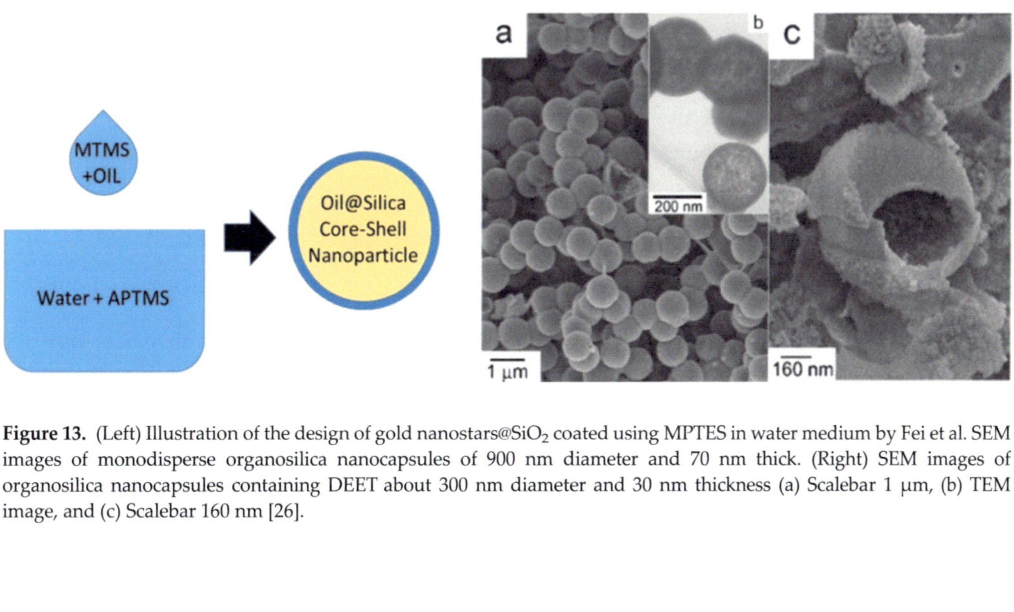

Figure 13. (Left) Illustration of the design of gold nanostars@SiO_2 coated using MPTES in water medium by Fei et al. SEM images of monodisperse organosilica nanocapsules of 900 nm diameter and 70 nm thick. (Right) SEM images of organosilica nanocapsules containing DEET about 300 nm diameter and 30 nm thickness (a) Scalebar 1 µm, (b) TEM image, and (c) Scalebar 160 nm [26].

Figure 14. (Left) Illustration of the design of silica-coated nonadecane using MTMS in water medium by Jin et al. (Right) SEM images of PCM microencapsulated spherical capsules: (a) an intact microcapsule and (b) a broken microcapsule of core-shell microstructure [28].

6. Conclusion

In this chapter, we summarize up-to-date the four major water-based silica-coating strategies for a variety of core nanomaterials. Currently, there are four successful water-based routes using different water-soluble silane precursors, namely sodium silicate, MPTMS, MPTES and MTMS. We reviewed these techniques in aqueous condition and their final core-shell morphologies. Their effective silica encapsulation leads to improved colloidal properties and creates new emerging materials based on silica-coated core-shell nanostructures. In particular, these simple large-scale aqueous synthesis techniques for silica-coated nanoparticles and their morphologies are important for paving new and practical applications. Four major water-based routes for silica-coating core-shell nanostructures are analyzed and their findings relating to sodium silicate, MPTMS, MPTES and MTMS are summarized below:

1. Sodium silicate as a precursor and its resultant shells are highly dependent on pH conditions and coating process requires a long period of time, lasting a few days.
2. Sodium silicate precursor is suitable for very thin silica-shell coatings (<2 nm), whose resultant silica-shell surface is often uneven, irregular and inhomogeneous.
3. Sodium silicate can be precipitated into silica-shell coatings under fully aqueous conditions (thin coat <6 nm), as well as under ethanol-water conditions (thick coating >10 nm).
4. Sodium silicate provide only –OH groups, which may demand extra functionalisation step to provide additional groups, such as –SH, $-NH_2$, –CHO, –COOH for conjugation.
5. MPTMS and MPTES precursors are effective in silica-coating metallic and semiconductor nanomaterials of different morphologies (nanospheres, nanorods, nanostars)
6. "Pre-hydrolysis" of MPTMS and MPTES is an essential step to successful silica coating.
7. MPTMS and MPTES precursors produce silica-coated core-shell nanomaterials that are highly monodispersed, spherical and uniform with smooth outer surfaces. Thickness can be easily controlled and thick dense shell can be coated within a few hours.
8. MPTMS and MPTES precursors provide intrinsic thiol –SH and silanol –OH functional groups, without any need for extra step of surface modifications.

http://dx.doi.org/10.5772/intechopen.74097

9. MPTMS and MPTES are proven effective for metals and semiconductors-based core-materials.

10. MTMS is proven effective in silica encapsulation of oil-based nano/microstructures.

11. MTMS is used in tandem with APTMS as a catalyst under oil/water emulsion.

This chapter reviewed a series of potentially reliable and reproducible synthesis techniques for high-quality silica-encapsulated core-shell nanomaterials. We compared and provided insights into the development of four water-based techniques to fabricate novel silica-encapsulated nanocomposites with unique core-shell architecture. These synthesis routes take place under mild reaction conditions, at ambient temperature and pressure, which are potentially applicable for large-scale production. The facile synthesis of silica shell under aqueous environment allows greater repeatability, facile steps for silanization, as compared with alcohol-based conventional Stober methods. They possess great potential to upscale batch production and present a biocompatible system that can be utilized for biomedical applications. The universality of these techniques were experimentally proven and demonstrated using a wide variety of core materials including Ag, Au, Pt, Fe_2O_3, CdTe, Gd, PCM, DEET and many others. In future, we hope to update the scientific community with more up-to-date research studies on the development of similar useful techniques for a simple one-pot aqueous synthesis of silica-encapsulated core-shell nanomaterials.

Acknowledgements

We acknowledge the support and funding of NRF POC Grant NRF2015NRF-POC001-025 and National University of Singapore for support and funding this work.

Author details

Kwok Wei Shah

Address all correspondence to: bdgskw@nus.edu.sg

National University of Singapore, Singapore

References

[1] Chaudhuri RG, Paria S. Core/shell nanoparticles: Classes, properties, synthesis mechanisms. Chemical Reviews. 2012;**112**:2373-2433. DOI: 10.1021/cr100449n

[2] Guerrero-Martınez A, Perez-Juste J, Liz-Marzán LM. Recent progress on silica coating of nanoparticles and related Nanomaterials. Advanced Materials. 2010;**22**:1182-1195. DOI: 10.1002/adma.200901263

[3] Liu S, Han M-Y. Silica-coated metal nanoparticles. Chemistry, an Asian Journal. 2010;**5**: 36-45. DOI: 10.1002/adfm.200400427

[4] Fernandez-Lopez C, Perez-Juste J, Pastoriza-Santos I, Liz-Marzán LM. Highly controlled silica coating of PEG-capped metal nanoparticles and preparation of SERS-encoded particles. Langmuir. 2009;**25**:24:13894-13899. DOI: 10.1021/la9016454

[5] Stober W, Fink A, Bohn E. Controlled growth of monodisperse silica spheres in the micron size range. Journal of Colloid Interface Science. 1968;**26**:62-69

[6] Liz-Marzán LM, Giersig M, Mulvaney P. Synthesis of nanosized gold-silica core-shell particles. Langmuir. 1996;**12**:4329-4335

[7] Shah KW. Nanostructured oxides and their coating on metal nanoparticles [thesis]. National University of Singapore; 2011

[8] Lee J, Choi JS, Park JH, Kim MH, Hong DW, Cho HC, Yang SH, Choi IS. Cytoprotective silica-coating of individual mammalian cells through bioinspired silicification. Angewandte Chemie, International Edition. 2014;**53**:8056-8059. DOI: 10.1002/anie.201402280

[9] Wolcott A, Gerion D, Visconte M, Sun J, Adam S, Chen S, Zhang JZ. Silica-coated CdTe quantum dots functionalized with thiols for bioconjugation to IgG proteins. The Journal of Physical Chemistry. B. 2006;**110**(11):5779-5789. DOI: 10.1021/jp057435z

[10] Liu S, Han M. Synthesis, functionalization, and bioconjugation of monodisperse, silica-coated gold nanoparticles: Robust bioprobes. Advanced Functional Materials. 2005;**15**: 961-967. DOI: 10.1002/adfm.200400427

[11] Zhou X, Kobayashi Y, Romanyuk V, Ochuchi N, Takeda M, Tsunekawa S, Kasuya A. Preparation of silica encapsulated CdSe quantum dots in aqueous solution with the improved optical properties. Applied Surface Science. 2005;**242**:281-286. DOI: 10.1016/j.apsusc.2004.08.022

[12] Hunyadi SE, Murphy CJ. Tunable one-dimensional silver-silica nanopeapod architectures. The Journal of Physical Chemistry. B. 2006;**110**:7226-7231. DOI: 10.1021/jp0603076

[13] Kobayashi Y, Sakuraba T. Silica-coating of metallic copper nanoparticles in aqueous solution. Colloids and Surfaces A: Physicochemical and Engineering Aspects. 2008;**317**: 756-759. DOI: 10.1016/j.colsurfa.2007.11.009

[14] Hu H, Wang Z, Pan L. Synthesis of monodisperse Fe_3O_4@silica core-shell microspheres and their application for removal of heavy metal ions from water. Journal of Alloys and Compounds. 2010;**492**:656-661. DOI: 10.1016/j.jallcom.2009.11.204

[15] Wang J, Zheng SR, Shao Y, Liu JL, Xu ZY, Zhu DQ. Amino-functionalized Fe_3O_4@SiO_2 core-shell magnetic nanomaterial as a novel adsorbent for aqueous heavy metals removal. Journal of Colloid and Interface Science. 2010;**349**:293-299. DOI: 10.1016/j.jcis.2010.05.010

[16] Li JF, Wang ZL, Tian ZQ. Surface analysis using shell-isolated nanoparticle enhanced Raman spectroscopy. Nature Protocols. 2013;**8**(1):52-65. DOI: 10.1038/nprot.2012.141

[17] Yuna Shin, JW Song, DC Kim, Taewook Kang. Facile preparation of ultrasmall void metallic nanogap from self-assembled gold-silica core-shell nanoparticles monolayer via kinetic control. Advanced Materials. 2015;**27**:4344-4350. DOI: 10.1002/adma.201501163

[18] Nakamura M. Nanostructured Oxides. Weinheim, Germany: Wiley-VCH Verlag GmbH & Co. KGaA; 2009. pp. 109-153. ISBN: 978-3-527-32152-0

[19] Niu D, Li YS, Ma Z, Hua Diao JLG, Chen HR, Zhao WR, Ruan ML, Zhang YL, Shi J. Preparation of uniform, water-soluble, and multifunctional nanocomposites with tunable sizes. Advanced Functional Materials. 2010;**20**:773-780. DOI: 10.1002/adfm.200901493

[20] Nakamura M, Ozaki S, Abe M, Matsumoto T, Ishimura K. One-pot synthesis and characterization of dual fluorescent thiol-organosilica nanoparticles as non-photoblinking quantum dots. Journal of Materials Chemistry. 2011;**21**:4689-4695. DOI: 10.1039/c0jm04259e

[21] Lee JY, Park W, Yib DK. Immunostimulatory effects of gold nanorod and silica-coated gold nanorod on RAW 264.7 mouse macrophages. Toxicology Letters. 2012;**209**:51-57 DOI: 10.1016/j.toxlet.2011.11.024

[22] Shah KW, Sreethawong T, Liu S-H, Zhang S-Y, Li ST, Han M-Y. Aqueous route to facile, efficient and functional silica coating of metal nanoparticles at room temperature. Nanoscale. 2014;**6**:11273-11281. DOI: 10.1039/c4nr03306j

[23] Shang M, Wang W, Zou H, Ren G. Coating Fe_3O_4 spheres with polypyrrole-Pd composites and their application as recyclable catalysts. Synthetic Metals. 2016;**221**:142-148. DOI: 10.1016/j.synthmet.2016.08.015

[24] Cui Y, Zheng X-S, Ren B, Wang R, Zhang J, Xia N-S, Tian Z-Q. Au@organosilica multifunctional nanoparticles for the multimodal imaging. Chemical Science. 2011;**2**:1463-1469. DOI: 10.1039/c1sc00242b

[25] Gao Y, DC Niu JLG, Shi J. Multifunctional gold nanostar-based nanocomposite: Synthesis and application for noninvasive MR-SERS imaging-guided photothermal ablation. Biomaterials. 2015;**60**:31-41. DOI: 10.1016/j.biomaterials.2015.05.004

[26] Fei B, Haifeng L, Wang RH, Xin JH. Monodisperse Organosilica microcapsules with functional groups by self-catalysis. Chemistry Letters. 2006;**35**(6):622-623. DOI: 10.1246/cl.2006.622

[27] Ottenbrite RM, Wall JS, Siddiqui JA. Self-catalyzed synthesis of organo-silica nanoparticles. Journal of the American Ceramic Society. 2000;**83**:3214-3215. DOI: 10.1111/j.1151-2916.2000.tb01709.x

[28] Jin Y, Lee W, Musina Z, Ding Y. A one-step method for producing microencapsulated phase change materials. Particuology. 2010;**8**:588-590. DOI:10.1016/j.partic.2010.07.009

Chapter 7

Delamination and Longitudinal Cracking in Multilayered Composite Nanostructured Coatings and Their Influence on Cutting Tool Wear Mechanism and Tool Life

Alexey Vereschaka, Sergey Grigoriev,
Nikolay Sitnikov, Gaik Oganyan,
Anatoliy Aksenenko and Andre Batako

Additional information is available at the end of the chapter

http://dx.doi.org/10.5772/intechopen.72257

Abstract

The wear and failure mechanism for multilayered nanostructured coatings has a number of significant differences from the one typical for monolithic single-layered coatings. In particular, while the strength of adhesion bonds at the "substrate-coating" boundary is important for monolithic coatings, then for multilayered nanostructured coatings, the strength of adhesion and cohesion bonds at interlayer boundaries and boundaries of separate nano-sublayers becomes of significant significance. Meanwhile, the delamination arising in the structure of multilayered nanostructured coatings can have both negative (leading to loss of coating uniformity and subsequent failure of coating) and positive influences (due to decrease of internal stresses and inhibition of transverse cracking). Various mechanisms of formation of longitudinal cracks and delaminations in coatings on rake tool faces, which vary based on the compositions and architectures of the coatings, are studied. In addition, the influence of internal defects, including embedded microdrops and pores, on the formation of cracks and delaminations and the failure of coatings is discussed. The importance of ensuring a balance of the basic properties of coatings to achieve high wear resistance and maximum tool life of coated metal-cutting tools is shown. The properties of coatings and the natures of their failures, as investigated during scratch testing and dry turning of steel C45, are provided.

Keywords: wear-resistant coatings, wear, crack, fracture, tool life, PVD coatings, delamination, nanoscale structures

1. Introduction

1.1. Background

Further increases in efficiency of machining and cutting speeds as well as tightening of reliability requirements associated with greater levels of automation of production result in the need to create new tool materials with enhanced performance characteristics. One way to improve the performance characteristics of tool materials is to enhance their surface properties by applying modified coatings [1]. In turn, the properties of modified coatings continue to be improved, and their architecture and elemental composition become more complicated. In particular, multilayered composite coatings, nanostructured and gradient coatings, and coatings with multicomponent elemental composition have been used extensively in recent years [2]. The use of a multilayered architecture of coatings and the use of nanostructured technology can significantly improve the performance characteristics of a new generation of coatings. However, along with the use of such coatings come new problems that did not occur with monolithic coatings of the first generation. In particular, problems arose concerning interlayer delamination and formation of specific longitudinal cracks in the structure of coating. A large number of studies examining problems of cracking have been conducted. The general assumption is that the formation of microcracks is associated with the displacement of dislocations [3–6]. A number of mechanisms for the formation of dislocation microcracks are well known [3, 4, 6]. In principle, those mechanisms provide for blocking of the progress of dislocation by some obstacle (e.g., a grain boundary, a boundary of nanolayers, or inclusion). If in some slip plane dislocations stop before a sufficiently powerful obstacle, then a cluster of dislocations is formed, and it causes a high concentration of stresses at the obstacle. This concentration of stresses results in formation of a dislocation microcrack. It should be noted that the problems of crack formation and delamination in the structures of multilayered coatings have not been studied as thoroughly as have other aspects of operation and wear of such coatings.

1.2. Literature review

Tabakov et al. [7, 8] considered mechanisms of cracking with respect to single-layer macroscale coatings on the basis of systems composed of TiN, TiCN, (Ti,Zr)N, and (Ti,Zr)CN. They discovered that coatings of a complex composition of (Ti,Zr)N and (Ti,Zr)CN are characterized by better resistance to intensive cracking. Tabakov et al. also considered multilayered coatings with macroscale structure: in particular, on the basis of systems composed of TiCN-(Ti,Zr)N-TiN, TiN-(Ti,Zr)N-TiN, TiCN-(Ti,Al)N-TiN, and TiCN-(Ti,Mo)N-TiN [9]. These studies proved that the introduction of zirconium nitride in the coating composition significantly reduces the tendency to cracking. The problems of cracking and brittle fracture of coatings consisting of Ti-TiN-(Ti,Cr,Al)N, Zr-(Zr,Cr)N-CrN, and Ti-TiN-(Ti,Cr,Al)N and Ti-(Al,Cr)N-(Ti,Al)N, Ti-(Al,Cr)N-(Ti,Cr,Al)N, and Zr-(Al,Cr)N-(Zr,Cr,Al)N also were addressed in papers [10–16]. A detailed review of existing papers in the field of crack formation in multilayered coatings, with classification of types of cracks and analysis of the mechanisms of their formation, is given in [17]. The topic of mathematical modeling of cracking in multilayered coatings with the use of an axis-symmetrical finite element method (FEM) model was considered by Skordaris et al. [18]. Wu et al. [19, 20] modeled cracking in single-layer coatings within the framework of linear elastic fracture mechanics (LEFM). M'Saoubi et al. [21] investigated the nature of wear, including brittle fracture and cracking of physical vapor deposition (PVD)-coated (TiN, (Ti,Si)N, (Ti,Al)N, and (Al,Cr)N) polycrystalline cubic boron nitride. Koseki et al. [22] examined the cutting performance of TiN-coated cutting tools. Defects (e.g., droplets, voids) in the coating were found to be the starting point of damage. The breakdown region is enlarged as the work material is caught in the damaged portion of the coating. Kumar and Curtin [23] considered the probable mechanisms of development and inhibition of cracks in microstructures: particularly at crack bridging by ductile ligaments, crack deflection by second-phase particles, microcrack formation, and stress-induced phase transformations. The same paper also includes an overview of methods for modeling the development of cracks using FEM and incorporating cohesive elements at the continuum level, as well as discrete dislocation methodology at the mesoscopic level, and coupled atomistic/continuum methods that transition atomic level information to the microscopic level. A large number of studies have been devoted to the investigation of causes and conditions for the formation of delaminations in multilayered composite macrostructures. To predict the occurrence of delaminations, the methods of layer-wise interface elements [24, 25], classical finite element analysis (FEA) [25, 26], and the virtual crack closure technique (VCCT) [27] are widely used. The issues concerning delamination of multilayered nanostructured coatings are also discussed in details in [28].

1.3. Mechanisms of crack formation

Starting from the theory of crack formation [3–5], the concentration of local tensile stresses σ_{ld} in the head of a series of edge dislocations caused by the action of a number of edge stresses τ can be determined using the following equation [29]:

$$\sigma_{ld} = \sqrt{\frac{d}{2x}}(\tau - \tau_i) \tag{1}$$

where $2d$ is the length of the slip band or the distance between the slip bands (the value may also correspond to crystalline grain diameter), x is the distance from the strip to the head of the cluster of dislocations, and τ_i is the stress of resistance to movement of dislocations (friction stress). If the local stress reaches the theoretical strength of the crystalline body σ_{theor}, determined by the equation

$$\sigma_{theor} = \sqrt{\frac{E\gamma}{a_0}} \tag{2}$$

where a_0 is the equilibrium distance between atoms, E is the modulus of elasticity, and γ is the plastic shear deformation, then conditions arise for the formation of a dislocation microcrack. Consequently, the criterion for the formation of a microcrack is as follows:

$$\sqrt{\frac{d}{2x}}(\tau - \tau_i) \geq \sqrt{\frac{E\gamma}{a_0}} \tag{3}$$

The number of positive or negative dislocations in a flat cluster near an obstacle can be expressed by an approximate formula (assuming $E \approx 2G$, where G is the shear modulus):

$$n \approx \frac{(\tau - \tau_i)}{bE} \tag{4}$$

Assuming the additional condition of $x \approx a_0$, from the joint solution of Eqs. (3) and (4), we obtain the condition necessary for the formation of a microcrack under the dislocation mechanism:

$$(\tau - \tau_i)nb = 2\gamma \tag{5}$$

Analysis of the given conditions for the formation of a dislocation microcrack results in the following conclusion. Local tensile stresses in the head of a number of dislocations are formed primarily because of tangential stresses τ and are not related in any way to tensile stresses (i.e., only shear stresses are crucial for the initiation of a microcrack). The defects already existing or emerging in the early stages of deformation of solid bodies result in the initiation and development of the failure processes. Various mechanisms of failure are realized depending on the structural and stress strain states of a solid body and also depending on the external medium [3, 6, 30]. The following are the most common mechanisms of microfailure of metals (**Figure 1**):

1. Viscous fracture (**Figure 1a**). This mechanism of failure is caused by the formation of micropores near inclusions or particles of the second phase, their growth, localization of microplastic deformation in the crosspieces between the pores, and, in the final stage, the fusion of micropores and the break of the bridges. The model and criterion for the formation of micropores are as follows: a micropore is formed when the cohesive stresses (bond stresses between the inclusion particles and the matrix) reach a critical stress. The existing models for the formation, growth, and fusion of micropores can be used to analyze micromechanisms of viscous fracture of a solid body in front of a crack tip. When the stresses in front of the inclusions (before a crack tip) reach critical values, the micropores are being formed. Further growth of micropores and localization of plastic deformation results in plastic blunting of crack tip, merging of micropores with crack tip, and subsequent growth of a viscous crack (**Figure 2**). In multilayered composite nanostructured coatings, such a mechanism of cracking can occur primarily in the formation of delaminations at interlayer boundaries, as well as at the boundaries of nano-sublayers.

2. Transcrystallite cleavage (**Figure 1b**). This mechanism is characterized by failure of a solid body (spread of a crack) along certain crystallographic planes. In polycrystalline bodies, the process of transcrystallite cleavage is realized not in one crystallographic plane, but through the distribution and subsequent integration of a multitude of microcracks of the cleavage that arises in a certain family of crystallographic grain planes. As a rule, transcrystallite cleavage is of a brittle nature, although plastic deformation processes are also possible. The described mechanism can be typical for the formation of cracks in monolithic single-layer coatings (TiN, ZrN, etc.), as well as in monolithic layers of multilayered coatings. In coatings with nano-sublayers, the development of cracks under the above mechanism is significantly constrained by the boundaries of nano-sublayers.

3. Intercrystallite fracture (**Figure 1c**). This mechanism consists of the initiation and propagation of microcracks along grain boundaries. This failure mechanism is related to the fact that the fracture energy necessary for propagation of a crack along the grain boundaries is lower than the corresponding energy of the transcrystallite cleavage. In the coatings,

intergranular fracture can occur in monolithic single-layer coatings (in particular, transverse cracks in the columnar crystalline structure of TiN are formed under the above mechanism). As a rule, several mechanisms of failure take place simultaneously, and as a consequence, a mixed type of failure occurs. Typically, three types of loading or displacement of the points of crack surfaces under the influence of an external load are considered [3, 4, 6]. The first type (type 1) includes the formation of normal detachment cracks, characterized by movement of the points of the crack surface under the action of a load in the direction perpendicular to the plane of the crack. In this case, the crack tends to open. Cracks of the transverse type (type 2) are cracks in which points of surfaces are displaced across the front of the crack (leading edge of the crack). Finally, longitudinal shear cracks (type 3) are characterized by displacement of the points of the crack surface along its front. It should be noted that these types of loading can be combined, thus forming complex types of loading. The conditions of operation of multilayered nanostructured modified coatings for a metal-cutting tool are most typically characterized by loading of type 1 (typical for the rake face of the tool, due to the constant formation and failure of adhesion bridges with the tool being machined) and type 2 (typical for the flank face of the tool, due to longitudinal compressive stresses and the resulting plastic deformation of a substrate). To describe the delamination process, it is also possible to use strain energy release rate (SERR), which represents energy dissipated during fracture per unit of newly created fracture surface area [31]. Delamination growth rates were correlated with the SERR by means of the Paris relation [32, 33]:

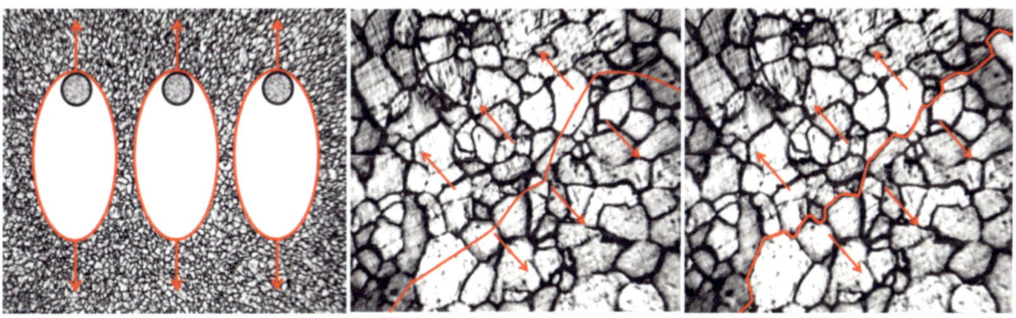

a b c

Figure 1. Mechanisms of failure: (a) viscous fracture, (b) transcrystallite cleavage, and (c) intercrystallite fracture.

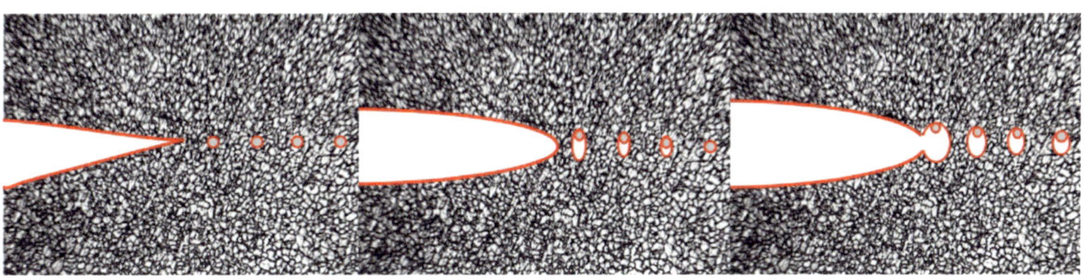

a b c

Figure 2. Pattern of the micromechanism of growth of a viscous crack: (a) inclusions at the crack tip, (b) growth of micropores in front of the crack tip, and (c) micropores merging with the crack tip.

$$\frac{dB_d}{dN} = Cf(G)^{n_p} \tag{6}$$

where B is delamination length (mm), N is the number of cycles, C is the Paris coefficient for delamination growth, G is the strain energy release rate (N/mm), n_p is the Paris exponent for delamination growth, and C and n are empirically determined parameters that depend on the materials and the temperature (and possibly other factors). As yet, no consensus has been reached on the correct form of $f(G)$. In particular, $f(G)$ can be represented as SERR at maximum fatigue load, G_{max}, and within the SERR range, ΔG. In turn, ΔG can be determined from the following formula [25]:

$$\Delta G = \left(\sqrt{G_{max}} - \sqrt{G_{min}}\right)^2 \tag{7}$$

The dependence (1) also can be represented by [26] as follows:

$$\frac{dB_d}{dN} = C(G_{max})^{n_p} \tag{8}$$

where G_{max} is the strain energy release rate at maximum fatigue load (N/mm).

Thus, the task of the present work is to study the process of delamination between layers of multilayer coatings and between nano-sublayers of nanostructured coatings. These processes have a significant impact on the overall performance of modifying coatings and products with such coatings (in particular, metal-cutting tools). An important feature of this work is that not only laboratory samples but also cutting tool samples that underwent cutting tests in real production conditions were considered. The peculiarities of delamination were investigated depending both on the elemental composition of the coatings and on their architecture (total coating thickness and thickness of the nanolayers).

2. Materials and methods

2.1. Deposition method

For deposition of nanoscale multilayered composite coatings (NMCC), a vacuum-arc VIT-2 unit [2], which was designed for the synthesis of coatings on substrates of various tool materials, was used. The unit was equipped with an arc evaporator with filtration of vapor-ion flow. In this study, the process, termed filtered cathodic vacuum-arc deposition (FCVAD) [10–16], was used for deposition of coatings on the tools to significantly reduce the formation of the droplet phase during formation of the coating. The use of the FCVAD process does not cause structural changes in carbide and provides the following:

- High adhesive strength of the coating in relation to the carbide substrate.
- Control of the level of the "healing" of energy impact on surface defects in carbide in the form of microcracks and micropores and formation of favorable residual compressive stresses in the surface layers of the carbide material.

- Formation of the nanoscale structure of the deposited coating layers (grain size, sublayer thickness) with high density due to the energy supplied to the deposited condensate and transformation of the kinetic energy of the bombarding ions into thermal energy in local surface volumes of carbide material at an extremely high rate of approximately 10^{14} K s^{-1}.

When choosing the composition of NMCC layers, in forming the coating of the three layered architecture [2, 10], the Hume-Rothery rule was used. This rule states that the difference in atomic dimensions in contacting compounds should not exceed 20% [34]. The parameters used at each stage of the deposition process of NMCC are shown in **Table 1**.

An uncoated carbide tool and a carbide tool with "reference" coating TiN, deposited via standard vacuum-arc technology of arc-PVD, were used as objects for comparative studies of tool life.

2.2. Microstructural studies

For microstructural studies of samples of carbide with coatings, a raster electron microscope FEI Quanta 600 FEG was used. The studies of chemical composition were conducted using the same raster electron microscope. To perform X-ray microanalysis, characteristic X-ray emissions resulting from electron bombardment of a sample were examined. The hardness (HV) of coatings was determined by measuring the indentation at low loads according to the method of Oliver and Pharr [35], which was conducted on a micro-indentometer microhardness tester (CSM Instruments) at a fixed load of 300 mN. The penetration depth of the indenter was monitored so that it did not exceed 10–20% of the coating thickness to limit the influence of the substrate. The adhesion characteristics were studied on a Nanovea scratch tester, which represents a diamond cone with apex angle of 120° and radius of top curvature of 100 μm. The tests were conducted with the load linearly increasing from 0.05 to 40 N. Crack length was 5 mm. Each sample was subjected to three trials. The obtained curves were used to determine two parameters: the first critical load, L_{C1}, at which the first cracks appeared in the coating, and the second critical load, L_{C2}, which caused the total failure of the coating.

2.3. Study of cutting properties

A study of the cutting properties of the tool made of carbide with developed NMCC was conducted using a lathe CU 500 MRD for longitudinal turning of steel C45 (HB 200). In the experiment, the cutters featured mechanical fastening of inserts made of carbide (WC + 15%

Process	p_N (Pa)	U (V)	I_{Al} (A)	I_{ZrNb} (A)	I_{Ti} (A)	I_{Cr} (A)
Pumping and heating of vacuum chamber	0.06	+20	120	80	65	75
Heating and cleaning of products with gaseous plasma	2.0	100 DC/900 AC f = 10 kHz, 2:1	80	—	—	—
Deposition of coating	0.36	−800 DC	160	75	55	70
Cooling of products	0.06	—	—	—	—	—

Note: I_{Ti} = current of titanium cathode, I_{Al} = current of aluminum cathode, I_{ZrNb} = current of zirconium-niobium cathode, I_{Cr} = current of chromium cathode, p_N = gas pressure in chamber, and U = voltage on substrate.

Table 1. Parameters of stages of the technological process of deposition of NMCC.

TiC + 6% Co) with square shapes (SNUN ISO 1832:2012) and with the following figures for the geometric parameters of the cutting part: $\gamma = -8°$, $\alpha = 6°$, $K = 45°$, $\lambda = 0$, and $R = 0.8$ mm. The study was performed for the following cutting modes: $f = 0.2$ mm/rev, ap = 1.0 mm, and vc = 250 m min^{-1}. Flank wear-land values (VB_c) were measured with a toolmaker's microscope MBS-10 as the arithmetic mean of four to five tests. A value of $VB_c = 0.4$ mm was taken as failure criterion. The study included statistical processing of tests of wear of cutting tools, sample mean value of wear, and sample mean square deviation of tool wear, which are random variables with different values in repeated experiments. Of note, during the experiments, outlying results were excluded. To exclude outlying results of the experiments, Irwin's criterion was used. To do that, the value of Irwin's criterion K_λ was defined, if the outlying result was the maximum value VB_{max}:

$$K_\lambda = (VB_c - VB_{max})/K_\sigma \tag{9}$$

and if the doubts were provoked by the wear value with minimum value VB_{min}:

$$K = (VB_c - VB_{min})/K_\sigma \tag{10}$$

The calculated value K_λ was compared to the critical value $K_{\lambda A}$, defined theoretically for a given level of significance level A and selection criterion n. If $K_\lambda < K_{\lambda A}$, then deviation of questionable value VB_c was considered as valid.

3. Results and discussion

3.1. Adhesion characteristics

The classical test that enables determination of the strength of the adhesive bond of a coating with a substrate by the scratch-test method also can be used for qualitative evaluation of the strength of the adhesive bond between individual coating layers and cohesive bond between nano-sublayers. The tests were conducted on a Nanovea scratch tester. The indenter was a diamond cone with an apex angle of 120° and radius of top curvature of 100 μm. The tests were performed with a load linearly increasing from 0.05 N to the final load (40 N). Crack length was 5 mm. Each sample was subjected to three trials. The obtained curves were used to determine two parameters: the first critical load L_{C1}, at which first cracks appeared in NMCC, and the second critical load L_{C2}, which caused the total failure of NMCC. Typical types of failure are presented in **Figure 3** (standard coating TiN), **Figure 4** (NMCC Zr-ZrN-(Nb,Zr,Ti,Al)N), and **Figure 6** (NMCC Ti-TiN-(Ti,Al)N). All investigated coatings showed a sufficiently high level of adhesion bonds with substrate. Numerous research efforts and the experience of the authors of this paper show that a scratch test does not have a unique correlation with the tool life of a coated tool [1]; the test allows only "rejecting" coatings with insufficient strength of adhesion bonds. However, this test enables a study of the nature of the coating failure, particularly from the point of view of delaminations that occur in its structure. Let us consider the nature of the failure of the single-layer monolithic TiN coating (**Figure 3**). A fairly smooth scribing groove is clear, with a clearly visible area of brittle fracture of the outer area of the coating. On the edges of the groove, cracks and splintered areas of the coating are visible. Patterns of failure of

http://dx.doi.org/10.5772/intechopen.72257

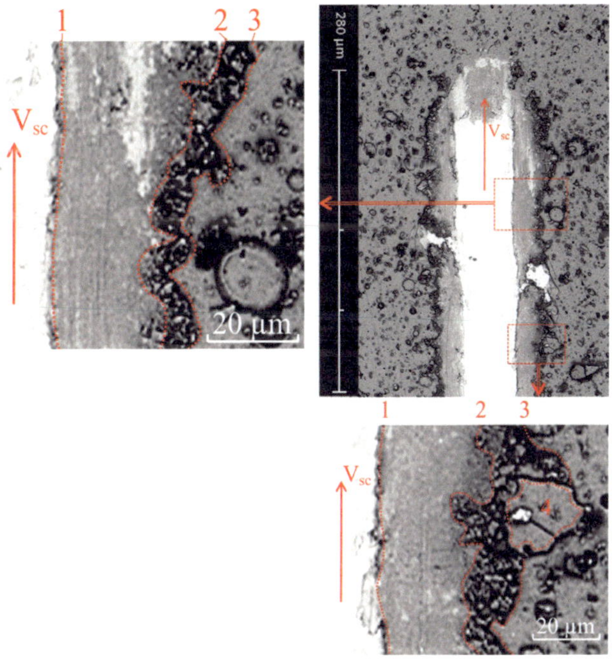

Figure 3. The nature of failure of coating TiN along a longitudinal crack, caused by a diamond indenter at critical (breaking) load [28]. Vsc, scribing direction. (1) "Substrate-coating" boundary, (2) boundary of the brittle fracture zone of the coating, (3) boundary of the scribing groove, and (4) splintered section of the coating.

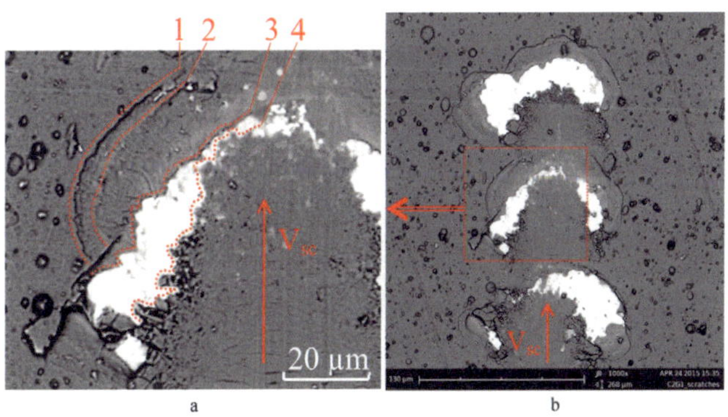

Figure 4. The nature of failure of NMCC Zr-ZrN-(Nb,Zr,Ti,Al)N along a longitudinal crack, caused by a diamond indenter at critical (breaking) load. Vsc, scribing direction [28]. (1) The boundary of the wedging spallation zone, (2) the delamination boundary between the intermediate and wear-resistant layers, (3) the boundary of "adhesion coating layer-substrate" delamination, and (4) the boundary of the coating zone, pressed by the tip of the scratch tester.

NMCC Zr-ZrN-(Nb,Zr,Ti,Al)N (**Figure 4**) and (NMCC Ti-TiN-(Ti,Al)N) (**Figure 5**) are characterized by a number of significant differences. The failures of those coatings occur under the mechanism of "wedging spallation." Meanwhile, NMCC Zr-ZrN-(Nb,Zr,Ti,Al)N shows extensive interlayer delaminations, whereas in NMCC Ti-TiN-(Ti,Al)N, similar delaminations are less pronounced, and delaminations between nano-sublayers also occur. Generally, this picture correlates with the nature of the failure of those coatings observed during cutting tests.

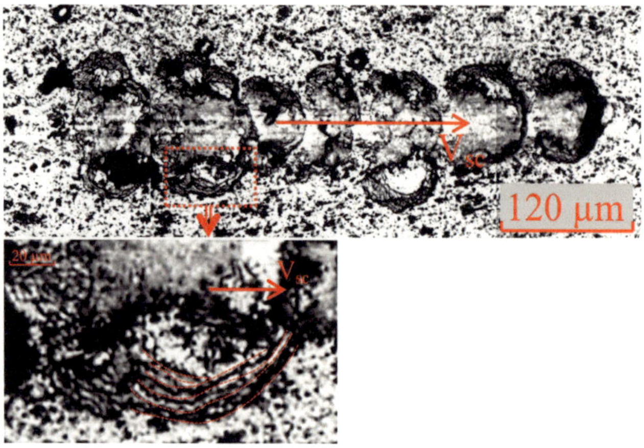

Figure 5. The nature of failure of NMCC Ti-TiN-(Ti,Al)N along a longitudinal crack, caused by a diamond indenter at critical (breaking) load [28]. Vsc, scribing direction.

The study of the scribing process for nanostructured coatings of large thickness (exceeding 10 μm) is of particular interest. In this case, it is possible to observe both coating failure caused by violation of adhesion bonds between layers and cohesive bonds between nano-sublayers and failure of a coating as a whole, when failure is not accompanied by delamination. Signs of failure of coating Ti-TiN-(Ti,Al)N (with coating thickness 13 μm) at scribing are shown in **Figures 6** and 7.

In particular, **Figure 6** shows both violation of the interlayer interface between layers TiN and (Ti,Al)N and persistence of strong adhesion bonds between nano-sublayers of layer (Ti,Al)N. At zoom in, it is possible to notice in **Figure 7** that in some cases, at critical loads, there is also failure of cohesive bonds between nano-sublayers, and that fact results in formation of a kind of "terraces," i.e., flat microsites with surface structure of a nano-sublayer. It is also possible to see signs of a tear-out of microdroplets embedded in the coating structure. **Figure 7** shows the "terrace-like" structure of failure zone of a nanostructured coating. A general structure of the coating under the study and the nature of cracking in it during the cutting tests are shown below, in **Figure 20**.

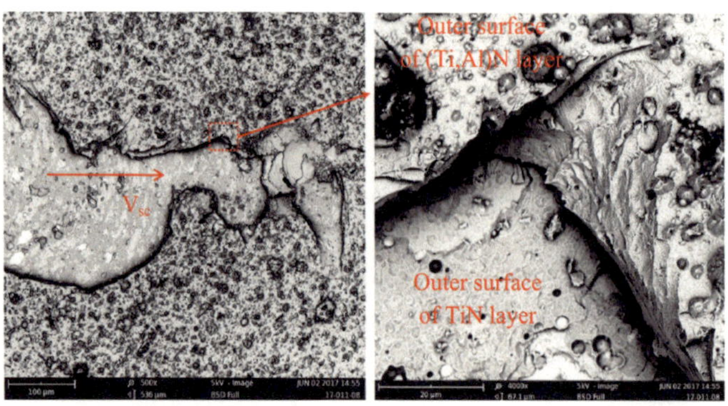

Figure 6. The nature of failure of NMCC Ti-TiN-(Ti,Al)N (coating thickness 13 μm) along a longitudinal crack, caused by a diamond indenter at critical (breaking) load [28].

http://dx.doi.org/10.5772/intechopen.72257

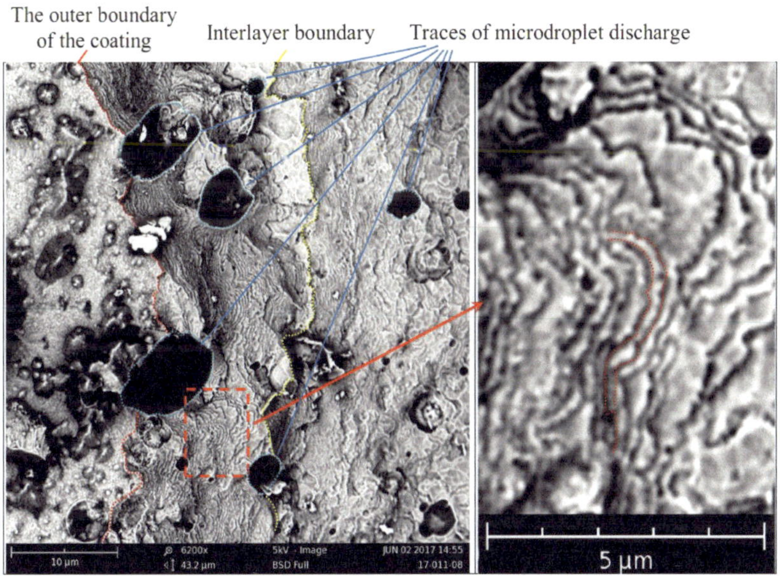

Figure 7. The nature of failure of NMCC Ti-TiN-(Ti,Al)N (coating thickness 13 μm) along a longitudinal crack, caused by a diamond indenter at critical (breaking) load [28].

3.2. Determination of basic properties of NMCC under cutting tests

This study was focused on the NMCC containing nitrides of Ti, Al, Cr, Zr, and Nb in its composition. For the detailed studies of various properties, NMCC were selected based on the following conditions:

- If earlier studies show significant increase in cutting properties and reliability of the tool [10–16].
- If the thermodynamic criterion ΔrG (Gibbs free energy change per mole of reaction) favored the formation of the NMCC.

To accomplish the research tasks, NMCC of various compositions were selected to meet the above conditions and were deposited using the FCVAD technology. The thicknesses of the coatings used in the studies were 2.4–5.0 μm. A wide range of thicknesses were selected on the basis of previous studies (in particular [10–16]), indicating the improvement in cutting performance with increase in coating thickness. The basic properties of the NMCC under study are presented in **Table 2**. Curves obtained by mathematical processing of the experimental data are shown in **Figure 8**.

The NMCC Ti-TiN-(Ti,Al)N shows better resistance for approximately 19 min of operation due to its high surface hardness; however, subsequently, the tool with such a coating begins to experience intensive wear. This fact can be related to the start of intense cracking and wear of this coating. As a result, the tool with NMCC Zr-ZrN-(Zr,Cr,Al)N showed better resistance, and it was characterized by a balanced combination of sufficiently high hardness and resistance to brittle fracture. Let us consider in detail the mechanism of cracking and failure of coatings, paying special attention to such aspects of those processes as longitudinal cracks and

#	Composition of NMCC	Tool life T_c (min) VB = 0.4 mm	Sublayer thickness (nm)	Total thickness (μm)	Adhesion, L_{C2} (N)	Hardness, HV (GPa)
1	Uncoated	8	—	—	—	18
2	TiN	18	—	2.8	31	30
3	Zr-ZrN-(Zr,Cr,Al, Nb)N	24	200	3.8	>40	34
4	Zr-ZrN-(Nb,Zr,Ti, Al)N	31	45–60	3.3	>40	34
5	Ti-TiN-(Ti,Al)N	28	65–90	5.0	>40	38
6	Zr-ZrN-(Zr,Cr,Al) N	37	15–45	3.4	39	36

Table 2. The basic properties of NMCC and periods of tool life of the carbide tools under study with the NMCC under study.

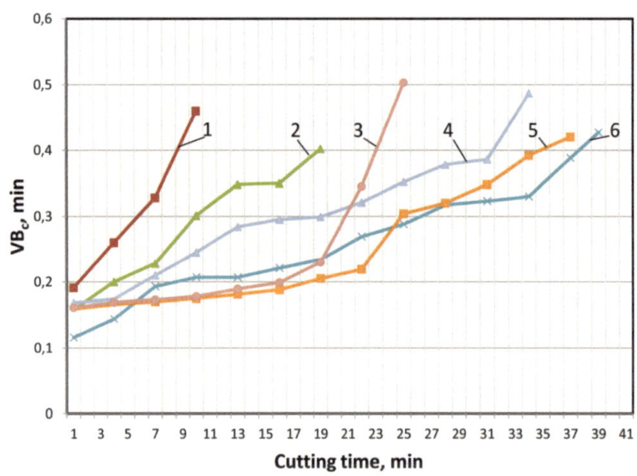

Figure 8. Dependence of wear VB on cutting time for dry turning of steel C45 at ap = 1.0 mm, f = 0.2 mm/rev, and vc = 250 m/min. (1) Uncoated, (2) TiN, (3) Zr-ZrN-(Zr,Cr,Al,Nb)N, (4) NMCC Zr-ZrN-(Zr,Cr,Al)N, (5) NMCC Ti-TiN-(Ti, Al)N, and (6) Zr-ZrN-(Zr,Cr,Al)N.

delaminations (interlayer delaminations and delaminations between nano-sublayers) form. Basic mechanism for the formation of longitudinal cracks and delaminations can be distinguished in a nanostructured multilayered coating because of the tearing force related to the adhesion interaction between the outer boundary of the coating and the material being machined (**Figure 9**), which has a prevailing fatigue characteristic and results in the formation of fatigue cracks due to the alternating processes of formation and failure of adhesion bridges in the system of "coating-material being machined." The considered mechanism is more typical for coatings on the rake face of the tool.

The action of the mechanism shown in **Figure 9** can result not only in the formation of longitudinal cracks and delaminations but also in the destruction of the surface layers of the coating and, consequently, in the deterioration of the tool life of the metal-cutting tool (**Figure 10**).

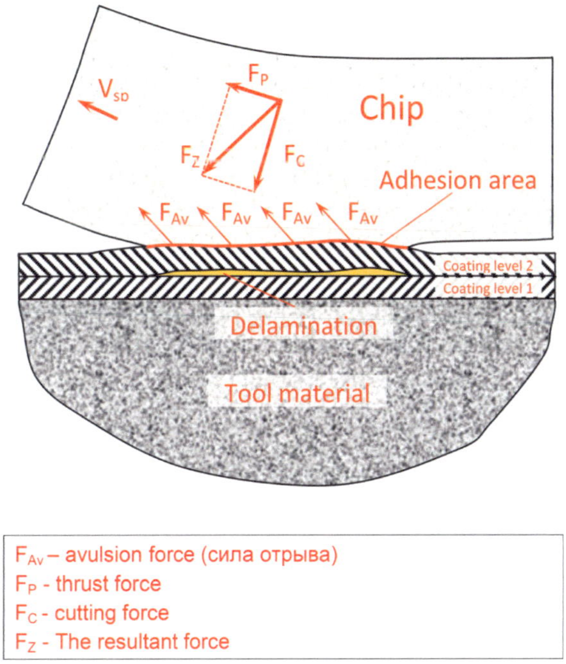

Figure 9. The mechanism of formation of longitudinal cracks and delaminations in a nanostructured multilayered coating during cutting due to the tearing force associated with adhesion interaction between the outer boundary of the coating and the material being machined [28].

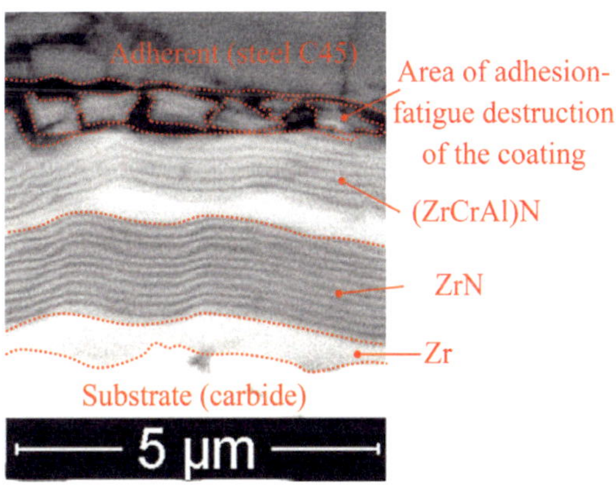

Figure 10. An example of the failure of the upper layer of NMCC Zr-ZrN-(Zr,Cr,Al)N because of the tearing force related to the adhesion interaction between the outer boundary of the coating and the material being machined (steel C45) [28].

An important distinctive feature of the development of longitudinal cracks in nanostructured coatings is the formation of bridges in the process of cracking due to the alternation of less plastic sublayers with more plastic ones in the coating structure. Such bridges inhibit the development of a crack by exerting a positive influence on coating crack resistance and, consequently, on the tool life of a cutting tool (**Figure 11**). This mechanism of inhibition of cracking is fairly close to the mechanism of action of bridges from a particle of a more plastic phase embedded in the brittle phase described, in particular, by Kumar and Curtin [23]. It should be noted that the studies of the propagation of longitudinal cracks in monolithic coatings revealed no such bridges. The strength of the bridges depends on the composition of the coating layers. In particular, in layers of (Zr,Cr,Al)N (**Figure 11a**), the bridges show significantly higher strength and ductility than in (Zr,Nb,Ti,Al)N (**Figure 11b**), where the bridges show a tendency to failure.

No such bridges are observed in NMCC Ti-TiN-(Ti,Al)N, and that may be connected with the high hardness and brittleness of the layer (Ti,Al)N. The failure of NMCC Ti-TiN-(Ti,Al)N often occurs in accordance with a pronounced "brittle fracture" scenario with the formation of a network of longitudinal and transverse cracks (**Figure 12**).

In the case of insufficiently strong adhesion bond between the coating layers or cohesive bonds between its nano-sublayers, delaminations of the classical form are formed between the layers of the coating or between its nano-sublayers. In particular, **Figures 13** and **14** show obvious delamination between the intermediate TiN layer and the wear-resistant (Ti,Al)N layer. In the structure of the coating presented in **Figure 14**, transverse cracks and delaminations also occur between nano-sublayers of the wear-resistant layer. In addition, it is possible to note a relatively positive role of delamination (1) as a factor of inhibition of transverse cracks (3). The transverse cracks (3) are decelerated at the boundary of the intermediate and wear-resistant layers, and they are not spreading in the intermediate TiN layer (**Figure 14**).

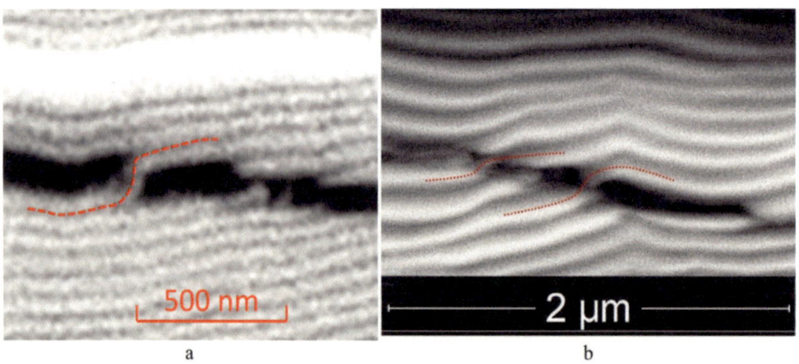

Figure 11. Deceleration of a longitudinal crack in NMCC Zr-ZrN-(Zr,Cr,Al)N (a) and Zr-ZrN-(Zr,Nb,Ti,Al)N (b) due the formation of bridges of more plastic nanolayers [17, 28].

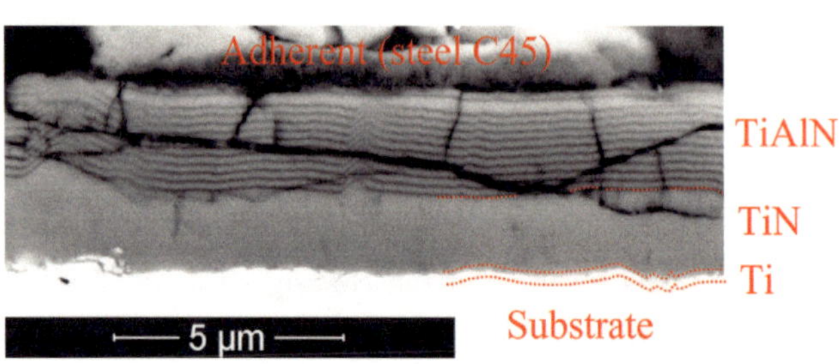

Figure 12. Failure of NMCC Ti-TiN-(Ti,Al)N with the formation of a network of longitudinal and transverse cracks [28].

http://dx.doi.org/10.5772/intechopen.72257

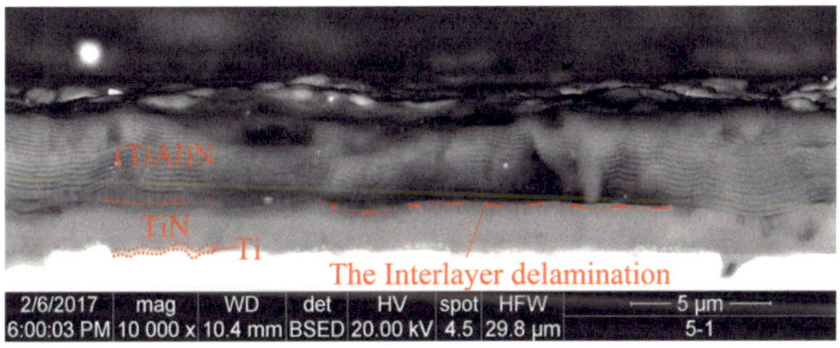

Figure 13. Interlayer delamination in the structure of NMCC Ti-TiN-(Ti,Al)N [28].

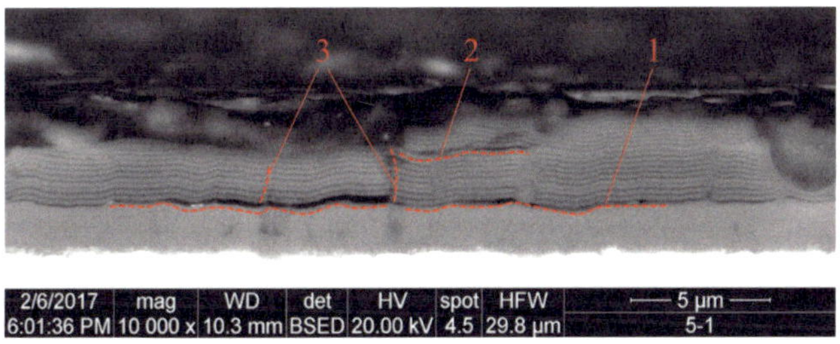

Figure 14. Interlayer delamination (1), delamination between nano-sublayers (2), and transverse cracks (3) in the structure of NMCC Ti-TiN-(Ti,Al)N [28].

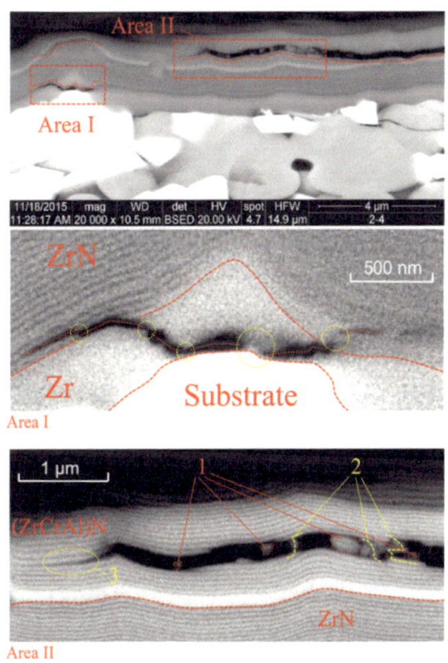

Figure 15. An example of formation of longitudinal cracks and delaminations in NMCC Zr-ZrN-(Zr,Cr,Al)N [28].

The patterns of formation of longitudinal cracks and delaminations often appear to be complex. In particular, **Figure 15** shows the mechanisms of cracking and delamination in NMCC Zr-ZrN-(Zr,Cr,Al)N. The area of this picture that is marked as AREA I contains a crack of a complex kind, combining delamination between the substrate and the adhesion layer Zr, passing into a transverse crack that cuts the adhesive layer, and turning into a series of delaminations between nano-sublayers of the intermediate coating layer. The initial factor stimulating the formation of this crack is microroughness of the substrate, formed by high carbide grain. In contrast, the area indicated as AREA II contains an example of extended delamination, reaching a width of 200–300 nm. The formation of this delamination resulted in (1) chipping of microcomponents of the coating, (2) formation of bridges of more plastic nanolayers, and (3) the crack development boundary.

Various defects in coatings (in particular, embedded microdrops and pores) can play an important role in the formation of longitudinal cracks and delamination. **Figure 16** shows how a crack reaches a macro droplet and forms branches. Meanwhile, one of the branches of the crack passes through a macro droplet, while the second crack branch traverses it along the contour. This photomicrograph reveals a separation of the material being machined from the coating; this separation indicates a low adhesive bond between the materials. Meanwhile, no separation of the coating from the tool material occurs due to a strong adhesive bond between them.

Another example of the effect of a microdroplet embedded in the structure of the coating on the formation of delaminations is shown in **Figure 17**. Here, a crack is formed directly above a microdroplet, and several parallel delaminations exist in the area adjacent to a microdroplet. These occurrences may be related to internal stresses arising during the coating deposition.

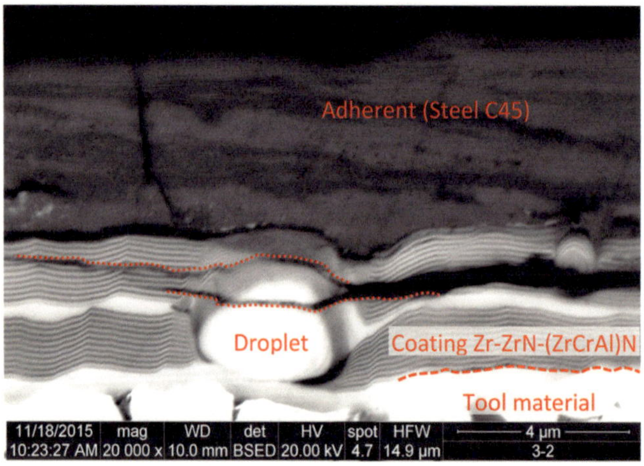

Figure 16. An example of development of a longitudinal crack in NMCC Zr-ZrN-(Zr,Cr,Al)N [28].

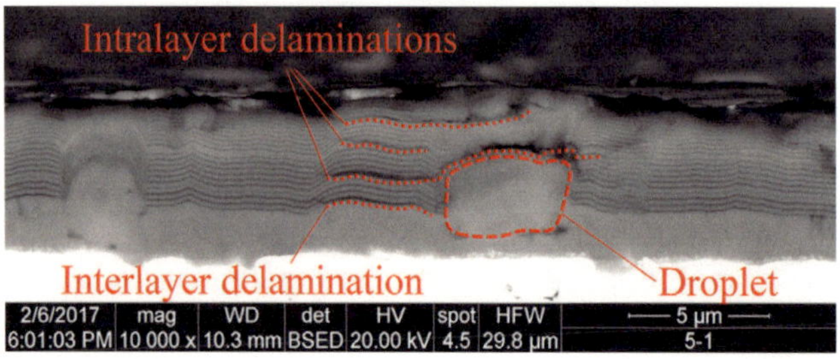

Figure 17. An example of development of a longitudinal crack in NMCC Ti-TiN-(Ti,Al)N [28].

Let us consider separately the process of delamination and failure in coating Zr-ZrN-(Zr, Cr, Al, Nb)N with thickness of sublayers of about 200 nm (**Figure 18a**). Due to thick sublayers, this coating cannot be called "nanostructured." This coating is characterized by a large number of delaminations, arising especially in the fracture zone adjacent to a wear crater. No bond bridges are formed between sublayers (**Figure 18** Area A), while delaminations are an important factor in failure of coating (**Figure 18c**).

Let us individually consider delaminations formed in NMCC of heavy thickness (usually exceeding 8 μm) because of heavy internal compressive stresses. Such delaminations can be formed with equal probability in the coating both on the rake and flank face of a tool. An example of formation of delaminations in "thick" NMCC is presented in **Figure 20**. It is possible to observe four clear delaminations located at approximately equal distance (about 20 nano-sublayers) from each other. Meanwhile, the delamination closest to the substrate (area A on **Figure 20**) passes exclusively along the boundary between the sublayers. At the same time, delaminations B, C, and D are rather longitudinal cracks because they also are characterized by breaks in the structures of nano-sublayers (**Figure 19**).

Various internal defects in "thick" NMCC (in particular, microdroplets embedded in the structure of the coating) become particularly important and result in local fracture of the coating because of the formation of multiple delaminations that weaken the coating structure and ultimately result in the formation of a transverse crack (**Figure 20**). As a result of the distortion of the coating structure associated with the curvature of the nano-sublayers because of rounding

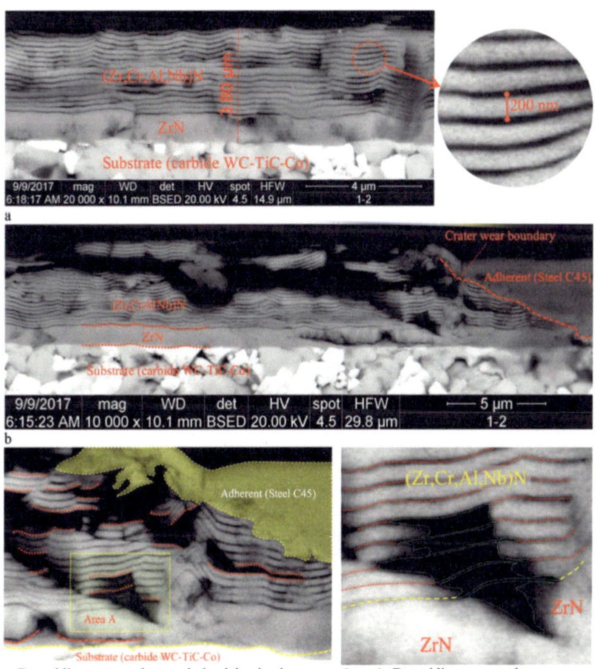

c. Dotted lines are used to mark the delamination zones.

Area A. Dotted lines are used to "reconstruct" the lost fragments of sublayers.

Figure 18. The process of delamination and failure in coating Zr-ZrN-(Zr,Cr,Al,Nb)N: general structure (a), destruction of the coating in the area of the crater wear boundary (b), boundary region "coating-adherent" (c).

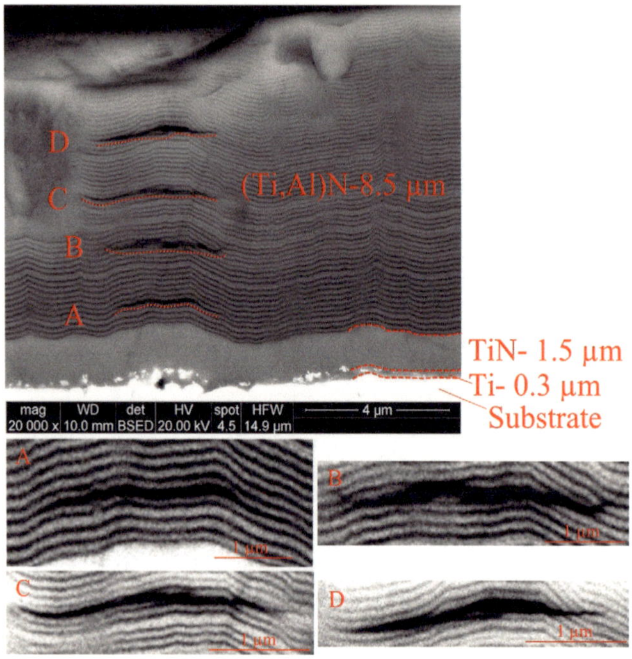

Figure 19. An example of the formation of delaminations in NMCC Ti-TiN-(Ti,Al)N (total thickness of the coating is 10.3 μm) [28].

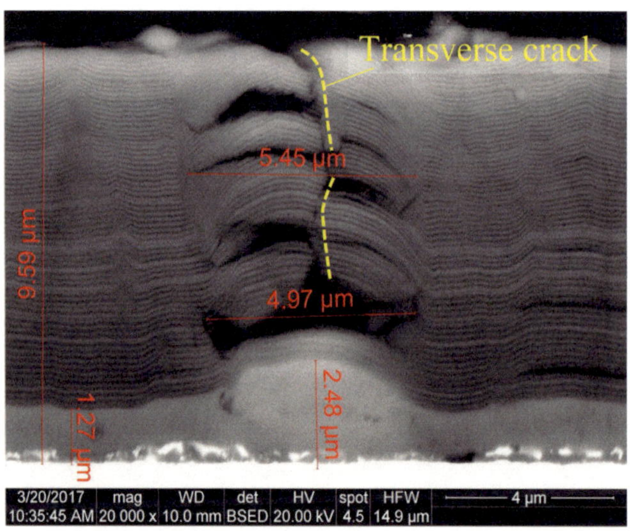

Figure 20. Formation of a transverse crack in the structure of NMCC Ti-TiN-(Ti,Al)N as a result of weakening of the structure of NMCC by multiple delaminations formed under the influence of internal stresses [28].

of an embedded microdroplet, internal stresses arise, which in turn result in the formation of corresponding delaminations and longitudinal cracks. Because (Ti,Al)N is a very hard, yet brittle compound, the chipping of fragments of nano-sublayers and formation of a transverse crack occur in the coating structure weakened by delaminations.

4. Conclusions

This study of the nature of the formation of longitudinal cracks and delaminations in multi-layered nanostructured coatings reveals the following:

1. Two important mechanisms result in formation of transverse cracks and delaminations:

   a. Tearing force associated with adhesion interaction between the outer boundary of the coating and the material being machined (typical for the rake face of the tool).

   b. Tearing force associated with plastic microdeformations in the surface layer of the tool substrate (more typical for the flank face of a tool).

2. The nature of the formation of longitudinal cracks and delaminations varies significantly for different coating compositions. In coatings with more plastic nanolayers, bridges can be formed, which inhibit the development of cracks. This can be clearly observed in NMCC Zr-ZrN-(Zr,Cr,Al)N and to a lesser extent in NMCC Zr-ZrN-(Zr,Nb,Ti,Al)N, while in coatings with more hard and brittle nanolayers (e.g., (Ti,Al)N), such bridges are not formed, and coatings are destructed under the mechanism of brittle failure.

3. Such coating defects as embedded microdrops and micropores can stimulate the development of longitudinal cracks and delaminations.

4. The following factors reduce the probability of formation of longitudinal cracks (delamination):

   a. Reduction of adhesion interaction between the outer boundary of the coating and the material being machined.

   b. Increase of adhesion bonds between coating layers and cohesive bonds between the nano-sublayers.

   c. Decrease in the level of plastic microdeformations of the tool substrate, in particular, through heat strengthening and/or diffusion saturation with alloying elements.

5. In general, coatings with thickness of sublayers of more than 100 nm do not form bond bridges which inhibit cracking. In such coatings, delamination develops more actively and leads to failure of coating structure.

6. NMCC of relatively large thickness (larger than 8 μm) may experience delamination during the deposition as a result of significant internal stresses. The presence of such delamination can contribute to brittle fracture of coatings, particularly NMCC based on a hard and brittle compound (Ti,Al)N.

Acknowledgements

This research was financed by the Ministry of Education and Science of the Russian Federation in the framework of the state order in the sphere of scientific activity (Leading researchers, project 16.9575.2017/6.7).

Author details

Alexey Vereschaka[1]*, Sergey Grigoriev[1], Nikolay Sitnikov[2], Gaik Oganyan[1], Anatoliy Aksenenko[1] and Andre Batako[3]

*Address all correspondence to: ecotech@rambler.ru

1 Moscow State Technological University STANKIN, Moscow, Russia

2 National Research Nuclear University MEPhI, Moscow, Russia

3 Liverpool John Moores University (LJMU), Liverpool, UK

References

[1] Vereschaka AS. Working Capacity of the Cutting Tool with Wear Resistant Coatings. Moscow: Mashinostroenie; 1993 (in Russian)

[2] Vereshchaka AA, Vereshchaka AS, Mgaloblishvili O, Morgan MN, Batako Nano-scale AD. Multilayered-composite coatings for the cutting tools. International Journal of Advanced Manufacturing Technology. 2014;**72**(1):303-317. DOI: 10.1007/s00170-014-5673-2

[3] Matvienko YG. The Models and Criteria of Fracture Mechanics. Moscow: FIZMATLIT; 2006 (in Russian)

[4] Broberg KB. Cracks and Fracture. San Diego: Academic Press; 1999

[5] Zehnder TA. Lecture Notes on Fracture Mechanics. Ithaca, New York: Cornel University; 2007

[6] Anderson TL. Fracture Mechanics: Fundamentals and Applications. New York: CRC Press LLC; 1995

[7] Tabakov VP. The influence of machining condition forming multilayer coatings for cutting tools. Key Engineering Materials. 2012;**496**:80-85

[8] Tabakov VP, Vereschaкa AS. Development of technological means for formation of multilayer composite coatings, providing increased wear resistance of carbide tools, for different machining condition. Key Engineering Materials. 2014;**581**:55-61

[9] Tabakov VP, Smirnov MY, Tsirkin AV. Productivity of End Mills with Multilayer Wear-resistant Coatings. UlSTU: Ulyanovsk; 2005 (in Russian)

[10] Vereschaka AA, Vereschaka AS, Bublikov JI, Aksenenko AY, Sitnikov NN. Study of properties of nanostructured multilayer composite coatings of Ti-TiN-(TiCrAl)N and Zr-ZrN-(ZrNbCrAl)N. Journal of Nano Research. 2016;**40**:90-98. DOI: 10.4028/www.scientific.net/JNanoR.40.90

[11] Vereschaka AA, Vereschaka AS, Batako AD, Hojaev OK, Mokritskii Y. Development and research of nanostructured multilayer composite coatings for tungsten-free carbides with extended area of technological applications. International Journal of Advanced Manufacturing Technology. 2016;**87**:3449-3457. DOI: 10.1007/s00170-016-8739-5

[12] Volkhonskii AO, Vereshchaka AA, Blinkov IV, Vereshchaka AS, Batako AD. Filtered cathodic vacuum arc deposition of nano-layered composite coatings for machining hard-to-cut materials. International Journal of Advanced Manufacturing Technology. 2016;**84**:1647-1660. DOI: 10.1007/s00170-015-7821-8

[13] Alexey AV, Sergey NG, Anatoly SV, Alexey YP, Batako AD. Nano scale multilayered composite coatings for cutting tools operating under heavy cutting conditions. Procedia CIRP. 2014;**14**:239-244. DOI: 10.1016/j.procir.2014.03.070

[14] Vereschaka AA, Volosova MA, Batako AD, Vereshchaka AS, Mokritskii BY. Development of wear-resistant coatings compounds for high-speed steel tool using a combined cathodic vacuum arc deposition. International Journal of Advanced Manufacturing Technology. 2016;**84**:1471-1482. DOI: 10.1007/s00170-015-7808-5

[15] Grigoriev SN, Vereshchaka AA. Methodology of formation of multi-layered coatings for carbide cutting tools. Mechanics & Industry. 2016;**17**:706. DOI: 10.1051/meca/2016065

[16] Vereschaka AA, Vereschaka AS, Batako ADL, Mokritskii BJ, Aksenenko AY, Sitnikov NN. Improvement of structure and quality of nano-scale multi-layered composite coatings, deposited by filtered cathodic vacuum arc deposition method. Nanomaterials and Nanotechnology. 2016;**7**:1-13. DOI: 10.1177/1847980416680805

[17] Vereschaka AA, Grigoriev SN. Study of cracking mechanisms in multi-layered composite nano-structured coatings. Wear. 2017;**378-379**:43-57. DOI: 10.1016/j.wear.2017.01.101

[18] Skordaris G, Bouzakis K-D, Charalampous P. A dynamic FEM simulation of the nano-impact test on mono- or multi-layered PVD coatings considering their graded strength properties determined by experimental–analytical procedures. Surface & Coatings Technology. 2015;**265**:53-61

[19] Wu X-F, Dzenis YA, Strabala KW. Free-edge stresses and progressive cracking in surface coatings of circular torsion bars. International Journal of Solids and Structures. 2008;**45**:2251-2264

[20] Wu X-F, Jenson RA, Zhao Y. Stress-function variational approach to the interfacial stresses and progressive cracking in surface coatings. Mechanics of Materials. 2014;**69**:195-203

[21] M'Saoubi R, Johansson MP, Andersson JM. Wear mechanisms of PVD-coated PCBN cutting tools. Wear. 2013;**302**:1219-1229

[22] Koseki S, Inoueb K, Usuki H. Damage of physical vapor deposition coatings of cutting tools during alloy 718 turning. Precision Engineering. 2016;**44**:41-54

[23] Kumar S, Curtin WA. Crack interaction with microstructure. Materials Today. 2007;**10**(9):34-44

[24] Hosseini-Toudeshky H, Hosseini S, Mohammadi B. Delamination buckling growth in laminated composites using layer wise-interface element. Composite Structures. 2010;**92**:1846-1856

[25] Pascoe JA, Rans CD, Alderliesten RC, Benedictus R. Fatigue disbonding of bonded repairs—An application of the strain energy approach. 27th ICAF Symposium; 5–7 June 2013; Jerusalem

[26] Pascoe JA, Rans CD, Benedictus R. Characterizing fatigue delamination growth behavior using specimens with multiple delaminations: The effect of unequal delamination lengths. Engineering Fracture Mechanics. 2013;**109**:150-160

[27] Rybicki E, Kanninen M. A finite element calculation of stress intensity factors by a modified crack closure integral. Engineering Fracture Mechanics. 1977;**9**(4):931-938

[28] Vereschaka AA, Grigoriev SN, Sitnikov NN, Batako A. Delamination and longitudinal cracking in multi-layered composite nano-structured coatings and their influence on cutting tool life. Wear. 2017;**390–391**:209-219. DOI: 10.1016/j.wear.2017.07.021

[29] Johnson KL. Contac Mechanics. Cambridge: Cambridge University Press; 1985

[30] Hertzberg RW. Deformation and Fracture Mechanics of Engineering Materials. New York: John Wiley & Sons; 1996

[31] Gent AN, Mars WV. Strength of elastomers. In: Mark JE, Erman B, Roland M, editors. The Science and Technology of Rubber. 4th ed. Boston: Academic Press; 2013. pp. 473-516

[32] Paris P, Gomez M, Anderson W. A rational analytic theory of fatigue. The Trend in Engineering. 1961;**13**:9-14

[33] Rans CD, Alderliesten RC, Benedictus R. Misinterpreting the results: How similitude can improve our understanding of fatigue delamination growth. Composites Science and Technology. 2011;**71**:230-238

[34] Hume-Rothery W. Atomic Theory for Students of Metallurgy. London: The Institute of Metals; 1969 (fifth reprint)

[35] Oliver WC, Pharr GMJ. An improved technique for determining hardness and elastic modulus using load and displacement sensing indentation. Journal of Materials Research. 1992;**7**:1564-1583

Chapter 8

Mechanical Properties of GO Nanostructures Prepared from Freeze-Drying Method

Yanhuai Ding, Hui Chen, Zheng Li, Huming Ren,
Xianqiong Tang, Jiuren Yin, Yong Jiang and
Ping Zhang

Additional information is available at the end of the chapter

http://dx.doi.org/10.5772/intechopen.71515

Abstract

Recently, 3D graphene oxide (GO) has attracted much attention due to its high specific surface area, multifunction, and facile preparation. Here, porous GO foams with extraordinary mechanical properties were prepared by using freeze-drying technique. The structure and mechanical properties of the GO foams have been characterized by X-ray diffraction, Fourier transform infrared spectroscopy, atomic force microscopy, and electronic universal testing machine. The unique structure endows the GO foams excellent elasticity, which can recover to its original shape even after compression hundreds of times. The density of GO foams has a significantly positive impact on the elastic modulus. Furthermore, the compressive strength of GO foams decreased linearly with decreasing relative humidity. A honeycomb model was constructed to investigate the effects of wall thickness, length, and included angle on the elastic modulus of GO foams. The structural evolution during the compression was revealed by finite element simulation.

Keywords: GO foams, mechanical properties, humidity sensitivity, Gibson honeycomb model, finite element simulation

1. Introduction

Recently, graphene and graphene oxide (GO) have been widely studied for various applications, such as sensors, field-effect transistors, Li-ion batteries, and polymer composites, due to their remarkable physical and chemical properties. As is well known, preparation of graphene monolayers by chemical reduction of GO sheets is considered as an ideal method to obtain a high yield of graphene. Actually, GO is a kind of functional nanomaterials used widely, which contains large number of hydrophilic groups such as hydroxyl, carboxyl, and carbonyl on both sides of the sheets. Since graphene was reckoned as a strong contender to carbon nanotubes, GO-based materials have attracted a great deal of attention due to their wide applications in gas sensors [1], Li-storage [2, 3], catalyst supports [4], and water purification [5]. GO nanosheets can be easily prepared by chemical exfoliation of graphite in bulk quantities. Due to its good dispersity and ease of post-functionalization, the researchers not only focus on the reduction of GO to graphene but also shift to explore various chemical and physical properties caused by chemical modification and structure tuning. Theoretical and experimental studies have demonstrated that GO exhibited size-dependent properties when its size down to the nanometer scale. GO sheets with different lateral size displayed different mechanical strength.

Though we view GO as potentially powerful and widely applicable, it is very difficult to recycle GO from the dispersion. To overcome this problem, various methods such as blending GO with polymer matrix, formation of GO sponge, and chemical reduced GO gels generated by hydrothermal method have been investigated. Recently, significant progress has been made in self-assembly of GO nanosheets into microporous or mesoporous networks such as GO foams. Several methods have been employed to prepare 3D GO foams such as self-assembly [6, 7], leavening process [8], electrochemical erosion [9], electrospinning method [10, 11], hydrothermal reaction [12], solvent evaporation method [13], and freeze-drying technique [14–16]. Among them, the freeze-drying technique has been intensively studied due to its low-cost, high efficiency, and high yield. The mechanical performance of GO foam is critical to its application under harsh environments; however, most researchers focused on the functionality of GO architectures. Here, 3D GO foams were prepared by freeze drying of GO aqueous dispersion. The relationship between structure and mechanical behavior of GO foams was investigated, and the effect of RH on the mechanical properties of GO foams was also studied. Simultaneously, the structural evolution of GO under uniaxial compression was simulated by finite element method.

2. Experimental details

2.1. Preparation of GO nanosheets

We used a modified Hummers' method to produce chemical exfoliated GO nanosheets with a thickness of 2–5 nm and lateral dimensions of 1–10 μm. In a typical process, graphite powders were oxidized by reacting with a mixture of $NaNO_3$ and concentrated H_2SO_4 in an ice bath, then $KMnO_4$ was added to the dispersion slowly. After fully oxidizing, the graphite oxide was washed with dilute HCl and deionized water pH of the wash solution was near neutral. The obtained graphite oxide was dispersed in deionized water and exfoliated through ultrasonication for 0.5 h.

2.2. Preparation of GO foams

GO foams were fabricated by freeze-drying method to form GO dispersion. The concentration of GO was set at 6.8 gL^{-1}. GO suspension was frozen into an ice cube in a refrigerator (−18°C) and then freeze dried with a condenser temperature of −20°C and inside pressure less than 20 Pa.

2.3. Tests of GO foams

The morphology, structure, and mechanical properties of GO foam were investigated by X-ray diffraction (XRD; Bruker D8 Advance), Fourier transform infrared spectroscopy (FTIR; Perkin-Elmer Spectrum 10), scanning electronic microscopy (SEM; JSM-6610LV), atomic force microscopy (AFM; Veeco Multimode Nanoscope 3D), and electronic universal testing machine (UTM; Instron 5943). The diameter and height of the specimens were 10 and 5 mm, respectively.

3. Results and discussion

AFM image, XRD pattern, and FTIR spectra of as-prepared GO foam are shown in **Figure 1**. Exfoliation of graphite oxide to GO was achieved by ultrasonication of graphite oxide. The exfoliated GO nanosheets are flat with an average thickness of about 2 nm, which means that the single-layer GO sheet was obtained, as shown in **Figure 1a**. XRD pattern of GO contains an intense 001 peak at $2\theta = 10°$ that corresponds to a d-spacing of approximately 0.8 nm. The disappearance of the native graphite peak at about 26° revealed the successful oxidation of the graphite powders. The increased interlayer spacing can weaken the van der Waals interactions between GO layers and make facile exfoliation possible. FTIR spectra of GO show that the spectral bands are corresponding to C-O stretching vibrations (1300–1000 cm^{-1}), C=O stretching vibrations from carbonyl and carboxylic groups (1720–1706 cm^{-1}), C=C stretching vibrations from unoxidized graphitic domains (1450–1680 cm^{-1}), C-H bending vibration (1465–1340 cm^{-1}), and O-H stretching vibrations (3430 cm^{-1}) [17, 18]. The presence of these functional groups on the GO surface leads to strong interaction of GO with water. In addition, the presence of the functional groups makes GO thermally unstable, as it undergoes pyrolysis at elevated temperatures.

Digital picture of a cylinder-like GO foam is presented in **Figure 2a**. The shape and size of the GO foam can be easily adjusted in the freeze-drying process. Foam-like structures with high porosity and flexibility have many important applications in actuators, catalytic supports, adsorption, and separation. The size of the foam can be easily adjusted by simply changing the initial concentration of the GO suspension. SEM image reveals that interconnected pore structure is formed where GO nanosheets traversed laterally and connected with other sheets. This honeycomb features endow GO foam extraordinary mechanical properties, which can undergo compression without collapsing.

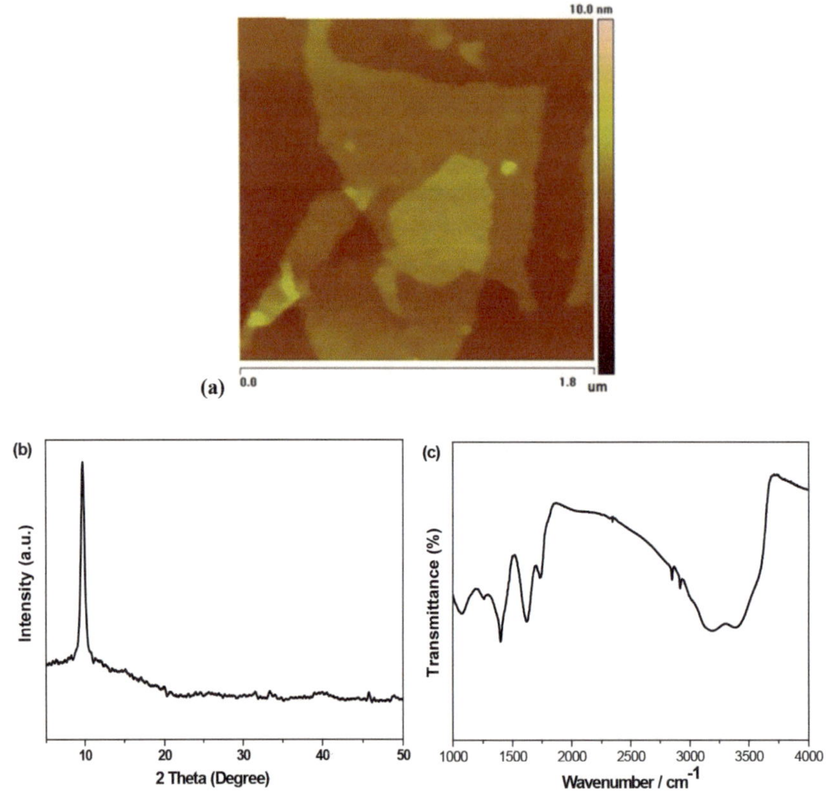

Figure 1. AFM image, XRD pattern, and FTIR spectra of as-prepared GO. (a) AFM image; (b) XRD; (c) FTIR.

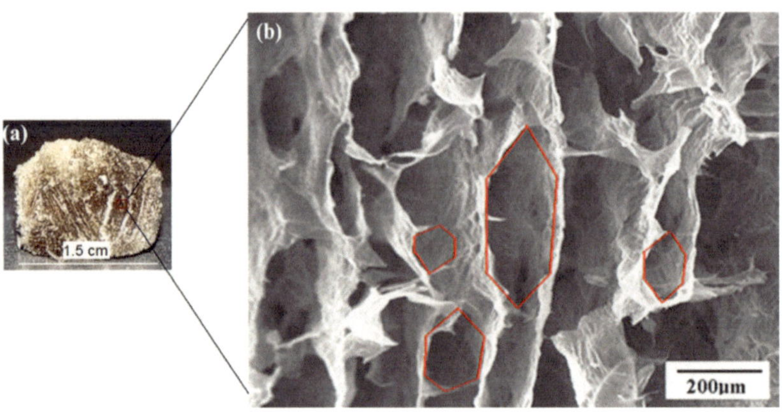

Figure 2. Microstructure of GO foams. The red line shows the honeycomb-like structure in GO foam.

GO foams were submitted to compression for 300 cycles, and each cycle involved an contact between sample and the indenter, a displacement to a prescribed value, and a retraction to the original position. The results of uniaxial compression experiments are shown in **Figure 3a**.

A compression load of 1.5 N was applied, and GO foams were pressed to 80% of the original length. After 300 cycles, the compression load was not changed, which can be ascribed to the well-ordered microstructures oriented along the compression direction. The load-displacement curves recorded during the compression cycling are shown in **Figure 3b**. All curves have a familiar form, in which displacement increases as applied load gradually increases. Once a maximum value of 1.5 N is achieved, the load decreases as the indenter is retracted. As we can see from the figure, no residual displacement is observed after the indenter was fully retracted. GO foams exhibited good elasticity, and it recovered to its original height even after 300 cycles.

The humidity sensitivity of the GO foam was characterized. As depicted in **Figure 4**, the maximum load compressed GO foam to its 80% of the original length is reduced with the increasing of RH. Surface properties of GO sheets were affected greatly by the RH due to its hydrophilicity, which caused a change in the mechanical behavior of GO foams.

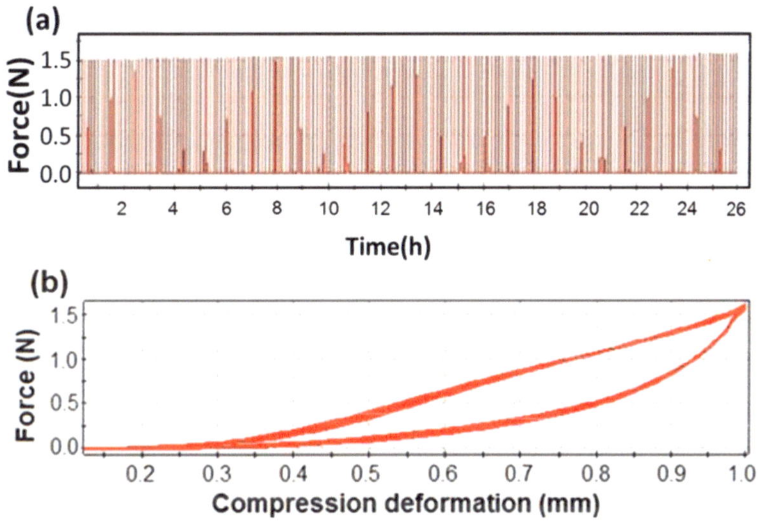

Figure 3. Uniaxial compression data of GO foams. (a) Uniaxial compression test. (b) Compression cycling.

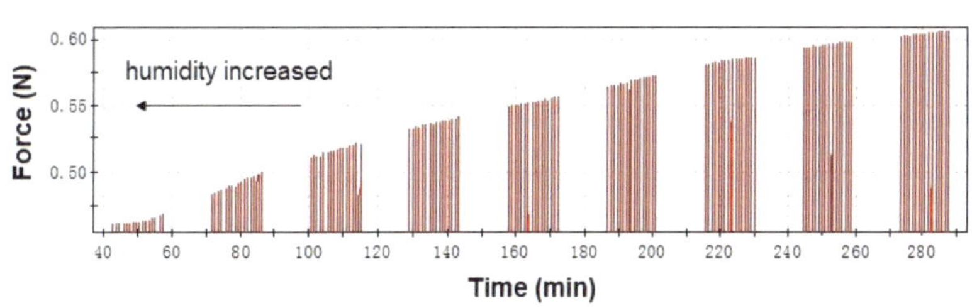

Figure 4. Load-displacement curves at different ambient humidity.

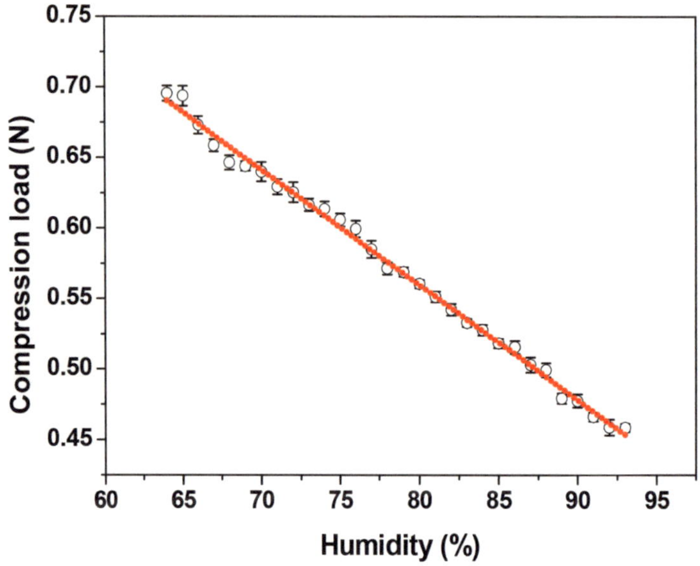

Figure 5. Relationship between the RH and mechanical properties of GO foams.

The relationship between the RH and mechanical properties of GO foams was shown in **Figure 5**. It is surprise that a linear relationship is observed between the maximum compression load and RH. GO foam exhibited very high sensitivity and good linearity with the RH in the range of 65–93%. Here, we give some discussions about the humidity sensing mechanism of the GO nanostructures. We believed that the large decrease in mechanical properties of GO foams was related to the adsorption of water molecules. As porous structure materials, GO foams have higher specific surface areas to absorb moisture easily. Thus, striking interaction between GO sheets was released due to the lubrication of water molecules. It may be the main reason for the decay of mechanical properties.

The pore size and porosity of GO foams were easily adjusted by changing the concentration of GO dispersions in the freeze-drying process. The effects of density and strain rate on the mechanical properties of GO foams are presented in **Figure 6**. The elastic modulus of GO foams with a density of 5.25, 10.93, and 15.45 mg cm^{-3} are 0.0635, 0.1715, and 0.3822 MPa, respectively. As expected, the elastic modulus was significantly affected by the density of the GO foam. GO foam shows a high resistance against the compression with increasing density. The yield stress displayed an increase as the strain rate increased due to the strain rate-dependent behavior of GO foams.

A honeycomb structure was constructed for the calculation of elastic modulus of GO foam. Two hypotheses were proposed: (1) GO foam suffers small deformation under uniaxial compression. (2) Honeycomb structure is remained during the compression. Then, the elastic modulus E and Poisson's ratio v could be expressed as:

http://dx.doi.org/10.5772/intechopen.71515

$$\left.\begin{aligned} E &= \frac{E_s\left(\frac{t}{l}\right)^3\left(\cos\theta+\frac{t}{l}\right)}{\left[\left(\frac{h}{l}+\sin\theta\right)\sin^2\theta\right]\cdot\left[1+(2.4+1.5\,\upsilon_s+\cot^2\theta)\left(\frac{t}{l}\right)^2\right]} \\ \upsilon_{12} &= \frac{\left(\cos\theta+\frac{t}{l}\right)\cos\theta\cdot\left[1+(1.4+1.5\,\upsilon_s)\left(\frac{t}{l}\right)^2\right]}{\left[\left(\frac{h}{l}+\sin\theta\right)\sin\theta\right]\cdot\left[1+(2.4+1.5\,\upsilon_s+\cot^2\theta)\left(\frac{t}{l}\right)^2\right]} \end{aligned}\right\} \quad (1)$$

Where t, l, and θ are the wall thickness, length, and included angle, respectively. The effects of wall thickness, length, and included angle on the mechanical properties of GO foams are shown in **Figure 7**. With an increase of h/l, the elastic modulus of GO foams decreased. However, the elastic modulus decreased with decreasing t/l. Additionally, the elastic modulus decreased with an increase in structural parameter θ.

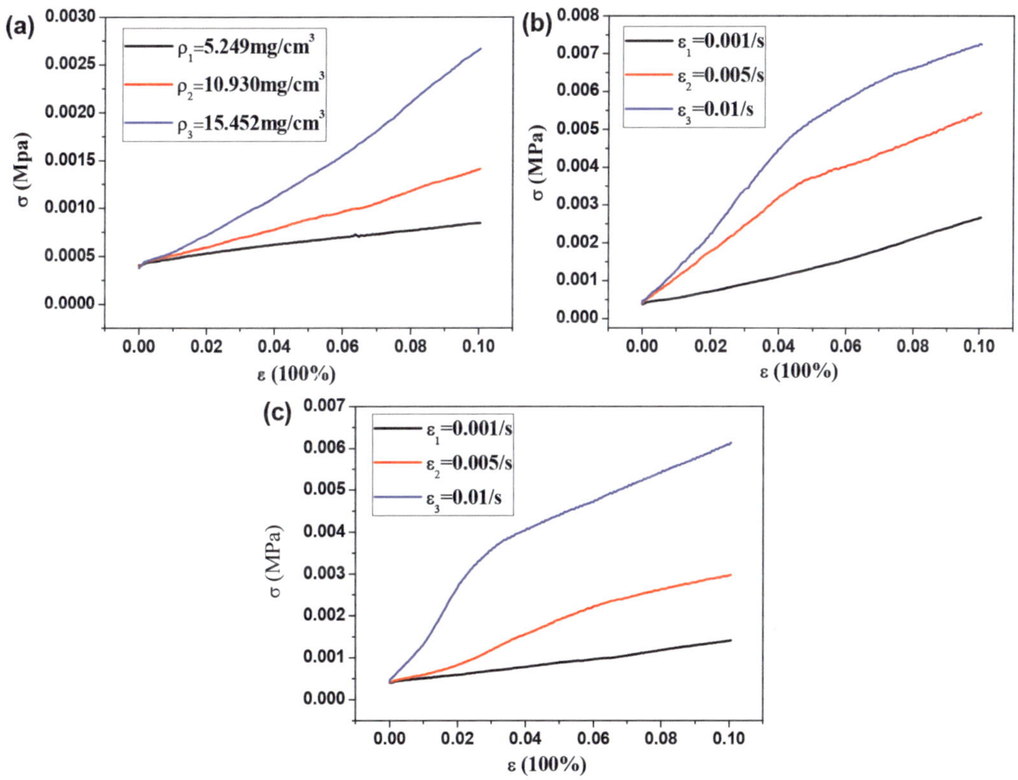

Figure 6. Effects of density and strain rate on the stress-strain behaviors of GO foams. (a) Loading rate 1 mm min^{-1}; (b) density 15.45 mg cm^{-3}; (c) density 10.93 mg cm^{-3}.

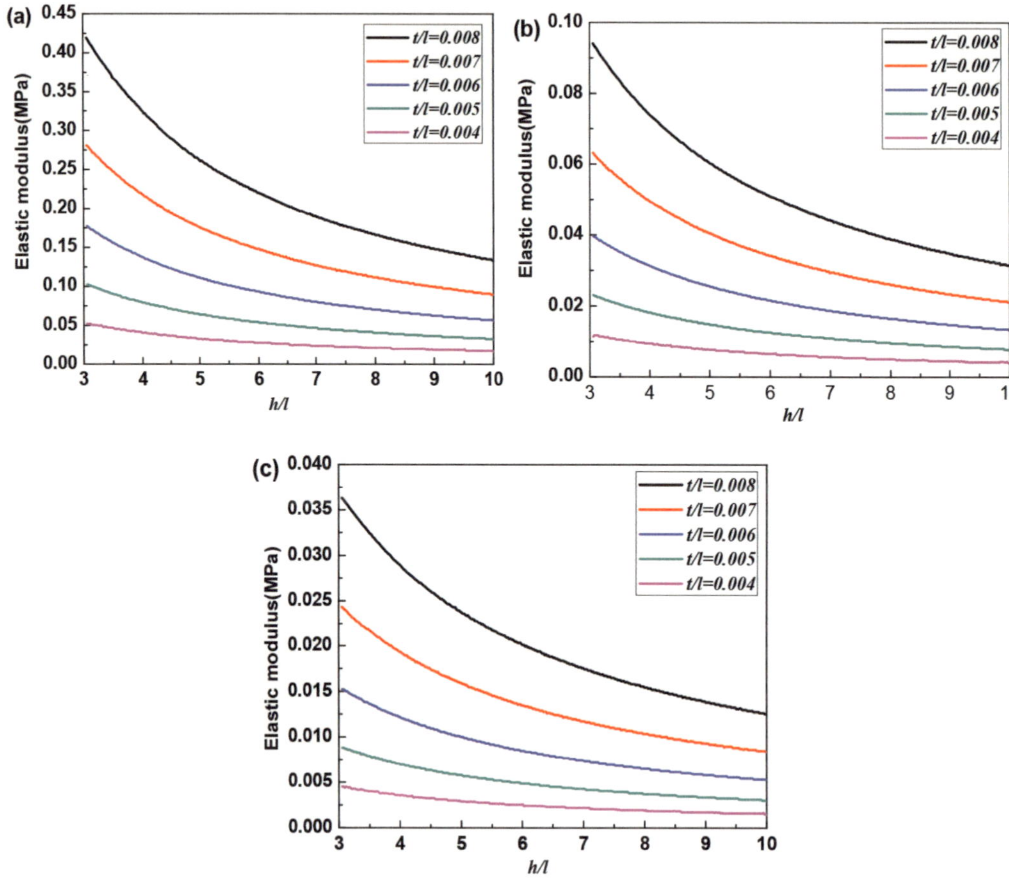

Figure 7. Effects of wall thickness (t), length (l), and included angle (θ) on the elastic modulus of GO foams. (a) $\theta = 15°$; (b) $\theta = 30°$; and (c) $\theta = 45°$.

The change of honeycomb-like GO foams before and after compression was simulated by FE method as shown in **Figure 8**. The FE model was created in Commercial software package (ANSYS 15.0). The bottom was fixed and the compression force was applied on the top of the model. The honeycomb structure can be described as a typical linear elastic material according to the experimental data. Then, the effect of cell size on the mechanical properties of honeycomb structure was calculated by FE method. The compressive stress of GO foams under small deformation can be released by the interconnected GO nanosheets. The strain was mainly concentrated at the middle of the long side when honeycomb structure undergone large deformation as shown in the strain nephogram. Buckling is the type of failure that has been observed in GO foam after compression. From the SEM image, the FE simulation has been proven to be effective in this case. The deformation of the honeycomb-like structure is mainly observed at the long side wall of the cell, which is consistent with the simulation results.

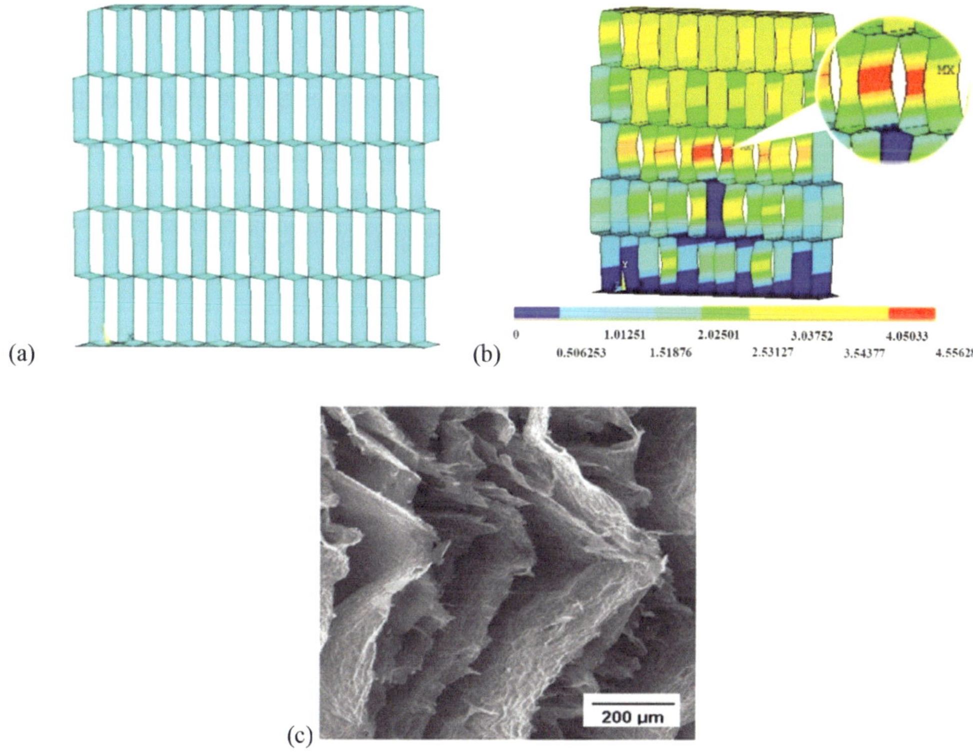

Figure 8. FE simulation and SEM image of honeycomb-like GO foams before and after compression. The insert in **Figure 8b** represents the strain concentrates on the middle of the long side.

4. Conclusions

Honeycomb-like structured GO foams were prepared by freeze-drying method. The GO network structure endows GO foam extraordinary mechanical properties. GO foams can recover to its original height even after 300 compression cycles. GO foams exhibits a linear relationship between the maximum compression load and RH. SEM characterization reveals that the deformation of the honeycomb-like structure under compression is mainly observed at the long side wall of the cell, which is consistent with the FE simulation results.

Acknowledgements

The financial support from the National Natural Science Foundation of China (No. 21376199 and 51,002,128) and Scientific Research Foundation of Hunan Provincial Education Department (No. 17A205) is greatly acknowledged.

Author details

Yanhuai Ding, Hui Chen, Zheng Li, Huming Ren, Xianqiong Tang, Jiuren Yin, Yong Jiang and Ping Zhang*

*Address all correspondence to: zhangp@xtu.edu.cn

Institute of Rheological Mechanics, Xiangtan University, Hunan, China

References

[1] Ming R, Ding Y, Chang F, He X, Feng J, Wang C, Zhang P. Humidity-dependant compression properties of graphene oxide foams prepared by freeze-drying technique. IET Micro & Nano Letters. 2013;**8**:66

[2] Compton OC, Nguyen ST. Graphene oxide, highly reduced graphene oxide, and graphene: Versatile building blocks for carbon-based materials. Small. 2010;**6**:711

[3] Vickery JL, Patil AJ, Mann S. Fabrication of graphene–polymer nanocomposites with higher-order three-dimensional architectures. Advanced Materials. 2009;**21**:2180

[4] Long Y, Zhang C, Wang X, Gao J, Wang W, Liu Y. Oxidation of SO_2 to SO_3 catalyzed by graphene oxide foams. Journal of Materials Chemistry. 2011;**21**:13934

[5] He Y, Liu Y, Wu T, Ma J, Wang X, Gong Q, Kong W, Xing F, Liu Y, Gao J. An environmentally friendly method for the fabrication of reduced graphene oxide foam with a super oil absorption capacity. Journal of Hazardous Materials. 2013;**260**:796

[6] Huang X, Qian K, Yang J, Zhang J, Li L, Yu C, Zhao D. Functional nanoporous graphene foams with controlled pore sizes. Advanced Materials. 2012;**24**:4419

[7] Chen W, Li S, Chen C, Yan L. Self-assembly and embedding of nanoparticles by in situ reduced graphene for preparation of a 3D graphene/nanoparticle aerogel. Advanced Materials. 2011;**23**:5679

[8] Niu Z, Chen J, Hng HH, Ma J, Chen X. A leavening strategy to prepare reduced graphene oxide foams. Advanced Materials. 2012;**24**:4144

[9] Favaro M, Agnoli S, Cattelan M, Moretto A, Durante C, Leonardi S, Kunze-Liebhäuser J, Schneider O, Gennaro A, Granozzi G. Shaping graphene oxide by electrochemistry: From foams to self-assembled molecular materials. Carbon. 2014;**77**:405

[10] Chen Y, Lu Z, Zhou L, Mai Y-W, Huang H. Triple-coaxial electrospun amorphous carbon nanotubes with hollow graphitic carbon nanospheres for high-performance Li ion batteries. Energy & Environmental Science. 2012;**5**:7898

[11] Chen Y, Li X, Zhou X, Yao H, Huang H, Mai Y-W, Zhou L. Hollow-tunneled graphitic carbon nanofibers through Ni-diffusion-induced graphitization as high-performance anode materials. Energy & Environmental Science. 2014;**7**:2689

[12] Xu Y, Sheng K, Li C, Shi G. Self-assembled graphene hydrogel via a one-step hydrothermal process. ACS Nano. 2010;**4**:4324

[13] Ding Y-H, Zhang P, Ren H-M, Zhuo Q, Yang Z-M, Jiang Y. Preparation of graphene/TiO_2 anode materials for lithium-ion batteries by a novel precipitation method. Materials Research Bulletin. 2011;**46**:2403

[14] Ye S, Feng J, Wu P. Highly elastic graphene oxide-epoxy composite aerogels via simple freeze-drying and subsequent routine curing. Journal of Materials Chemistry A. 2013;**1**:3495

[15] Mi X, Huang G, Xie W, Wang W, Liu Y, Gao JP. Preparation of graphene oxide aerogel and its adsorption for Cu^{2+} ions. Journal of Carbon. 2012;**50**:4856

[16] Mohandes F, Salavati-Niasari M. Freeze-drying synthesis, characterization and in vitro bioactivity of chitosan/graphene oxide/hydroxyapatite nanocomposite. RSC Advances. 2014;**4**:25993

[17] Guo H-L, Wang X-F, Qian Q-Y, Wang F-B, Xia X-H. A green approach to the synthesis of graphene nanosheets. ACS Nano. 2009;**3**:2653-2659

[18] Cote LJ, Cruz-Silva R, Huang J. Flash reduction and patterning of graphite oxide and its polymer composite. Journal of the American Chemical Society. 2009;**131**:11027-11032

Section 2

Applications

Chapter 9

Convective Transport Characteristics of Nanofluids in Light-Weight Metal Foams with High Porosity

Huijin Xu, Zhanbin Xing, Fuqiang Wang and Changying Zhao

Additional information is available at the end of the chapter

http://dx.doi.org/10.5772/intechopen.72291

Abstract

Metal foams can be well used as ideal materials for various efficient heat transfer devices due to light weight, high specific, and high thermal conductivity. Nanofluids have higher thermal conductivities than traditional fluid, so it can be used as an efficient heat transfer characteristics medium. This paper focuses on heat transfer of nanofluid, metal foam and the combination of the two. The physical properties of nanofluid and metal foam are summarized. The characteristics of flow and heat transfer are introduced. This work creates a close connection between scientific research and practical applications of this dual heat transfer enhancement method.

Keywords: metal foam, nanofluid, heat transfer, forced convection, natural convection, phase change

1. Introduction

Metal foam owns the advantages of light weight, high specific surface area, high thermal conductivity and relatively high permeability. Owing to recent advances in manufacturing technologies, metal foam becomes commercially available. The metal foam can be well used as an ideal material for manufacturing efficient heat transfer devices: heat exchangers, heat sinks, solar collectors and catalyst reformers. The practical structure of metal foam (copper) is shown in **Figure 1** and the schematic diagram of convective heat transfer through metal foam is shown in **Figure 2**. From **Figures 1** and **2**, the high specific surface area of metal foam, and lots of pores can be found in metal foams. When fluid flowing through metal foam, the high specific surface area can provide a large surface area to the heat transfer. So the metal foam is a very good extended surface of heat transfer. Nanofluid is a special kind of engineered colloids made of a base fluid and nanoparticles (1–100 nm). Nanofluid owns the higher thermal conductivity and single-phase heat transfer coefficient than the base fluid does, so it can be

Figure 1. The structure of convective heat transfer through metal foams.

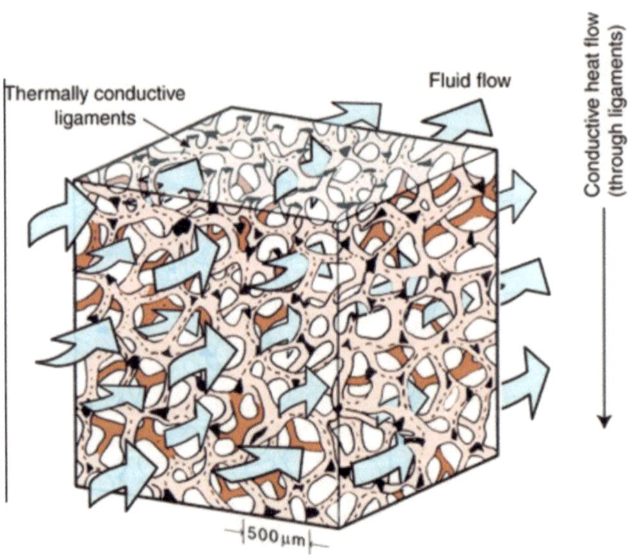

Figure 2. The schematic of convective heat transfer through metal foams.

used as an efficient heat transfer medium. The micrograph of the nanofluid is shown in **Figures 3** and **4** shows the schematic diagram of nanoparticles. Nanoparticle is very small and microscopic effects are very obvious. With the addition of nanoparticles, the physical properties of the nanofluid are changed, which made the nanofluid beneficial to heat transfer and attracts widespread attention of scholars. A tremendous number of investigations on the nanofluid can be found in literatures. Furthermore, a lot of experimental researches were conducted on the convective heat transfer of nanofluids, most of which showed that the nanofluid is able to enhance the convective heat transfer. The advantage of the nanofluid and that of the porous foam can be combined as one to further enhance the heat transfer of thermal equipment. For nanofluids flowing through metal foams, some studies have been reported.

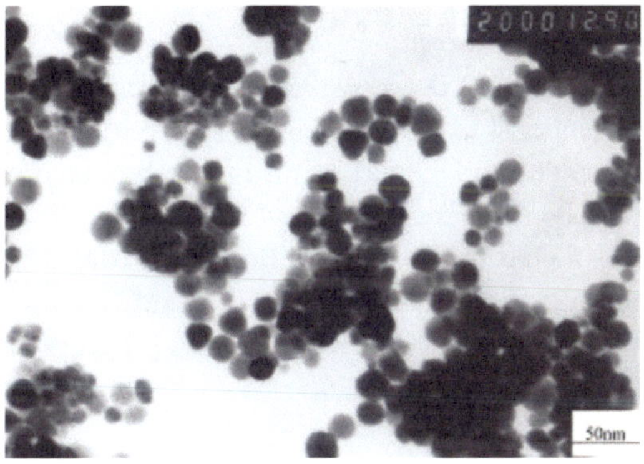

Figure 3. Nanofluid.

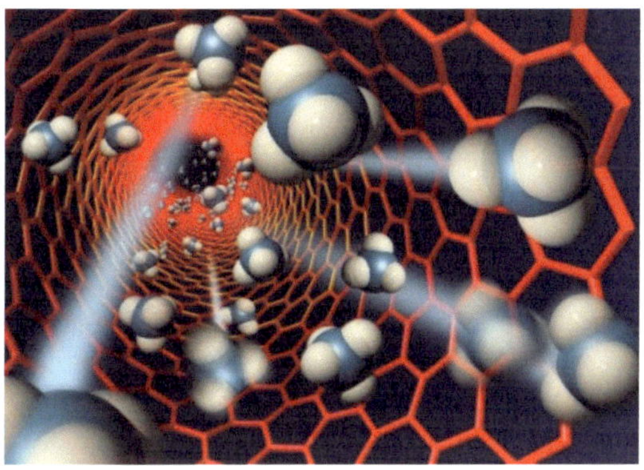

Figure 4. Nanoparticles.

In this chapter, flow and thermal transport of nanofluids in metal foams are presented. The recent advances for metal foams, nanofluids, and the combination of nanofluids and metal foams are reviewed. The performance of forced convection, natural convection, and phase change heat transfer of nanofluids in metal foams are analyzed. The contents and the brief introduction for this chapter are shown in the following. Although there is great application potential of nanofluids in thermal science, little attention has been paid to the effect of non-uniform nanoparticle concentration on convective heat transfer of nanofluid in metal foam based on local thermal non-equilibrium (LTNE) model. Hence, laminar convective heat transfer of nanofluid in metal foam with fully developed hydraulic and thermal fields is discussed. The flow and heat transfer characteristics of nanofluids to this case are discussed as well.

2. Thermal transport in metal foams

Over the last several decades, flow and heat transfer in metal foam have been studied experimentally, theoretically or numerically by many scholars. In this section, the properties and characteristics of metallic porous media are firstly presented and the recent research progress on thermal transport in porous media is reviewed.

2.1. Pressure drop and permeability

The Darcy model is the first model to describe the percolation theory of porous media. In 1856, Darcy proposed a linear relationship between seepage velocity and pressure drop. Although the theory is simple and easy to understand, but it is very limited. Forchheimer modified the Darcy model by adding the inertia terms associated with the velocity square the equation, but still cannot be applied to the turbulent region. Brinkman considered the effective viscous dissipation term and modified the Darcy model, and found that the results are closer to the molecular diffusion at large porosity [1, 2]. In 1952, Ergun [3] proposed the empirical formula for permeability of porous medium:

$$K = \frac{d_f^2 \varepsilon^3}{150(1-\varepsilon)^2} \tag{1}$$

In 1998, Calmidi and Mahajan [4] proposed the empirical formula of the metal foam permeability based on the experiment:

$$\frac{K}{d_p^2} = 0.00073(1-\varepsilon)^{-0.224}\left(\frac{1.18}{1-e^{-(1-\varepsilon)/0.04}}\sqrt{\frac{1-\varepsilon}{3\pi}}\right)^{-1.11}. \tag{2}$$

Bhattacharya and Mahajan [5] and Plessis et al. [6] proposed an empirical formula of the permeability coefficient and the inertia coefficient by using a foam sample with a pore size of 45–100 PPI and a porosity of 0.973–0.978, and employed water and glycerol as the liquid phase. Many scholars studied the pressure drop and the permeability of the flow in porous media, and the formulas were given in **Table 1**.

2.2. Effective thermal conductivity

The effective thermal conductivity is of great significance for the study of heat transfer mechanism in porous media. Maxwell [10] firstly studied the effective thermal conductivity of

Time	Researcher	Empirical formula	Equation numbers
1994	Plessis et al. [6]	$\frac{K}{{d_l}^2} = \frac{\varepsilon^2}{36\chi(\chi-1)}, F = \frac{2.05\chi(\chi-1)}{\varepsilon^2(3-\chi)} \cdot \frac{\sqrt{K}}{d_l}$	(3)
2000	Paek et al. [7]	$\frac{\Delta p\sqrt{K}}{L\rho u^2} = f = \frac{1}{Re_K} + 0.105 = \frac{\mu}{\rho u\sqrt{K}} + 0.105$	(4)
2006	Liu et al. [8]	$f = 22\frac{1-\varepsilon}{Re_{d_p}} + 0.22, 30 < Re_{d_p} < 300$ $f = 0.22, Re_{d_p}$	(5)
2002	Fourie and Plessis [9]	$f_{Fourie} = (3-\chi)(\chi-1)\frac{C_D\chi^{1.5}}{24\varepsilon^3}$ $C_D = 1 + (10/Re^{0.667}) = 1 + 10(\rho ud(\chi-1)/2\mu\varepsilon)^{-0.667}$	(6)

Table 1. Models for predicting pressure drop and permeability of flow in porous media.

porous media immersion heat transfer. After that, many scholars studied the effective thermal conductivity of porous media. Most studies of effective thermal conductivity are mainly focused on the volume fraction of each component:

$$k_e = \varepsilon k_f + (1-\varepsilon)k_s \tag{7}$$

The above model is mainly concentrated on sand, cylindrical, spherical packing bed and fiber insulation blanket, but the estimation of the effective thermal conductivity of metal foam has a large deviation from the experimental result [11]. Calmidi and Mahajan [4] respectively measured the effective thermal conductivity of ERG aluminum foam at the low temperature (ignoring radiation heat transfer) by using air and water as the flow phase. Boomsma and Poulikakos [12] proposed an efficient thermal conductivity model for predicting the three-dimensional ideal cellular structure of metal foam, which is in good agreement with the experimental data of Calmidi and Mahajan [4]. Bhattacharya and Mahajan [5] used the circular cylinder model to modify the Calmidi's correlation. Hadim and North [13] generalized the correlation coefficient of the thermal conductivity model proposed by Wakao et al. [14], and make it applicable to calculate the effective thermal diffusivity and the stagnation thermal conductivity of sintered porous media. The formulas were given in **Table 2**.

Time	Researcher	Empirical formula	Equation numbers
1992	Calmidi and Mahajan [4]	$k_{fe} = \left\{\frac{2}{\sqrt{3}}\left[\frac{r\left(\frac{b}{L}\right)}{k_f+\left(1+\frac{b}{L}\right)\frac{k_f}{3}} + \frac{(1-r)\frac{b}{L}}{k_f+\left(\frac{b}{L}\right)\frac{2}{3}\left(-k_f\right)} + \frac{\frac{\sqrt{3}}{2}-\frac{b}{L}}{k_f+\left(\frac{b}{L}\right)\frac{4r}{3\sqrt{3}}\left(-k_f\right)}\right]\right\}^{-1}$	(8)
		$k_{se} = \left\{\frac{2}{\sqrt{3}}\left[\frac{r\left(\frac{b}{L}\right)}{\left(1+\frac{b}{L}\right)\frac{k_s}{3}} + \frac{(1-r)\frac{b}{L}}{\frac{2}{3}\left(\frac{b}{L}\right)k_s} + \frac{\frac{\sqrt{3}}{2}-\frac{b}{L}}{\frac{4r}{3\sqrt{3}}\left(\frac{b}{L}\right)k_s}\right]\right\}^{-1}$	(9)
		$\frac{b}{L} = \frac{-r+\sqrt{r^2+\frac{2}{\sqrt{3}}(1-\varepsilon)\left[2-r\left(1+\frac{4}{\sqrt{3}}\right)\right]}}{\frac{2}{3}\left[2-r\left(1+\frac{4}{\sqrt{3}}\right)\right]}, r = 0.09$	(10)
		$k_f = \alpha_l k_l + \alpha_v k_v$	(11)
2001	Boomsma and Poulikakos [12]	$\frac{k_{ec}}{k_s} = \frac{1}{\sqrt{2}}\left\{\frac{4\lambda}{2e^2+\lambda\pi(1-e)} + \frac{3e-2\lambda}{e^2} + \frac{\left(\sqrt{2}-2e\right)^2}{2\pi\lambda^2\left(1-2e\sqrt{2}\right)}\right\}^{-1}$	(12)
		$\mathrm{e} = 0.339, \lambda = \sqrt{\frac{\sqrt{2}\left(2-\frac{5\sqrt{2}}{8}e^3-2\varepsilon\right)}{\pi\left(3-4\sqrt{2}e-e\right)}}$	
2002	Bhattacharya and Mahajan [5]	$k_{fe} = \left[\frac{2}{\sqrt{3}}\left(\frac{t/L}{k_f+\left(\left(k_s-k_f\right)/3\right)} + \frac{\frac{\sqrt{3}}{2}-t/L}{k_f}\right)\right]^{-1}$	(13)
		$\frac{t}{L} = \frac{-\sqrt{3}-\sqrt{3+(1-\varepsilon)\left(\sqrt{3}-5\right)}}{1+\frac{1}{\sqrt{3}}-8/3}$	(14)
1982	Wakao [14]	$(k_e)_x = \varepsilon k_f + (1-\varepsilon)k_s + 0.5 Pr Re_d\left(\frac{u}{u_m}\right)k_f$	(15)
		$(k_e)_y = \varepsilon k_f + (1-\varepsilon)k_s + 0.1 Pr Re_d\left(\frac{u}{u_m}\right)k_f$	(16)

Table 2. Models for predicting effective thermal conductivity of porous media.

2.3. Convective heat transfer coefficient

Lu et al. [15, 16] studied the forced convection characteristics of shell-and-tube heat exchangers filled with high porosity metal foams. Qu et al. [17] experimentally studied the natural convection of air in a open-call copper foam, and found that there is a turning point in the Grashof number for small porosity (ε = 0.9). Guo [18] numerically simulated the laminar forced-convection heat transfer in a porous medium flat plate channel with constant heat flux and analyzed the flow and heat transfer performance. Fand et al. [19] immersed the porous medium in water or silicone oil with the porous medium randomly stacked by glass spheres. Many researchers studied convective heat transfer of flow in porous media, the formulas for predicting Nusselt number were given at **Table 3**.

Because of the difference between the thermal conductivity of the fluid and that of the metal foam, the heat is diffused at a different rate between the two phases. So some researchers hold that the solid and fluid phases have different temperatures, namely LTNE model. Convective heat transfer performance in metal foams was numerically investigated based on the local thermal equilibrium (LTE) model and the LTNE model and the velocity and temperature fields was obtained.

The steady forced convective heat transfer in a tube fully filled with metal foam is numerically considered under the boundary condition of a uniform temperature. Effects of porosity on mean Nusselt number with LTE/LTNE models are shown in **Figure 5**. The LTE and LTNE Nusselt numbers are both decreased with an increase in foam porosity. The relative deviation is reduced by increasing porosity, due to the greatly decreased solid effective thermal resistance. When porosity is greater than 95%, the relative deviation between LTE/LTNE Nusselt numbers is lower than 20%. For ε >95%, the LTE model can be treated as a rapid estimation tool for thermal performance of metal foams.

Difference between solid and fluid thermal conductivities is the most significant quantity for metal foam LTNE effect. **Figure 6** presents the effects of thermal conductivity ratio on mean Nusselt numbers with LTE/LTNE models. The Nu difference for LTE/LTNE models is reduced when thermal conductivity ratio is increased, which is attributed to that k_{fe}/k_{se} is increased

Time	Researcher	Empirical formula	Equation numbers
1982	Fand et al. [19]	$NuPr^{0.0877} = 0.618Ra^{0.698} + 8.54 \times 10^6 Ge \times \text{sech}Ra, (0.001 < Re_{max} = 3)$	(17)
		$NuPr^{0.0877} = 0.766Ra^{0.374}\left(\frac{C_1 d}{C_2}\right)^{0.173} (3 < Re_{max} = 100)$	(18)
2003	Boomsma et al. [20]	$Nu = \frac{q}{A_{con}\left(T_{pl} - T_{c,inlet}\right)}\frac{D_{hyd}}{k_c} = \frac{mc\left(T_{c,outlet} - T_{c,inlet}\right)}{A_{con}\left(T_{pl} - T_{c,inlet}\right)}\frac{D_{hyd}}{k_c}$	(19)
2007	Arisetty et al. [21]	$Nu = \frac{2Rh}{k_f} = \frac{2Rq_w}{k_f\left(T_w - T_{f,b}\right)}$	(20)
2000	Calmidi and Mahajan [22]	$Nu_{sf} = \frac{h_{sf} d_f}{kf} = C_T Re_{d_f}^{0.5} Pr^{0.37} = C_T\left(\frac{V d_f}{\varepsilon \upsilon}\right)^{0.5} Pr^{0.37}$	(21)
2005	Brito and Rodríguez [23]	$Nu = 1.1Re^{0.43}Pr^{1/3}, 10 < Re < 100$	(22)
		$Nu = 1.2Re^{0.43}Pr^{1/3}, 20 < Re < 240$	(23)

Table 3. Models for predicting Nusselt number of heat transfer in porous media.

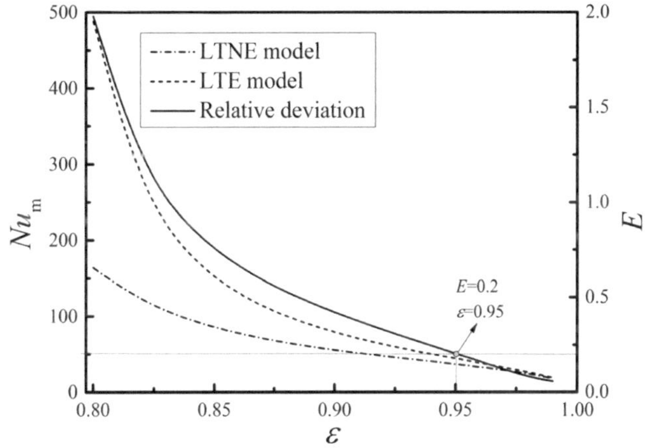

Figure 5. Effect of porosity on mean Nusselt number.

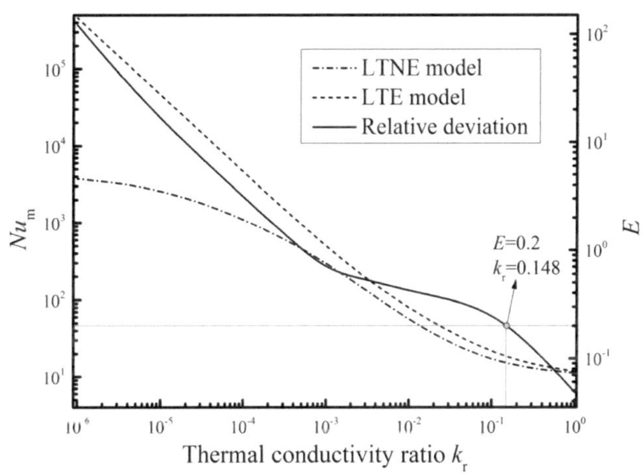

Figure 6. Effect of thermal conductivity ratio on mean Nusselt number with LTE/LTNE models.

with an increase in thermal conductivity ratio. When fluid thermal conductivity equals to solid thermal conductivity ($k_r = 1$), LTE Nusselt number coincides with LTNE Nusselt number, in which condition with the LTNE effect is negligible and the LTE assumption holds. In addition, the relative deviation gradually decreases to zero with an increase in thermal conductivity ratio. The relative deviation is reduced to 20% when thermal conductivity ratio is increased to 0.148. The LTE model can be roughly used for prediction of heat transfer in porous foams instead of LTNE model in the range $k_f/k_s > 0.148$.

3. Transport phenomena in nanofluids

The concept of nanofluid, by adding nanoparticles into a base fluid, is firstly proposed in 1995 [24]. Since then, lots of work has been done on the transport phenomena of nanofluid. In this work, the basic features of nanofluid are comprehensively presented. Nanofluid is a new type heat exchange medium which is made by mixing highly conductive nanoparticles and the traditional heat transfer fluid. Due to the addition of nanoparticles, the density, the thermal conductivity and the viscosity of nanofluid are obviously different from those of traditional media, and can be used as a more efficient heat exchange medium.

3.1. Thermal conductivity

Lee et al. [25] have measured the thermal conductivity of four nanofluids: copper oxide and water, copper oxide and ethylene glycol, alumina and water, alumina and ethylene glycol. Li and Xuan [26] analyzed the mechanisms of nanofluids to improve the thermal conductivity. Xie et al. [27] measured the thermal conductivity of the alumina nanoparticle suspension. The influence of pH value of suspension, specific surface area of the dispersed system, crystallization of solid phase and thermal conductivity of the base fluid on the nanofluid thermal conductivity was studied. Eastman et al. [28] measured the thermal conductivity of the copper nanofluid and found that the thermal conductivity of nanoparticles is increased obviously. Guo [29] used KD-2 thermal analyzer to measure the thermal conductivity of the nanofluid. Using the temperature oscillation technique, Das et al. [30] prove that the thermal conductivity of copper oxide/water and alumina/water increases with an increase in the temperature and a decrease in the particle size. Patel et al. [31] have also obtained similar conclusions through experiments. Ebrahimnia-Bajestan et al. [32] applied nanofluids to the solar system, and studied the laminar convection heat transfer of the TiO_2/water nanofluid in a tube by experimental and numerical methods. Many researchers proposed models of nanofluids thermal conductivity base on experimental study [33–47], and the formulas were given in **Table 4**.

3.2. Viscosity and friction factor

At present, there is no suitable theory to predict the viscosity of nanofluids accurately. Einstein [48] proved that the relative viscosity of the suspension is a simple function of the volume fraction of suspended particles. Scholars revised the formula in different aspects, and put forward their correction models respectively [39, 49]. With a least-square curve fitting, Maïga et al. [50] proposed a correlation based on some experimental data available in the open literature. Shafahi et al. [51] indicated that the nanofluid viscosity is a function of the temperature and proposed the correlations. Scholars proposed correlations of different nanoparticle types. The formulas of nanofluids viscosity were given in **Table 5**.

Xuan et al. [34, 53] found that the friction factor of nanofluids is almost the same as that of water at the same velocity, and is independent of the volume fraction of nanoparticles. Therefore, the friction factor of nanofluids is calculated with a single-phase model:

$$\lambda_{\mathrm{nf}} = \frac{\Delta P_{\mathrm{nf}} d}{L} \frac{2g}{u_{\mathrm{m}}^2} \tag{24}$$

3.3. Convective heat transfer

As a new type heat exchanging medium, the nanofluid has a very pronounced enhancement effect on the convective heat transfer. Scholars have carried out a series of studies on the

Time	Researcher	Empirical formula	Equation numbers
1873	Maxwell [33]	$k_{nf} = \frac{k_{np}+2k_{bf}+2(k_{np}-k_{bf})\phi}{k_{np}+2k_{bf}-(k_{np}-k_{bf})\phi} k_{bf}$	(25)
1999	Lee et al. [25]	$k_{nf} = \frac{k_{np}+(n-1)k_{bf}-(n-1)(k_{bf}-k_{np})\phi}{k_{np}+(n-1)k_{bf}+(k_{bf}-k_{np})\phi} k_{bf}, n = 3/\psi$	(26)
2011	Lee et al. [35]	$\frac{k_{nf}}{k_{bf}} = 1 + 4.4\left(\frac{T}{T_{fr}}\right)^{10}\left(\frac{k_{np}}{k_{bf}}\right)^{0.03} Pr^{0.66} Re^{0.4}$	(27)
2003	Wang et al. [37]	$k_{nf} = \frac{(1-\phi)+3\phi\int_0^\infty \frac{k_{cl}(r)n(r)}{k_{cl}(r)+2k_{bf}}dr}{(1-\phi)+3\phi\int_0^\infty \frac{k_{cl}n(r)}{k_{cl}(r)+2k_{bf}}dr} k_{bf}$	(28)
2004	Yu et al. [38]	$k_{nf} = \frac{k_{np}+2k_{bf}+2(k_{np}-k_{bf})(1+\beta)^3\phi}{k_{np}+2k_{bf}-(k_{np}-k_{bf})(1+\beta)^3\phi} k_{bf}$	(29)
2013	Hadadian [39]	$\frac{k_{nf}}{k_{bf}} = \frac{k_{np}+2k_{bf}-2\phi(k_{bf}-k_{np})}{k_{np}+2k_{bf}+\phi(k_{bf}-k_{np})} + \frac{\rho_{np}\phi c_{p,np}}{2k_{bf}}\sqrt{\frac{k_B T}{3\pi r_{cl}\mu_{bf}}}$	(30)
2005	Xue and Xu [40]	$\left(1-\frac{\phi}{\alpha}\right)\frac{k_{nf}-k_{bf}}{2k_{nf}+k_{bf}} + \frac{\phi(k_{nf}-k_{lr})(2k_{lr}+k_{np})-\alpha(k_{np}-k_{lr})(2k_{lr}+k_{nf})}{\alpha(2k_{nf}+k_{bf})(2k_{lr}+k_{np})-2\alpha(k_{np}-k_{lr})(k_{lr}+k_{nf})} = 0, \alpha = \left(\frac{d_{np}}{d_{np}+t}\right)^3$	(31)
2002	Keblinski et al. [42]	$Re_{d_{np}} = \frac{C_{RM} d_{np}}{\upsilon}$	(32)
2005	Xue [44]	$k_{nf} = k_{bf}\frac{1-\phi+2\phi\frac{k_{np}}{k_{np}-k_{bf}}ln\frac{k_{bf}+k_{np}}{2k_{bf}}}{1-\phi+2\phi\frac{k_{bf}}{k_{np}-k_{bf}}ln\frac{k_{bf}+k_{np}}{2k_{bf}}}$	(33)
2016	Esfe et al. [45]	$\frac{k_{nf}}{k_{bf}} = 0.991 + 0.276T\phi + 77.6\phi^2 + 3641.2T\phi^2 + \frac{0.00217}{\sin(T-\phi)} - 6.01\times10^{-6}T^2 - 3647.1T\phi\sin\phi$	(34)
2005	Chon et al. [46]	$\frac{k_{nf}}{k_{bf}} = 1 + 64.7\phi^{0.746}\left(\frac{d_{bf}}{d_{np}}\right)^{0.369}\left(\frac{k_{np}}{k_{bf}}\right)^{0.7476} Pr^{0.9955} Re^{1.2321}$	(35)
2011	Khanafer and Vafai [47]	$\frac{k_{nf}}{k_{bf}} = 1 + 1.011\phi + 2.438\phi\left(\frac{47}{d_{np}}\right) - 0.025\phi\left(\frac{k_{np}}{0.613}\right)$	(36)

Table 4. Models for predicting nanofluid thermal conductivity.

Time	Researcher	Empirical formula	Equation numbers
1906	Einstein [48]	$\mu_{nf} = (1+2.5\phi)\mu_{bf}$	(37)
2005	Maïga et al. [50]	$\mu_{nf} = (1+1.73\phi+123\phi^2)\mu_{bf}$	(38)
2010	Shafahi et al. [51]	$\mu_{nf} = 2.9\times10^{-7}T^2 - 2.0\times10^{-4}T + 3.4\times10^{-2}$ for $\phi = 1\%$ $\mu_{nf} = 3.4\times10^{-7}T^2 - 2.3\times10^{-4}T + 3.9\times10^{-2}$ for $\phi = 4\%$	(39)
2013	Yang et al. [52]	$\mu_{nf} = \mu_{bf}(1+39.11\phi+533.9\phi^2)$ for alumina $\mu_{nf} = \mu_{bf}(1+5.45\phi+108.2\phi^2)$ for titania	(40)

Table 5. Models for predicting nanofluid viscocity.

convective heat transfer of nanofluids. Xuan et al. [34, 53] established an experimental system to measure the convective heat transfer coefficient of nanofluid and the laminar flow and turbulent flow friction factors in the channel. In nanofluid, the nanoparticles undergo thermophoretic motion with in the temperature gradient field. Researchers have taken more and more attention to the thermophoretic motion of nanoparticles [54, 55]. For the heating of the side wall in a rectangular channel, Berkovski-Polevikov's coefficients have good agreement with the experimental data with length-width ratio between 1 and 10, and MacGregor-Emery's coefficient has good agreement with the experimental data with length-width ratio greater than 10 [56]. Maïga et al. [50] considered the influence of the nanoparticle volume fraction and the Reynolds number on the average convective heat transfer coefficient of water-based nanofluid. Sakai et al. [57] improved the Buongiorno model for the convective heat transfer of nanofluids, so that it can be applied to continuity equations, momentum equations and energy equations without the effect of nanoparticle volume fraction distribution. Jia and Wang [58] improved the Eubank-Proctor model and fitted out a coefficient of mixture flow considering natural convection. Yang et al. [59] made two kinds nanofluids using graphite nanoparticles, and measured the laminar-flow convective heat transfer coefficient in a horizontal tube heat exchanger. Formulas for Nusselt number of nanofluids convection were given in **Table 6**.

Buongiorno [60] proposed a mathematical model on the non-uniform volume traction of nanoparticles. He assumed incompressible flow, no chemical reactions, negligible external forces, dilute mixture, negligible viscous dissipation, negligible radiative heat transfer, and LTE between nanoparticles and the base fluid [52, 60].

The forced convective heat transfer of the nanofluid in a plain tube at the full development section was studied by the numerical method. **Figure 7** is the effect of nanoparticle volume fraction on the Nusselt number and k_w/k_B (k_w is the thermal conductivity of the wall and k_B is the average thermal conductivity). It can be seen that the Nusselt number increases at first and then decreases with an increase in the nanoparticle volume fraction, and there is a maximum value of the Nusselt number with a suitable nanoparticle volume fraction. The increase in the nanoparticle volume traction leads to a more uniform velocity, which is beneficial to the convective heat transfer. In addition, the high nanoparticle volume fraction can increases the thermal conductivity, so an increase in the nanoparticle volume traction is beneficial to the convective heat transfer. The ratio of the wall thermal conductivity to the average thermal conductivity decreases with an increase in the nanoparticle volume fraction. When the ration of k_w/k_B decreases, the Nusselt number is reduced. The effect of nanoparticle volume traction is

Time	Researcher	Empirical formula	Equation numbers
2000	Xuan and Roetzel [53]	$Nu_{\mathrm{nf}} = 0.4328\left(1.0 + 11.285\phi^{0.754} Pe_{\mathrm{np}}^{0.218}\right) Re_{\mathrm{nf}}^{0.333} Pr_{\mathrm{nf}}^{0.4}$ for laminar flow $Nu_{\mathrm{nf}} = 0.0059\left(1.0 + 7.6286\phi^{0.6886} Pe_{\mathrm{np}}^{0.001}\right) Re_{\mathrm{nf}}^{0.9238} Pr_{\mathrm{nf}}^{0.4}$ for turbulent flow	(41)
2011	Corcione [56]	$Nu = 0.18\left(\frac{PrRa}{0.2+Pr}\right)^{0.29}\left(\frac{H}{W}\right)^{0.13}, \quad 1 \le \frac{H}{W} \le 2, 10^{-3} \le Pr \le 10^5, 10^3 \le \frac{PrRa}{0.2+Pr}\left(\frac{H}{W}\right)^{-3}$ $Nu = 0.22\left(\frac{PrRa}{0.2+Pr}\right)^{0.28}\left(\frac{H}{W}\right)^{-0.09}, \ 2 \le \frac{H}{W} \le 10, Pr \le 10^5, Ra \le 10^{13}$	(42)
2005	Maïga et al. [50]	$Nu = 0.086Re^{0.55}Pr^{0.5}$ for constant heat flux $Nu = 0.28Re^{0.35}Pr^{0.36}$ for constant wall temperature	(43)
2015	Jia and Wang [58]	$Nu_{\mathrm{nf}}\left(\frac{\mu_{\mathrm{wnf}}}{\mu_{\mathrm{nf}}}\right)^{0.14} = 2.11\left[Gz_{\mathrm{nf}} + 0.574\left(Gr_{\mathrm{nf}}Pr_{\mathrm{nf}}\frac{d}{L}\right)^{0.75}\right]^{1/3}, Gz_{\mathrm{nf}} = \frac{d}{L}Re_{\mathrm{nf}}Pr_{\mathrm{nf}}$	(44)
2005	Yang et al. [59]	$\frac{1}{U} = \frac{1}{h_{nf}\left(\frac{A_i}{A_o}\right)} + \frac{D_o}{2k} ln \frac{D_o}{D_i} + \frac{1}{h_o} = \frac{A_o \Delta T_{lm}}{Q}$ $\Omega = Nu_{\mathrm{nf}} Pr_{\mathrm{nf}}^{-1/3}\left(\frac{L}{D}\right)^{1/3}\left(\frac{\mu_{\mathrm{nf}}}{\mu_{\mathrm{w}}}\right)^{-0.14}$	(45)

Table 6. Models for predicting Nusselt number of nanofluid convective heat transfer.

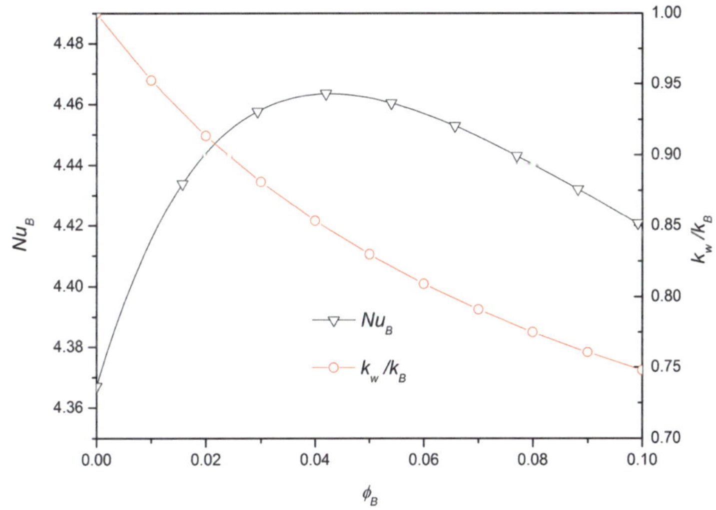

Figure 7. Effect of nanoparticle volume fraction on mean Nusselt number.

more obvious with a larger k_w/k_B, and the effect of k_w/k_B is more obvious with larger nanoparticle volume traction.

Figure 8 is the relationship between the Nusselt number and N_{BT}. N_{BT} is a dimensionless parameter related to the Brown motion and the thermophoretic motion. It can be found that the nanofluid is unbenefited for heat transfer with low N_{BT} (<0.2). There is a maximum Nusselt number when N_{BT} is from about 0.4 to 0.5. Then the Nusselt number decreases with an increase in N_{BT}. When the N_{BT} is close to 10 or greater, the Nusselt number tends to be constant. The enhancement via the Brown diffusion motion causes nanoparticle to disturb the

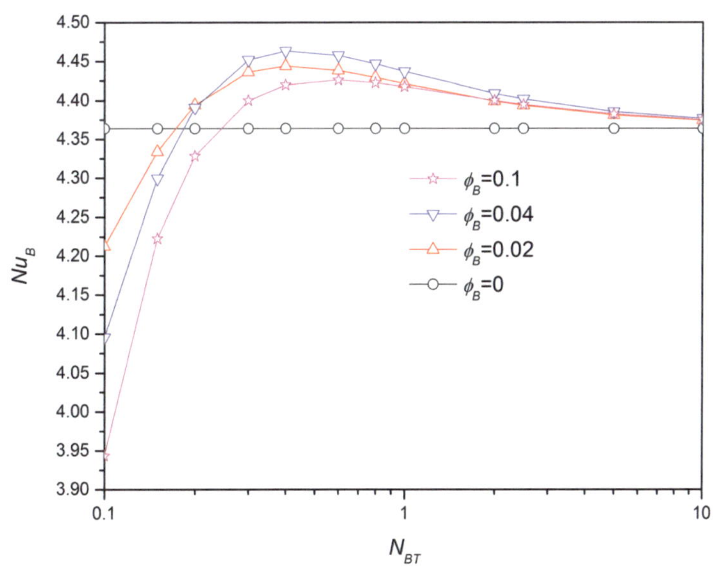

Figure 8. Effect of N_{BT} on mean Nusselt number.

flow more effectively, causing local turbulence to enhance the heat transfer between nanoparticles and the base liquid. Nanoparticles will move to the cold region (wall) by thermophoresis diffusion. For large nanoparticle aggregating, nanoparticle of other areas is too small, so it has little heat transfer enhancement with too large N_{BT}.

4. Convection of nanofluids in metal foams

Even though metal foams own excellent thermal performance, poor heat conduction ability of most heat transfer fluids restricts further heat transfer improvement in metal foams, for which the combination of the metal foam and the nanofluid with highly conductive nanoparticles is a promising solution. In this chapter, the transport characteristics of nanofluids flowing through metal foams. In this chapter, the recent advances on the forced convection and natural convection of nanofluids in porous foams will be firstly reviewed and the latest research concerns from the perspective of fundamental research will be put forward.

4.1. Experimental data

Cheng [61] tested the heat transfer performance of the heat pipe with different nanofluid volume fractions and liquid filling rates, and also tested the heat transfer performance of the screen suction core heat pipe. Hajipour et al. [62] studied the mixed convection of alumina/water nanofluid in a vertical square channel partially filled with open metal foams under the constant wall heat flux using the experimental and numerical method. Goodarzi et al. [63] studied the laminar and turbulent mixing flow and heat transfer of Cu/water nanofluids in a shallow rectangular cavity using a two-phase mixture model. Mao [64] studied the generation, fusion and detachment of boiling bubbles on the smooth plate and foam metal surface. Nazari et al. [65] studied the influence of the interaction between nanofluid and porous medium of extended surface on the heat exchanger thermal performance, and the forced convection of alumina/water nanofluids in a circular tube filled with metal foams was studied experimentally with isothermal boundary conditions.

4.2. Modeling the forced convective heat transfer

Matin and Pop [66] studied the force convection heat transfer of nanofluids in a horizontal porous medium channel at fully developed section with constant heat flux. Xu et al. [67] investigated the dual heat transfer enhancement of nanofluids flowing in a metal foam channel by numerical method based on the local non thermal equilibrium model. Mahdi et al. [68] summarized the influence of the porosity, permeability, inertial coefficient and effective heat exchange coefficient of porous media, and also studied the effect of thermodynamic parameters of nanofluids. Sivasankaran and Narrein [69] proposed a numerical simulation of laminar pulsating heat transfer and hydraulic characteristics of alumina/water nanofluid in a three-dimensional spiral microchannel radiator, using the modified viscosity equation and the two-phase mixing model.

In Xu et al. [70], velocity and temperature fields are numerically obtained. The effects of some key parameters on flow and heat transfer of nanofluid in porous media are analyzed. For the nanofluid flowing through metal foams, the nanoparticle volume fraction is a most important parameter, the effect of which on pressure drop is shown in **Figure 9**. As can be seen, with the increase in volume fraction, the pressure drop per unit length gradually increases and the increasing amplitude for pressure drop also increases. This is attributed to that the dynamic viscosity and the density of nanofluid are increased sharply with the increase in volume fraction. **Figure 10** shows the effect of nanoparticle volume fraction on heat transfer for two different nanoparticles (Al_2O_3 and TiO_3). As the nanoparticle volume fraction increases, Nusselt number gradually increases but the increasing amplitude is reduced. This is attributed to that the thermal conductivity increasing amplitude is decreased with an increase in

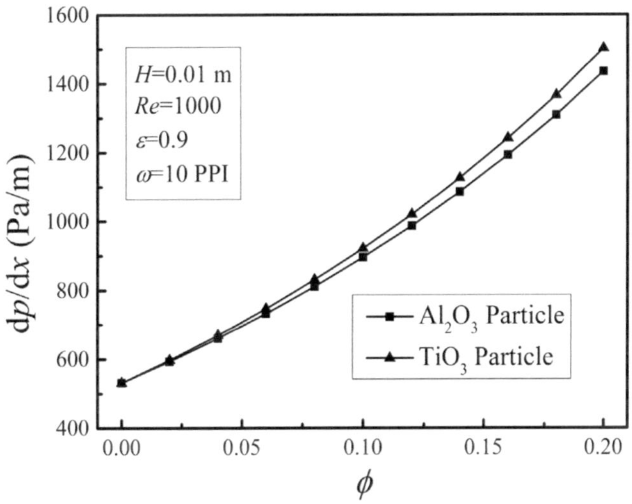

Figure 9. Effects of nanoparticle volume fraction on the overall pressure drop.

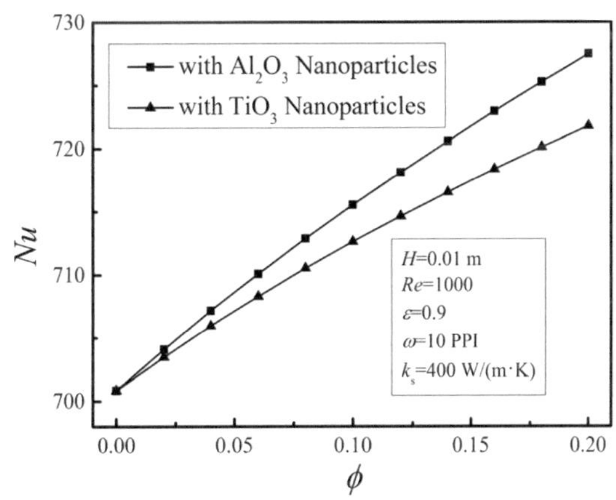

Figure 10. Effects of nanoparticle volume fraction on heat transfer.

nanoparticle volume fraction. Due to thermal conductivity of Al_2O_3 is higher than that of TiO_3, Nusselt number of Al_2O_3 is higher than that of TiO_3 as shown in **Figure 10**. From **Figure 10**, the maximum heat transfer augmentation of nanofluid is about 3.8% for Al_2O_3 and 3.0% for TiO_3, which is very useful for further improving thermal performance of metal foam heat exchangers and heat sinks, especially for high heat-flux applications.

4.3. Modeling the natural convective heat transfer

Sun and Pop [71] studied the steady natural convection of the water-based nanofluid in a right triangle shell filled a porous medium using the numerical method. It is found that the average Nusselt number can be increased by increasing the nanoparticle volume fraction under a low Rayleigh number, but the average Nusselt number decreases with an increase in the nanoparticle volume fraction under a high Rayleigh number. Sherement [72] established a Buongiorno mathematical model for the three-dimensional natural convection of nanofluids in porous media, and considered that the heterogeneous models of nanoparticles are more suitable. Bhadauria and Agarwal [73] proposed a detailed model of the nanofluid saturated porous layer.

A lattice Boltzmann (LB) model for the nanofluid natural convection in a porous medium was established by using the volume-averaging method. **Figures 11** and **12** show the velocity and temperature distributions for $Ra = 10^4$ and 10^6, respectively. The other parameters are set to $Da = 0.0001$, $\varepsilon = 0.6$, and $\phi = 0.5\%$. For $Ra = 10^4$, the temperature is very uniform in the y direction, and the heat is transferred mainly by heat conduction. Streamlines are nearly in parallel with the gravitational direction. In **Figure 12**, the high temperature region at the upper left side and the low temperature region the bottom right side respectively diffuse heat with a

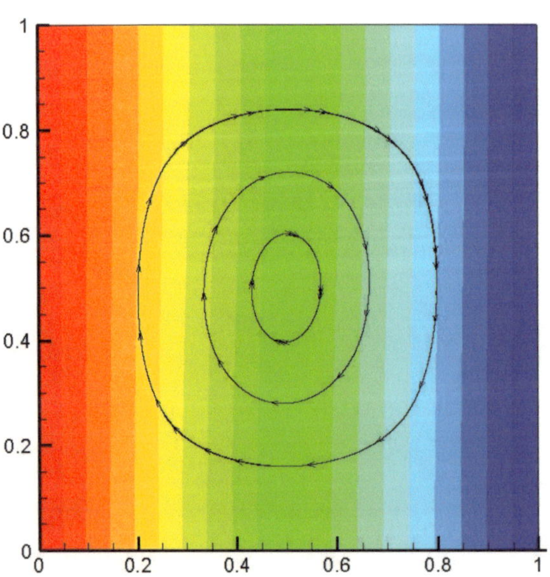

Figure 11. The distributions of the flow and the temperature for $Ra = 10^4$.

http://dx.doi.org/10.5772/intechopen.72291

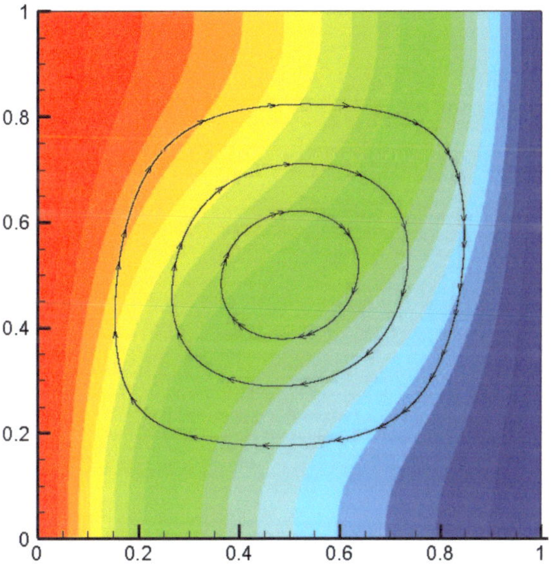

Figure 12. The distributions of the flow and the temperature for $Ra = 10^6$.

quicker speed than other region. The change of the heat transfer regime from the heat conduction to the natural convection is clearly shown in **Figures 11** and **12**.

Figure 13 shows the effect of the Rayleigh number on the average Nusselt number with $Da = 10^{-3}$ and $Da = 10^{-4}$. With an increase in the Rayleigh number, the increasing amplitude of Nu_{ave} is mild for the small Rayleigh number, and Nu_{ave} increases sharply for large Rayleigh numbers. The effect of Ra on the flow and Nu_{ave} is more obvious for the high Darcy number.

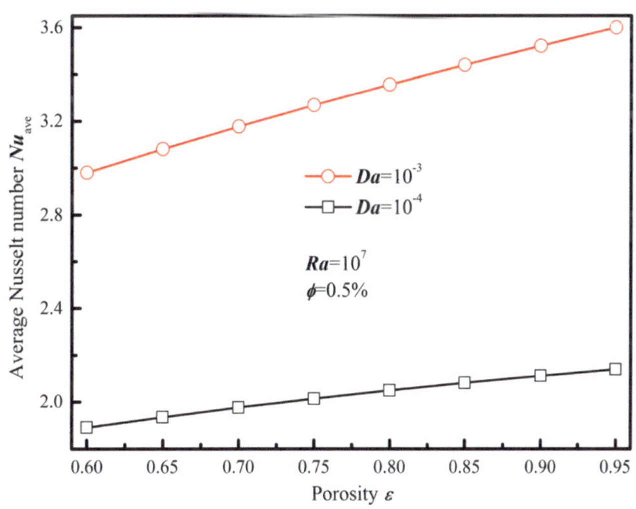

Figure 13. Effect of the Rayleigh number on the heat transfer.

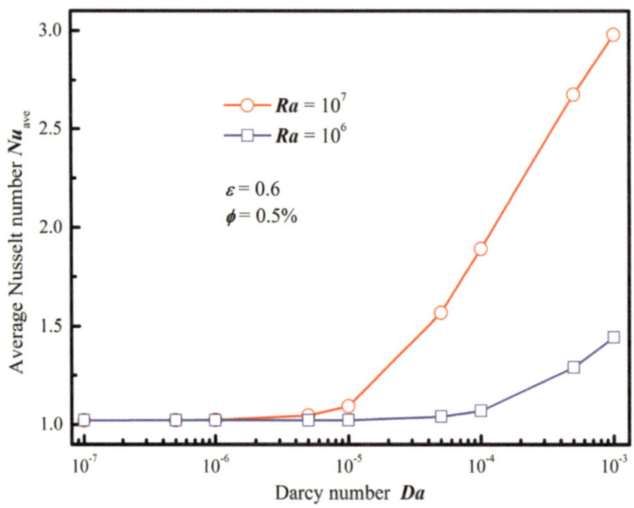

Figure 14. Effect of the Darcy number on the heat transfer.

Figure 14 shows the detailed relationship between the Darcy number and the average Nusselt number. The average Nusselt number almost does not change when *Da* is less than 10^{-5}. As *Da* increases from 10^{-7} to 10^{-6}, Nu_{ave} only increases by 0.03% for $Ra = 10^6$ and by 0.14% for $Ra = 10^7$. Natural convection almost can be ignored and there is only heat conduction when the Darcy number is less than 10^{-5}.

Figure 15 shows the effect of the nanoparticle volume traction on Nu_{ave} with different thermal conductivities of nanoparticles (k_p). Physical characteristics of nanofluids differ for different nanoparticle volume tractions. The thermal conductivity of the nanofluid increases with an

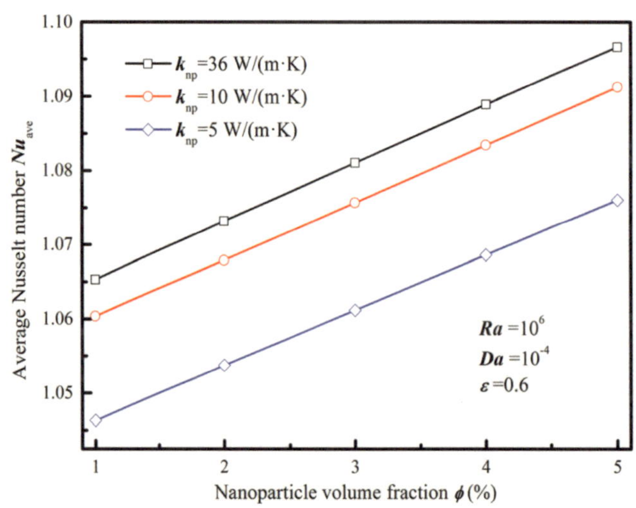

Figure 15. Effect of the nanoparticle concentration on the heat transfer.

increase in the nanoparticle volume traction, and it is beneficial for promoting the heat transfer, which leads to the increased average Nusselt number.

5. Phase change heat transfer

Phase change heat transfer of nanofluid in porous foams is a relatively new theme. In this chapter, the basic scientific problems for this topic will be firstly presented and then the recent research advances will be reviewed. The future research points will also be discussed.

5.1. Liquid-gas phase change heat transfer

Boiling heat transfer is used in a variety of industrial processes and applications, such as refrigeration, power generation, heat exchangers, cooling of high-power electronics components and cooling of nuclear reactors [74]. The use of nanofluids for boiling heat transfer enhancement is a promising solution that is currently being explored by many researchers for pool boiling applications.

Lee and Mudawar [75] have undertaken an experimental study to explore the benefits of using alumina/water nanofluid for microchannel cooling applications. They revealed the enhancement of the heat transfer coefficient for single-phase laminar flow. However, in the two-phase regime, the nanofluids caused the surface deposition in micro-channels, and large agglomerates of nanoparticles were formed. Kim et al. [76] investigated the subcooled flow boiling using dilute alumina, zinc oxide and diamond water-based nanofluids. Kim et al. [77] studied the pool boiling by experiment with water-based nanofluids containing Al_2O_3, ZrO_2 and SiO_2 nanoparticles. An irregular porous structure was formed at the surface. You et al. [78] measured the CHF in pool boiling using a flat, square copper heater submerged into nanofluids at a sub atmospheric pressure of 2.89 psia. Nanoparticle deposition was observed by Bang and Chang [79], who also measured a CHF enhancement of 50% with alumina-water nanofluids on a stainless steel plate. Zhu et al. [80] developed a boiling heat transfer coefficient correlation of the refrigerant/lubricating oil mixture on the surface of the metal foam surface.

Several researchers have noticed the nano-deposition at the heater surface, which can alter the surface area, the surface wettability and the bubble nucleation. The nucleation site density, the bubble departure diameter and the bubble frequency are all affected by the nanofluid boiling. It was found by several researchers [77, 78] that bubble diameters increase during boiling with nanofluids, but the nucleation site density decreases with the addition of nanoparticles into the base fluid.

5.2. Liquid-solid phase change heat transfer

Liquid-solid phase change in porous media is frequently encountered in lots of natural and engineering systems. Over the past several decades, this problem has been extensively investigated analytically, experimentally and numerically [81]. Thermal management systems based on latent heat storage of phase change materials (PCMs) can be widely used. Many researches are focused on demonstrating the performance improvement over pure PCM-based thermal management systems and the free and forced-convection heat transfer phenomena inside the porous media [82, 83].

Hong and Herling [84] experimentally studied the effect of surface area density on the performance of paraffin-infiltrated aluminum foams with pore sizes from 500 to 2 mm. Lafdi et al. [85] also conducted an experimental study with paraffin-infiltrated aluminum foams and found that both pore size and porosity affected the performance of the system. Tian and Zhao [86] performed similar experiments with copper foams. With the advancement of fabrication techniques for microcellular metal foams [87], the effect of pore size and porosity becomes more interesting due to the extremely large surface area enabled by metal foams with small pore size. Numerical models were also developed to predict the temperature profile of PCM metal foam systems. These models had an origin in Boomsma and Poulikakos [12], where the effective thermal conductivity (k_e) of an infiltrated porous metal foam was estimated based on a tetrakaidecahedron pore model. Tao et al. [88] investigated the latent heat storage (LHS) performance of metal foams/paraffin composite phase change material (CPCM) using lattice Boltzmann method. Gao et al. [89] also used the lattice Boltzmann method to simulate solid-liquid phase change with natural convection in porous media under LTNE conditions.

A new sort of nanofluid phase change material (PCM) is developed by suspending a small amount of nanoparticles in melting paraffin by Wu et al. [90]. Zheng et al. [91] found that Ag/1-Tetradecanol showed remarkably high thermal conductivity and reasonably high phase change enthalpy. Khodadadi et al. [92] numerically simulated the solidification of Cu/H_2O nanofluids in a vertical square enclosure. Guo [93] numerically obtained that room with alumina/paraffin as PCM ceiling is a good way of saving the required cool energy in summer. Wu et al. [94] investigated the effects of Cu nanoparticles on the thermal conductivity and the phase change heat transfer of Cu/paraffin PCM by the Hot Disk thermal constants analyzer and infrared monitoring methods respectively. The results show that adding nanoparticles is an efficient way to enhance the phase change heat transfer of PCM.

6. Summary

Metal foams and nanofluids are greatly potential for the application of practical thermal applications since they are beneficial for heat transfer enhancement. A review of previous study for different convective flow and heat transfer regimes about the metal foam and the nanofluid is presented in this article. The effects of several parameters in metal foam and nanofluid properties, thermal boundary conditions, and flow and heat transfer characteristics were analyzed. Previous studies have shown that nanofluid and metal foam can enhance heat transfer. Some suggestions for future works should be paid attention to, as turbulent flow of nanofluids flow in metal foams, new models for the heat transfer of nanofluids in metal foams, the micro effect of nanofluid, the non-Newtonian effect of nanofluids, and the slip effect of nanofluid in metal foams.

Acknowledgements

This work is supported by the National Natural Science Foundation of China (No. 51406238), the Fundamental Research Funds for the Central Universities (No. 17CX02047), the Foundation for Outstanding Young Scientist in Shandong Province (No. BS2014NJ009), and the Postdoctoral Science Foundation of China (No. 2015 M570363).

Author details

Huijin Xu[1,2]*, Zhanbin Xing[2], Fuqiang Wang[3] and Changying Zhao[1]

*Address all correspondence to: hjxu1015@gmail.com

1 China-UK Low Carbon College, Shanghai Jiao Tong University, Shanghai, China

2 Department of Energy and Power Engineering, College of Pipeline and Civil Engineering, China University of Petroleum (Huadong), Qingdao, China

3 School of Automobile Engineering, Harbin Institute of Technology at Weihai, Weihai, China

References

[1] Zheng KC, Wen Z, Wang ZS, Guo-Feng L, Liu XL, Wu WF. Review on forced convection heat transfer in porous media (in Chinese). Acta Physica Sinica. 2012;**61**(1):1-11

[2] Amhalhel G, Furmański P. Problems of modeling flow and heat transfer in porous media. Biulctyn Instytutu Techniki Cieplnej Politechniki Warsza Wskiej. Journal of Power Technologies. 1997;**85**:55

[3] Ergun S. Fluid flow through packed columns. Chemical Engineering Progress. 1952;**48**(2): 89-94

[4] Calmidi VV, Mahajan RL. The effective thermal conductivity of high porosity fibrous metal foams. Journal of Heat Transfer. 1992;**121**(2):466-471

[5] Bhattacharya AC, Mahajan R. Thermophysical properties of high porosity metal foams. International Communications in Heat and Mass Transfer. 2002;**45**:1017-1031

[6] Plessis PD, Montillet A, Comiti J, Legrand J. Pressure drop prediction for flow through high porosity metallic foams. Chemical Engineering Science. 1994;**49**:3545-3553

[7] Paek JW, Kang BH, Kim SY, Hyun JM. Effective thermal conductivity and permeability of aluminum foam materials. International Journal of Thermophysics. 2000;**21**(2):453-464

[8] Liu JF, WT W, Chiu WC, Hsieh WH. Measurement and correlation of friction characteristic of flow through foam matrixes. Experimental Thermal and Fluid Science. 2006;**30**(4): 329-336

[9] Fourie JG, Plessis JPD. Pressure drop modelling in cellular metallic foams. Chemical Engineering Science. 2002;**57**(14):2781-2789

[10] Maxwell JC. A Treatise on Electricity and Magnetism. Dover, New York: Oxford University Press; 1954

[11] Zhao CY. Review on thermal transport in high porosity cellular metal foams with open cells. International Journal of Heat and Mass Transfer. 2015;**55**(13):3618-3632

[12] Boomsma K, Poulikakos D. On the effective thermal conductivity of a three-dimensionally structured fluid-saturated metal foam. International Journal of Heat and Mass Transfer. 2001;**44**(4):827-836

[13] Hadim H, North M. Forced convection in a sintered porous channel with inlet and outlet slots. International Journal of Thermal Sciences. 1999;**44**(1):33-42

[14] Wakao N, Kagei SD. Heat and mass transfer in packed beds. AICHE Journal. 1982;**1**(2): 193-199

[15] Lu W, Zhao CY, Tassou SA. Thermal analysis on metal-foam filled heat exchangers. Part I: Metal-foam filled pipes. International Journal of Heat and Mass Transfer. 2006;**49**(15): 2751-2761

[16] Zhao CY, Lu W, Tassou SA. Thermal analysis on metal-foam filled heat exchangers. Part II: Tube heat exchangers. International Journal of Heat and Mass Transfer. 2006;**49**(15): 2762-2770

[17] Qu ZG, Xu ZG, Tao WQ, Lu TJ. Experimental study of natural convective heat transfer in horizontally-positioned cellular metal foams with open cells (in Chinese). Journal of Xi'an Jiao Tong University. 2009;**43**(1):4

[18] Guo LY. The Optimal Analysis and Effect of Flow and Heat Transfer in a Parallel-Plate Channel Filled with Porous Media (in Chinese). Lanzhou: Lanzhou University of Technology; 2013

[19] Fand RM, Steinberger TE, Cheng P. Natural convection heat transfer from a horizontal cylinder embedded in a porous medium. International Journal of Heat and Mass Transfer. 1986;**29**:119-133

[20] Boomsma K, Poulikakos D, Zwick F. Metal foams as compact high performance heat exchangers. Mechanics of Materials. 2003;**35**(12):1161-1176

[21] Arisetty S, Prasad AK, Advani SG. Metal foams as flow field and gas diffusion layer in direct methanol fuel cells. Journal of Power Sources. 2007;**165**(1):49-57

[22] Calmidi VV, Mahajan RL. Forced convection in high porosity metal foams. Journal of Heat Transfer. 2000;**122**(3):557-565

[23] Brito J, Rodríguez W. Heat transfer characterization of metallic foams. Industrial and Engineering Chemistry Research. 2005;**44**(24):9078-9085

[24] Choi SUS. Enhancing thermal conductivity of fluids with nanoparticle, developments and applications of non-Newtonian flows. Applied Physics A: Materials Science and Processing. 1995;**231**(66):99-105

[25] Lee SP, Choi S, Li S, Eastman JA. Measuring thermal conductivity of fluids containing oxide nanoparticles. Journal of Heat Transfer. 1999;**121**(2):280-289

[26] Li Q, Xuan YM. A preliminary analysis of the intensified thermal-conductivity mechanism of nano-fluids (in Chinese). Journal of Engineering for Thermal Energy and Power. 2002;**17**(6):568-571

[27] Xie HQ, Wang JC, Xi TG, Liu Y, Ai F. Thermal conductivity enhancement of suspensions containing nanosized alumina particles. Journal of Applied Physics. 2002;**91**(7): 4568-4572

[28] Eastman JA, Choi SUS, Li S, Yu W, Thompson LJ. Anomalously increased effective thermal conductivities of ethylene glycol-based nanofluids containing copper nanoparticles. Applied Physics Letters. 2001;**78**(6):718-720

[29] Guo SS. The Researeh on thermophys1eal Characteristic of Nanofluids (in Chinese). Hangzhou: Zhejiang University; 2006

[30] Das SK, Putra N, Thiesen P, Raetzel W. Temperature dependence of thermal conductivity enhancement for nanofluids. Journal of Heat Transfer. 2003;**125**(4):567-574

[31] Patel HE, Das SK, Sundararajan T, Nair AS. Thermal conductivities of naked and monolayer protected metal nanoparticle based nanofluids: Manifestation of anomalous enhancement and chemical effects. Applied Physics Letters. 2003;**83**(14):2931-2933

[32] Ebrahimnia-Bajestan E, Moghadam MC, Niazmand H, et al. Experimental and numerical investigation of nanofluids heat transfer characteristics for application in solar heat exchangers. International Journal of Heat and Mass Transfer. 2016;**92**(1):1041-1052

[33] Maxwell JC. Treatise on Electricity and Magnetism. Oxford: Clarendon Press; 1873

[34] Xuan Y, Li Q. Investigation on convective heat transfer and flow features of nanofluids. Journal of Heat Transfer. 2003;**125**(1):151-155

[35] Lee JH, Lee SH, Choi C, Jang S, Choi S. A review of thermal conductivity data, mechanisms and models for nanofluids. International Journal of Micro-nano Scale Transport. 2011;**1**(4):269-322

[36] Corcione M. Empirical correlating equations for predicting the effective thermal conductivity and dynamic viscosity of nanofluids. Energy Conversion and Management. 2011; **52**:789-793

[37] Wang BX, Zhou LP, Peng XF. A fractal model for predicting the effective thermal conductivity of liquid with suspension of nanoparticles. International Journal of Heat and Mass Transfer. 2003;**46**(14):2665-2672

[38] Yu W, Choi SUS. The role of interfacial layers in the enhanced thermal conductivity of nanofluids: A renovated Maxwell model. Journal of Nanoparticle Research. 2004;**6**(4): 355-361

[39] Hadadian M, Samiee S, Ahmadzadeh H, et al. Nanofluids for heat transfer enhancement–A review. Physical Chemistry Research. 2013;**1**(1):1-33

[40] Xue Q, Xu WM. A model of thermal conductivity of nanofluids with interfacial shells. Materials Chemistry and Physics. 2005;**90**(2):298-301

[41] Jang SP, Choi SUS. Role of Brownian motion in the enhanced thermal conductivity of nanofluids. Applied Physics Letters. 2004;**84**(21):4316-4318

[42] Keblinski P, Phillpot SR, Choi SUS, Eastman JA. Mechanisms of heat flow in suspensions of nano-sized particles (nanofluids). International Journal of Heat and Mass Transfer. 2002;**45**(4):855-863

[43] Gao L, Zhou XF. Differential effective medium theory for thermal conductivity in nanofluids. Physics Letters A. 2006;**348**(3):355-360

[44] Xue QZ. Model for thermal conductivity of carbon nanotube-based composites. Physica B Condensed Matter. 2005;**368**(1–4):302-307

[45] Esfe MH, Afrand M, Yan WM, Akbari M. Applicability of artificial neural network and nonlinear regression to predict thermal conductivity modeling of Al_2O_3-water nanofluids using experimental data. International Journal of Heat and Mass Transfer. 2016;**66**:246

[46] Chon CH, Kihm KD, Lee SP, Choi SUS. Empirical correlation finding the role of temperature and particle size for nanofluid (Al_2O_3) thermal conductivity enhancement. Applied Physics Letters. 2005;**87**(15):435

[47] Khanafer K, Vafai K. A critical synthesis of thermalphysical characteristics of nanofluids. International Journal of Heat and Mass Transfer. 2011;**54**(19):4410-4428

[48] Einstein A. Eineneuebestimmung der moleküldimensionen. Annalen der Physik. 1906;**324**: 289-306

[49] Brinkman HC. The viscosity of concentrated suspensions and solutions. Journal of Chemical Physics. 1952;**20**(4):571-581

[50] Maïga SEB, Palm SJ, Nguyen CT, Roy G, Galanis N. Heat transfer enhancement by using nanofluids in forced convection flows. International Journal of Heat and Fluid Flow. 2005;**26**(4):530-546

[51] Shafahi M, Bianco V, Vafai K, Manca O. Thermal performance of flat-shaped heat pipes using nanofluids. International Journal of Heat and Mass Transfer. 2010;**53**(7):1438-1445

[52] Yang C, Li W, Sano Y, Mochizuki M, Nakayama A. On the anomalous convective heat transfer enhancement in nanofluids: A theoretical answer to the nanofluids controversy. Journal of Heat Transfer. 2013;**135**(5):054504

[53] Xuan Y, Roetzel W. Conceptions for heat transfer correlation of nanofluids. International Journal of Heat and Mass Transfer. 2000;**43**(19):3701-3707

[54] Semenov NS. Mechanism of particle thermophoresis in pure solvents. Philosophical Magazine. 2003;**83**(17):2199-2208

[55] Fu HL. Theoretical Study on Thermophysical Properties and Migration Properties of Nanofluids (in Chinese). Suzhou: Soochow University; 2013

[56] Corcione M. Heat transfer in nanoparticle suspensions. International Journal of Heat and Mass Transfer. 2011;**32**:65-77

[57] Sakai F, Li W, Nakayama A. A Rigorous Derivation and its Applications of Volume Averaged Transport Equations for Heat Transfer in Nanofluid Saturated Metal Foams. The International Heat Transfer Conference; 2014

[58] Jia T, Wang RX. Convective heat transfer characteristics of MWNTs water-based nanofluid (in Chinese). Journal of Refrigeration. 2015;**36**(1):35-39

[59] Yang Y, Zhang ZG, Grulke EA, Anderson WB, Wu G. Heat transfer properties of nanoparticle-in-fluid dispersions (nanofluids) in laminar flow. International Journal of Heat and Mass Transfer. 2005;**48**(6):1107-1106

[60] Buongiorno. Convective transport in nanofluids. Journal of Heat Transfer. 2006;**128**(3): 240-250

[61] Cheng QF. Experimental Research on Heat Transfer Characteristics of Nanofluids in Foam Metal Heat Pipe (in Chinese). Zhenjiang: Jiangsu University; 2013

[62] Hajipour M, Dehkordi AM. Mixed-convection flow of Al_2O_3–H_2O nanofluid in a channel partially filled with porous metal foam: Experimental and numerical study. Experimental Thermal and Fluid Science. 2014;**53**(2):49-56

[63] Goodarzi M, Safaei MR, Vafai K, Ahmadi G, Dahari M. Investigation of nanofluid mixed convection in a shallow cavity using a two-phase mixture model. International Journal of Thermal Sciences. 2014;**75**:204-220

[64] Mao YB. The heat transfer characteristics of metal foams in nanofluids pool boiling (in Chinese). Refrigeration. 2015;**34**(4):6-10

[65] Nazari M, Ashouri M, Kayhani MH, Tamayol A. Experimental study of convective heat transfer of a nanofluid through a pipe filled with metal foam. International Journal of Thermal Sciences. 2015;**83**:33-39

[66] Matin MH, Pop I. Forced convection heat and mass transfer flow of a nanofluid through a porous channel with a first order chemical reaction on the wall. International Communications in Heat and Mass Transfer. 2013;**46**(8):134-141

[67] Xu HJ, Gong L, Huang SB, Qu ZG, Xu MH. Heat transfer enhancement of nanofluids in metal foams (in Chinese). Journal of Engineering Thermophysics. 2014;**35**(8):1586-1590

[68] Mahdi RA, Mohammed HA, Munisamy KM, Saeid NH. Review of convection heat transfer and fluid flow in porous media with nanofluid. Renewable and Sustainable Energy Reviews. 2015;**41**:715-734

[69] Sivasankaran S, Narrein K. Numerical investigation of two-phase laminar pulsating nanofluid flow in helical microchannel filled with a porous medium. International Communications in Heat and Mass Transfer. 2016;**75**:86-91

[70] Xu HJ, Gong L, Huang SB, Xu MH. Flow and heat transfer characteristics of nanofluid flowing through metal foams. International Journal of Heat and Mass Transfer. 2015; **83**(83):399-407

[71] Sun Q, Pop I. Free convection in a triangle cavity filled with a porous medium saturated with nanofluids with flush mounted heater on the wall. International Journal of Thermal Sciences. 2011;**50**(11):5141-5153

[72] Sherement MA, Pop I, Rahman MM. Three-dimensional natural convection in a porous enclosure filled with a nanofluid using Buongiorno's mathematical model. International Journal of Heat and Mass Transfer. 2015;**82**:396-405

[73] Bhadauria BS, Agarwal S. Convective transport in a nanofluid saturated porous layer with thermal non equilibrium model. Transport in Porous Media. 2011;**88**:107-131

[74] Barber J, Brutin D, Tadrist L. A review on boiling heat transfer enhancement with nanofluids. Nanoscale Research Letters. 2011;**6**(1):1-16

[75] Lee J, Mudawar I. Assessment of the effectiveness of nanofluids for single-phase and two-phase heat transfer in micro-channels. International Journal of Heat and Mass Transfer. 2007;**50**(3):452-463

[76] Kim SJ, Mckrell T, Buongiorno J, Hu LW. Subcooled flow boiling heat transfer of dilute alumina, zinc oxide, and diamond nanofluids at atmospheric pressure. Nuclear Engineering and Design. 2010;**240**(5):1186-1194

[77] Kim SJ, Bang IC, Buongiorno J, Hu LW. Effects of nanoparticle deposition on surface wettability influencing boiling heat transfer in nanofluids. Applied Physics Letters. 2006;**89**(15):718

[78] You SM, Kim JH, Kim KH. Effect of nanoparticles on critical heat flux of water in pool boiling heat transfer. Applied Physics Letters. 2003;**83**(16):3374-3376

[79] Bang IC, Chang SH. Boiling heat transfer performance and phenomena of Al_2O_3-water nanofluids from a plain surface in a pool. International Journal of Heat and Mass Transfer. 2005;**48**(12):2407-2419

[80] Zhu Y, HT H, Ding GL, Peng H, Huang XC, Zhuang DW, et al. Nucleate pool boiling heat transfer characteristics of refrigerant/oil mixture on metal foam covers (in Chinese). CIESC Journal. 2011;**62**(2):329-335

[81] Sundarram SS, Li W. The effect of pore size and porosity on thermal management performance of phase change material infiltrated microcellular metal foams. Applied Thermal Engineering. 2014;**64**:147-154

[82] Zhao CY, Lu W, Tian Y. Heat transfer enhancement for thermal energy storage using metal foams embedded within phase change materials (PCMs). Solar Energy. 2010;**84**(8):1402-1412

[83] Siahpush A, O'Brien J, Crepeau J. Phase change heat transfer enhancement using copper porous foam. Journal of Heat Transfer. 2008;**130**(8):318-323

[84] Hong ST, Herling DR. Effects of surface area density of aluminum foams on thermal conductivity of aluminum foam-phase change material composites. Advanced Engineering Materials. 2010;**9**(7):554-557

[85] Lafdi K, Mesalhy O, Shaikh S. Experimental study on the influence of foam porosity and pore size on the melting of phase change materials. Applied Physics Letters. 2007;**102**(8):083549

[86] Tian Y, Zhao CY. A numerical investigation of heat transfer in phase change materials (PCMs) embedded in porous metals. Energy. 2011;**36**(9):5539-5546

[87] Sundarram SS, Jiang W, Li W. Fabrication of small pore-size nickel foams using Electroless plating of solid-state foamed immiscible polymer blends. Journal of Manufacturing Science and Engineering. 2015;**136**(2):021002

[88] Tao YB, You Y, He YL. Lattice Boltzmann simulation on phase change heat transfer in metal foams/paraffin composite phase change material. Applied Thermal Engineering. 2016;**93**:476-485

[89] Gao D, Tian FB, Chen Z, Zhang D. An improved lattice Boltzmann method for solid-liquid phase change in porous media under local thermal non-equilibrium conditions. International Journal of Heat and Mass Transfer. 2017;**110**:58-62

[90] Wu S, Zhu D, Zhang X, Huang J. Preparation and melting/freezing characteristics of Cu/paraffin Nanofluid as phase-change material (PCM). Energy and Fuels. 2010;**24**(3):1894-1898

[91] Zeng JL, Cao Z, Yang DW, Sun LX, Zhang L. Thermal conductivity enhancement of Ag nanowires on an organic phase change material. Journal of Thermal Analysis and Calorimetry. 2010;**101**(1):385-389

[92] Khodadadi JM, Hosseinizadeh SF. Nanoparticle-enhanced phase change materials (NEPCM) with great potential for improved thermal energy storage. International Communications in Heat and Mass Transfer. 2007;**34**(5):534-543

[93] Guo CX. Application study of nanoparticle-enhanced phase change material in ceiling board. Advanced Materials Research. 2011;**150**:723-726

[94] Wu SY, Wang H, Xiao S, Zhu DS. An investigation of melting/freezing characteristics of nanoparticle-enhanced phase change materials. Journal of Thermal Analysis and Calorimetry. 2012;**110**(3):1127-1131

Chapter 10

Nanotechnologies in Cultural Heritage - Materials and Instruments for Diagnosis and Treatment

Rodica-Mariana Ion, Sanda-Maria Doncea and Daniela Țurcanu-Caruțiu

Additional information is available at the end of the chapter

http://dx.doi.org/10.5772/intechopen.71950

Abstract

This chapter aims to evaluate the nanomaterials that can be used to diagnostic, conservation and restoration of different artifacts and monuments and that can contribute to solving the problems which could appear during weathering processes of them. The nanotechnology, as a new and revolutionary area in science, can improve the traditional methods currently used for restoration and preservation in cultural heritage and can contribute to the creation of new highly specialized methods for diagnostic and treatment of different artifacts or even monuments. With a smaller size, a higher penetrability, viscosity, thermal and magnetic properties, in comparison with the traditional materials, the nanomaterials can contribute to solve the problems deriving from specific phenomena that could appear during the intervention and to identify the potential newly formed products in the treated materials. In this chapter, some aspects about the nanomaterials used for conservation and restoration of stone and paper artifacts are evidenced and discussed.

Keywords: nanomaterial, nanotechnology, cultural heritage, hydroxyapatite

1. Introduction

The term "preservation" may have different meanings depending on the field in which it is applied [1]. The knowledge of the preservation of artworks on different artifacts is not limited to historical and semiotic analyses. Preservation nowadays requires an interdisciplinary team with a solid knowledge of materials science, chemistry, physics, microbiology, art history and nanotechnology in order to contribute and offer solutions to prevent the natural aging of some artifacts (paper documents, stone, paints, etc.) and to provide useful and basic predictions about the degradation of the cultural patrimony and can come up with viable solutions [2, 3].

Generally speaking, the term "nano" defines an extremely small entity in the order of 10^{-9} m [4].

The applications of nanotechnologies in the field of preservation of cultural heritage could include:

- Diagnosis of the damaged surfaces to obtain physicochemical and structural information on the materials that form the historic heritage and for identifying the surface damaging type: alveolarization, fractures and so on [5].
- New instruments and diagnostic methods, in order to make an informed decision on the materials to use during the further phases of restoration [6].
- Innovative methods for cleaning surfaces affected by polluting substances and black crusts [7].
- Treatment of surfaces (protection, waterproofing, self-cleaning) [8].
- New products for consolidation and protection of natural and artificial stones (compatible products) [9].
- Materials and innovative processes against raising of humidity and against sulfation for the conservation and/or restoration interventions on art manufactures of different nature [10].

The present chapter means to complete a survey on the use of nanotechnology for the preservation and restoration of the stone monuments and different damaged paper and stone artifacts. The scientific principles behind numerous nanomaterials on different types of common movable and fixed artistic substrates are discussed. Compared to traditional methods, these new tools have the benefit of considerably less impact on both the operators and the environment. Different types of nanoparticles currently used to produce conservation treatments with enhanced material properties are discussed.

1.1. Nanomaterials for restoration and conservation of cultural heritage

Nanostructures represent a stage of matter between agglomerated molecules and structures and are typically characterized by a large surface area that affects their physicochemical properties. Innovative applications are of nanostructures and are based on two types of unique properties associated with nanostructures [11]:

1. New optical properties due to the generation of quantum effects.
2. Changes in reactivity and mechanical properties due to small physical dimensions and a large area of the specific surface. The advantages of small granule sizes in comparison with the agglomerated materials include: a low sintering temperature, super-elasticity, improved diffusion, improved dielectric and tribological properties.

One of the directions for the use of nanoparticles, nowadays, is the preservation of cultural heritage. The application of nanotechnologies to different artifacts has recently proved the huge potential of this science to apply in the field of preservation of universal cultural heritage [12, 13]. Solid nanodispersions, micelles, gels or microemulsions offer new viable solutions for the restoration and preservation of works of art. Some recently concerns are related to the synthesis and the application of nanoparticles of Ca and Mg or hydroxyapatite to paper and stone preservation [14]. In this chapter, the most used methods of nanoparticle synthesis and some of their recent applications for the preservation of artifacts are presented. The novelty of this area resides in a type of cultural heritage material (stone, paper) and starts with the main degradation paths and discussing protocols for the application of innovative nanomaterial-based tools for cleaning, consolidation or deacidification, which represent the majority of the case studies encountered in restoration and conservation procedures.

2. Artifacts conservation and restoration

2.1. Stone

2.1.1. Influence of pollutants

The main types of atmospheric pollutants which could affect the stony monuments are shown in **Table 1**.

Usually, calcium carbonate can take three forms of polymorphs: calcite, aragonite and vaterite [15]. It is known that in the calcite variety (calcium carbonate variety detected by X-ray diffraction (XRD)) dissolution is affected by the presence of foreign substances Mg^{2+} (from $MgSO_4{\cdot}7H_2O$, called epsomite), which is considered one of the major cations in seawater and

Nr.crt.	Pollutant	Source	Effect on stone
1	Carbon dioxide and carbon monoxide	Internal combustion engines	Acid rain, soil pollution
2	Nitrogen and sulfur oxides	Combustion of fossil fuels by motor vehicles or thermoelectric power plants	Acid rain, soil pollution
3	Volatile organic compounds (hydrocarbons and hydrocarbon derivatives)	Combustion of fossil fuels, the evaporation of fuels, solvents used in various industrial processes	Acid rain, soil pollution
4	Suspended particulates—very small particles similar with gas may contain iron oxides, heavy metals (lead, cadmium, manganese, chromium), asbestos fibers or other pollutants	Combustion of fossil fuels, human activity	Acid rain, soil pollution

Table 1. Pollutant source and effects on stone.

groundwater. Also, SO_4^{2-} (from atmospheric pollution responsible for calcite conversion in gypsum) and NO_3 (responsible for solubilizing the stone in the wall) are the most damaging pollutants or different artifacts [16]. Processes such as crystallization and salt dissolution contribute to new pores, which are responsible for an accelerated damaging process through the microcracks generated in the stone. The dissolution rate significantly increases in the presence of NaCl solutions, due to electrostatic reasons. Calcium carbonate ($CaCO_3$) can be found in soils, rocks and sediments. Among the minerals, calcium carbonate is one of the most sensitive ones to weathering [17]. Due to rapid weather destruction, small amounts of $CaCO_3$ can dominate the geochemical behavior of aquatic systems. Given the sensitivity of this rock, it is necessary to study the influence of climatic and environmental factors on this rock [18].

The pollutant effects on buildings and monuments degradation require a multidisciplinary approach using conventional and unconventional methods to achieve a good understanding of the mechanisms and consequences of such pollution [6]. It has been established that the hydraulic properties of stones and their traction resistance are the most important parameters that control the stone resistance to decomposition, and these parameters are widely used to estimate their durability.

The effects of the environment on some monuments (Basarabi-Murfatlar Churches) have been assessed through various analytical investigations: thermal analysis, XRD, EDXRF and ion chromatography. When the temperature is less than 120°C, a weight loss due to the absorbed water is recorded, especially in the absence of hydrated salts. In the temperature range 200–600°C, a loss of chemically bonded water is registered, but no other compounds subjected to weight loss. After 600°C, CO2 loss is observed, due to the decomposition of carbonates, higher for the most degraded stones [19].

Stone surface alteration (determined by scanning electron microscopy (SEM) analysis coupled with diffused X-ray spectroscopy (SEM-EDX)) could be classified in several types:

a. Surface damage caused by sulfur and calcium and by the calcite reaction with the sulfur oxides present in the medium.

b. Alteration of the surface caused by deposits, when no chemical reactions occurred. These deposits are composed of dust from the anthropogenic particles. These degraded layers are mainly caused by the epigenetic formation of gypsum. In areas with high traffic, the sulfur oxide content is considerably higher [20]. There are two different mechanisms of destruction in the process of damaging the stones under the action of environmental conditions:

- Chemical alteration of calcite and precipitation on gypsum that catches atmospheric particles on the surface.
- Physical phenomenon of deposition of atmospheric particles on the surface.

Kruhl and Nega [21] investigated the fragmented form of quartz particles and found that the size of the fragments decreases as temperature increases during suture formation, while Bernal et al. [22] concluded that the processes of degradation cause the fragmentation of the surface size. Daniele and Taglieri [13] studied the morphology of quartz aggregates in granite and showed that the fragmented dimensions of the quartz aggregates are different depending on the type of granite. The sedimentary rocks are those originating:

a. Sedimentation and cementation under natural pressure of particles of natural rocks eroded by wind, rain, sunbathing, conglomerate and sandstone.

b. Precipitation from natural solutions—gypsum, limestone, dolomite and travertine; sedimentation and consolidation of residues of dead organisms (shells, shells)—diatomite, limestone and chalk. The limestone is derived from marine sediments and fossilized organic lakes (such as crinoids or brachiopods), consisting of calcium carbonate or calcite. Sandstones are the result of sedimentation of sand together with silica or calcium carbonate; the silica in the tiles may have the same adverse effect encountered at the granite. Both sedimentary rocks—limestone and sandstone—are easier to cut than granite, which has made them quite often used in the masonry of historical buildings [23, 24]. On the Romanian territory, there are many masonry made of siliceous sandstone or calcareous sandstone, the first being those with higher compressive strength. The mineralogical composition and crystal structure of the stone monuments before and after the conservation treatment can be determined by X-ray diffraction (XRD) [25, 26]. A sample of $CaCO_3$ (**Figure 1**) has a series of diffraction peaks at 2ϑ: 29.5, 39.5, 47.6, 48.58, etc., corresponding to the calcite phase, which shows that in the water the precipitated calcium carbonate crystals were formed mainly of calcite.

In **Figure 2**, there are Fourier transform infrared (FTIR) peaks corresponding to the vaterite phase generated inside and outside of the damaged limestone surfaces.

Very clear bands could be observed in FTIR spectra, which show significant differences between not-treated and treated limestones, especially at the bands assigned to carbonate groups and to sulfate bands (**Figure 3**).

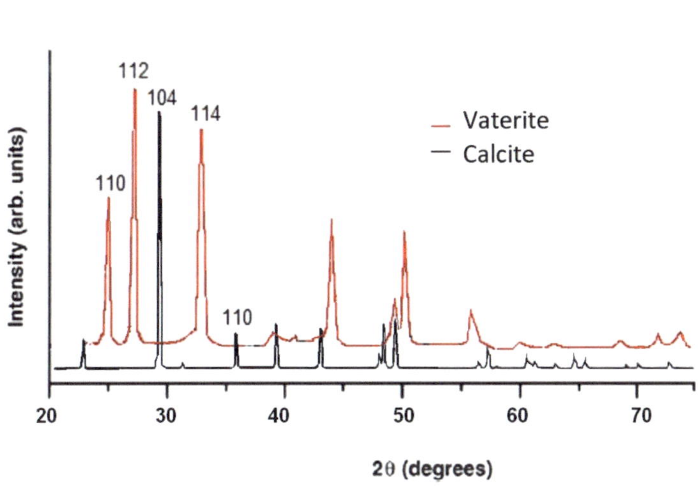

Figure 1. X-ray diffraction of calcite.

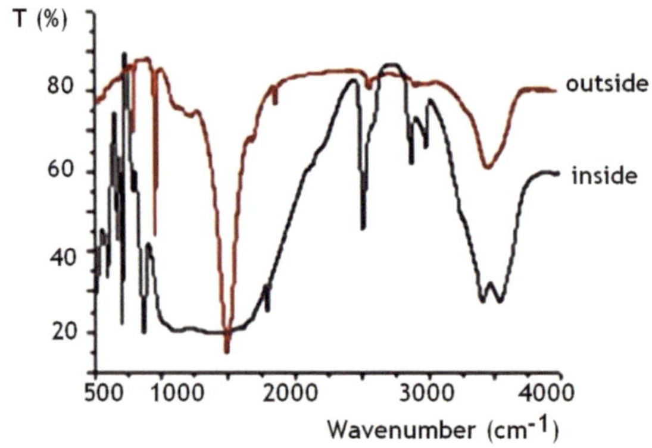

Figure 2. FTIR spectra of inside and outside treated limestone surface.

2.2. Causes of natural stone degradation

The natural stone is subjected to a slow and continuous process of deterioration (known as alteration— or degradation, or even "decay"), which is a phenomenon caused by physico-chemical causes other than mechanical actions: moisture; crystallization of salts in the mass of the material; deposition of pollutants on the surface of the rock, acting through chemical and/or biological processes; high temperature variations during day and night (strong heat cycles - temperature drop) or accidental fire action; and erosion due to strong winds (process without significant mechanical stress) [25–27].

Moisture is one of the main factors that attack stones through the capillary phenomenon, and limestone, dolomite and marble are most affected by this external factor [28]. During 10 years of exposure to a 100 cm column of atmospheric precipitation, calcareous rocks can reduce their thickness by 0.2 mm. Moisture can also affect the rocks that contain the tiny black, feldspar and tremolite. In the case of granite, the moisture from the pores can lead to the splitting of the stone into layers of 1–3 mm thick, through the expansion-contraction cycles (important to the reversible frost-thaw phenomena, because the freezing volume increases with 9%).

Salt crystallization is one of the main phenomena of the destruction of porous materials, which occurs by penetrating the material pores by the aqueous solution containing dissolved

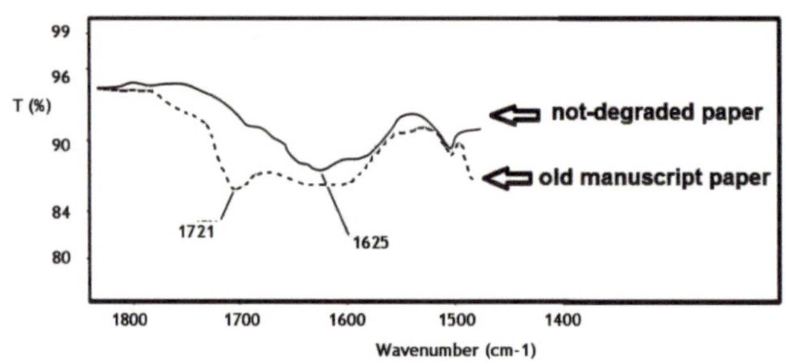

Figure 3. FTIR spectra of not-treated and treated limestones.

Nr.crt.	Factor	Source	Effect on stone
1	Moisture	Rain, freeze-thaw cycles, groundwater,	Construction, surface water mitigating through the shoulder, ditches or imbibing through in the paved surface of the pavement
2	Salt crystallization	Mirabilite (Na2SO4• 10H2O) and thenardite (Na2SO4)	Severe alveolar weathering, granular disintegration, efflorescences and subflorescences in buildings, monuments and quarry
3	Freeze-thaw cycles	Weather seasons	Severe alveolar weathering, granular disintegration, efflorescences and subflorescences in buildings, monuments and quarry
4	Black crusts - soiling on buildings. This aids the production of gypsum, which crystallizes on the surface	Exposition to moisture in the form of fog, mist or dew. Diesel engines are one of the most damaging particle sources	Ongoing decay mechanisms, such as salt weathering

Table 2. Causes of natural stone degradation.

pollutants, such as mineral salts, combustion gases, powdered plant residues and microorganisms, **Table 2**. The tensions in the surfaces that delineate the pores due to the salts crystallized here, known as "stone efflorescence," represents a great danger for the durability of a historical structure. The most common salts in the efflorescence phenomenon of masonry are sulfates, carbonates and nitrates (sodium, magnesium, calcium, potassium), generated from the atmospheric pollutants (SO_2, NO_2 and CO_2), which are transformed in (HNO_3, H_2SO_4 and H_2CO_3) in contact with the mortars [24, 29].

Freeze-thaw cycles. If the mortar pores are filled with water, a pressure will be exerted on the pore walls of the mortar due to the increase in the volume (by 9%) of the frozen water. Thus, prolonged freeze-thaw cycles will progressively degrade mortars with moisture. The most sensitive to this phenomenon is lime mortars [30].

2.3. Type of inorganic nano-consolidants

The nanomaterials are adequate materials for the architectural heritage, due to their consolidation and protection capacity of damaged building materials. The nanoparticles are able of self-cleaning coatings for a preventive protection system for historical surfaces, preserving the original aspect of treated elements, decreasing the deposition of pollutants and soiling and reducing the onset of external degradation processes due to soiling phenomena. The nanoparticles must have the following attributes: termal stability, biologically and chemically inertness, nontoxic, low cost, stable toward visible or near UV light, good adaptability to various environment and good adsorption in solar spectrum. In addition, these treatments can also have water repellent properties, which favor this self-cleaning action. On the other hand, the presence of soluble salts is recognized as an important decay agent of stone heritage.

Thus, in the last few years, the study of the application of nanoparticles as a desulfating agent for stone, mortars and wall paintings is being carried out [31].

The principle of the inorganic materials is to create an insoluble "paste" that fills the pores of the stone. There is a large area of such consolidants, as it is shown in **Table 3**.

Some inorganic materials, such as calcium hydroxide $Ca(OH)_2$, magnesium hydroxide $Mg(OH)_2$, barium hydroxide $Ba(OH)_2$, strontium hydroxide $Sr(OH)_2$ and hydroxyapatite (HAp), have already been used as consolidants for different damaged carbonate stones [32]. Calcium hydroxide has been used as nanoparticles (130–300 nm), dispersed in alcohols, as nanosols (50–250 nm), as "pastelike" in ethanol and as calcium hydroxide microparticles (1–3 mm) [33]. But, due to their low porosity, high moisture content of the substrate, oversaturation of the material and quick evaporation of the solvent, calcium hydroxide is not an optimal

Nr.crt.	Consolidant	Effect
1	Alcalo-silicates	Deposition of silica in the limestone mass
2	Silico-fluoride	Silicone tiles can form a cemented crust on the outside, a layer of higher hardness. In general, these solutions are no longer recommended
3	Alkaline hydroxides	The consolidation effect is still low, requiring repeat treatment to make it more effective. The process remains relatively uneconomical
4	Strontium and barium hydroxides	These solutions seem to be more effective than calcium-based, but experts believe the problem of the durability of the treated stone, the application to historical structures
5	Inorganic builders (zinc stearate and aluminum stearate, aluminum sulfate, phosphoric acid, phosphate and hydrofluoric acid)	Superficial penetration into the pores of the stone. Danger of microorganism population
6	Alcosilanes (or alkoxylans). Increased mechanical strength has also been reported with approx. 20% of the silicon tiles treated, which is already performing	The best materials with a reinforcing function A deep penetration into the pores of silky tiles. The penetration of the alkoxylans in the stone occurs at a depth of 20–25 mm, which means much more than the inorganic builders
7	Acrylic polymers (methyl-methacrylate, methyl-acrylate, ethyl-methacrylate and butyl-methacrylate).	All of these builders have increased the resistance of the treated stone layer, but unfortunately they are unstable in color under the action of the already mentioned agents
8	Vinyl polymers (polyvinyl chloride, polyvinylchloride-chlorinate, polyvinyl acetate)	Polyvinyl acetate may cause a glossy-glassy appearance on the surface of the stone. On the other hand, if the polymers were insufficiently diluted in the solvent, layers were formed which could represent a screen for retaining the moisture and salts in the stone, that is, exactly the opposite of what was intended by applying the waterproofing treatment
9	The polyurethanes	Treatment is very effective, but warmth and light produce opposite effects

Table 3. Consolidants used for stone and their effects.

option for this monument [34]. This is the reason for finding other optimal materials. Also, a reduced penetration depth and a limited solubility of lime in water are causing chromatic alteration of stone surface [35, 36]. Except the metallic hydroxides above mentioned, for stone consolidation could be used hydroxyapatite (HAp) [37, 38]. It is a natural mineral form of calcium apatite having the formula $Ca_{10}(PO_4)_6(OH)_2$ and has the ability to readily accept in its structure numerous substitution ions for both Ca^{2+} and PO_4^{3} ions, with the chemical composition altering and the morphological structure [39]. The OH^- ion may be replaced by fluoride, chloride or carbonate ions, producing fluorouracil or chlorapatite [40].

3. Paper

3.1. Considerations on the methods of conserving the historical paper

Hydrolysis of acid-catalyzed cellulose is the main source of paper degradation. It is well known that the degradation process resides in the manufacture of low quality paper. At the beginning of the eighteenth century, papermaking technology changed and the paper began to be made of increasingly reactive materials (wood pulp) and acidic substances (rosin, used for sizing, chlorine for bleaching and so on). After a long period of exposure to environmental conditions (e.g., temperature, humidity, light), these substances accelerate chemical degradation of the paper. The global effect is the rapid decrease in the resistance to degradation of paper documents, especially on paper made since the eighteenth century [41–43]. There is a general consensus on the inevitable treatment of deacidification in paper preservation. Deacidification involves a complete neutralization of the paper and, in most cases, the introduction of an alkaline reservoir that opposes the acidity assault in the environment (e.g., pollution) [44]. In this context, many studies have been developed addressing acidity elimination processes [45]. The best methods of deacidification are based on nonaqueous solvents. Less polar fluids minimize the risk of ink solubility. Among these most important deacidification methods are the Wei T'o method and the Bookkeeper method [46].

Recently, a new method has been proposed based on alcohols dispersed in calcium hydroxide nanoparticles that give good results in deacidification of the paper. The above-mentioned studies have also been extended to the treatment of paper with magnesium hydroxide nanoparticles (brucellosis) because it has been shown that magnesium reduces the rate of oxidative degradation of the cellulose substrate due to exposure to light.

Magnesium hydroxide nanoparticles, $Mg(OH)_2$ (brucite), can be obtained by hydrating Mg (MgO), precipitating magnesium salts with an alkaline solution and electrolyzing an aqueous solution of a bounce magnesium.

$Mg(OH)_2$ is usually accomplished by the first two methods [47]. When magnesium hydroxide powder is used as a precursor for the synthesis of magnesium oxide, the size of the hydroxide particles, the shape and degree of agglomeration of these are key parameters for many applications. An example is the sintering stage in the ceramic manufacturing process. Magnesium-based alkaline compounds are of great importance in the preservation of cultural heritage. Such an area of interest is also the treatments for deacidification of acidic paper in order to preserve it. Both the Wei T'o and Bookkeeper methods are based on the use of magnesium compounds that generate $Mg(OH)_2$ "in situ" after hydrolysis. Unfortunately, some studies have demonstrated a limited homogeneous distribution of alkaline reservoirs [48]. The purpose of the new studies was to synthesize magnesium nanoparticles following the procedure similar to that of calcium hydroxide, which yielded very good results. Unlike the Wei T'o and Bookkeeper methods, these nanoparticles once deposited on the paper immediately acquire the role of deacidifier or buffer. Synthetic pathways for $Mg(OH)_2$ nanoparticles have an important role both theoretically and practically. Several papers have shown that the precipitation of metallic hydroxides in their salts is strongly affected by variation in synthesis parameters. These include high-temperature reactions, reagent concentration and aging time. In particular, it was demonstrated that temperatures above 100°C promote the formation of nanoparticles in nonaqueous media. Some studies also reported significant effects of organic solvents related to the shape and size of precipitated particles. The obtained particles were characterized for determination of their chemical nature by: FT-TR spectroscopy, specific surface area (SA) measurement of dry powder, X-ray diffraction (XRD) characterization and shape and size characterization by SEM and TEM electronic microscopy with scanning or electron transmission. Applying deacidification of nanoparticle paper is compared to Wei T'o [49].

In some cases, during the degradation process, it is possible to identify some odorous compounds such as vanillin or vanilic acid, well identified by gas chromatography, **Figure 4**.

3.1.1. Synthesis of nanoparticles of alkaline hydroxides

This process can be considerably stopped or slowed down by deacidification treatment. The excellent deacidification agents are $Mg(OH)_2$ and $Ca(OH)_2$ because they provide very good

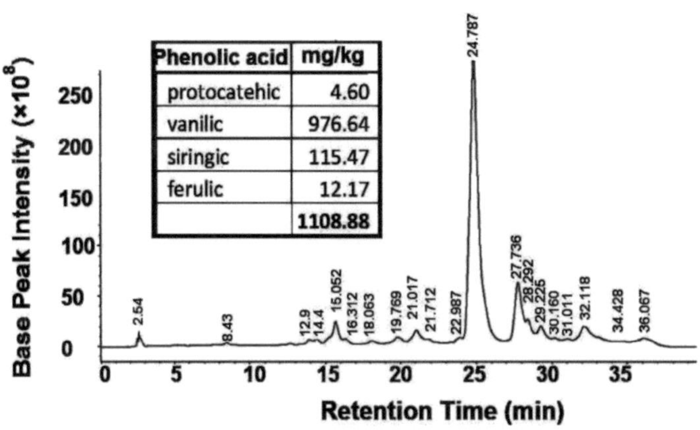

Figure 4. GC-MS chromatogram of the damaged paper.

physicochemical compatibility with the support, and after conversion into carbonates, they work as alkaline reservoirs without producing any undesirable side effect [50].

Nanoparticles dispersed in alcohols can be applied to the paper by spraying, impregnating or dipping [51]. This method produces in situ hydroxides and requires dispersants to stabilize the system. The solvents used are volatile, environmentally friendly and with low surface tension so that they act as carrier solids for solid particles and ensure the homogeneity and penetration of nanoparticles into the depth of cellulose fibers [52].

The synthesis of $Ca(OH)_2$ and $Mg(OH)_2$ nanoparticles was made in heterogeneous medium, and CaO and MgO reagents were used. The oxides in turn were obtained from the corresponding carbonates, which were initially milled and then milled to a size of 100 μm. Then, the carbonates were calcined at 1000°C. Both carbonates were of analytical purity.

The process consists of suspending in a mixture of isopropyl alcohol and water in each of the two stoichiometric proportions of the respective oxides according to the equations given above. The hydrolysis reaction was carried out at 80°C for about 30 min (using deionized water).

3.1.2. *Treatment of paper document with suspension of alkaline hydroxide nanoparticles*

The resulting particles were used to treat paper with HAp in isopropanol, and the historical paper was manually sprayed. The pulverized sheets were from a book printed in Romania, in Bucharest, from a private collection from the end of the nineteenth century. Untreated paper was taken as a reference. The paper was then treated with the nanoparticles of $Ca(OH)_2$ or $Mg(OH)_2$ and examined by scanning electron microscopy (SEM) [37, 38]. Initially untreated, unwritten, unprinted and uncolored paper was investigated by SEM and AFM, too, as a reference (**Figure 5**).

The cellular microarchitecture of cellulose was investigated by SEM. The micrographs obtained for this sample revealed a densely packed cellulose fiber network which, on a microscopic scale, inside the sheet of paper appears randomly oriented without having a majority

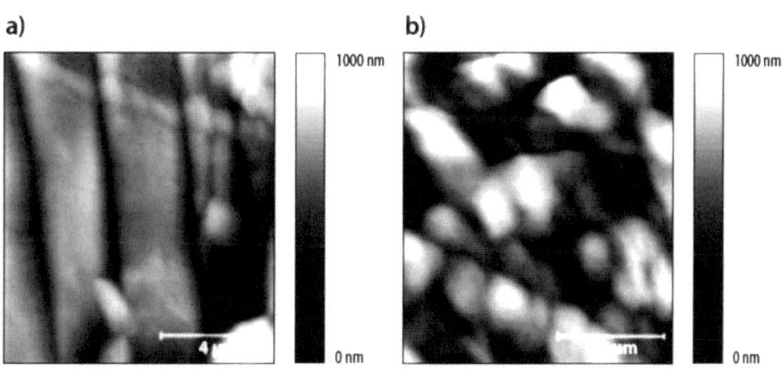

Figure 5. SEM images for not-damaged and damaged cellulose substrates.

direction of microfiber orientation. The fibers are homogeneous and seem to come from plant fibers, perhaps cotton or linen. The size of the fibers is different, some of them are whole, others are broken. Some fibers have inlay that might be salt crystals. The presence of mineral crystals in paper can be considered as a consequence of how it is made (**Figure 6**). The presence of luminous areas on the image is a consequence of either the presence of a thicker part of the glue material or a rupture of the paper [53].

On a macroscopic examination of the sprayed sample of nanoparticles of $Mg(OH)_2$, no negative influence on the optical parameters of the paper is observed. Some white deposits of $Mg(OH)_2$ were formed on the paper surface more pronounced in this case than in the case of $Ca(OH)_2$, although both consolidants have the same concentration and volume applied to the same type of paper. Not all $Mg(OH)_2$ nanoparticles neutralize the acidity of the paper and the unreacted quantity will be carbonated over time in the presence of atmospheric CO_2.

3.1.3. Hydroxyapatite nanoparticles: synthesis and characterization

Hydroxyapatite (HAp) was obtained by the modified precipitation chemical method, and the synthesized substance was analyzed by spectral techniques: atomic force microscopy (AFM), scanning electron microscopy (SEM) (**Figure 7**) X-ray diffraction (XRD) and spectroscopy in Fourier transform infrared (FTIR).

As synthesis reagents were used, calcium nitrate tetrahydrate, $Ca(NO_3)_2{\cdot}4H_2O$, and dibasic ammonium phosphate, $(NH_4)_2HPO_4$, were previously dissolved each in deionized water. Then, the solution of $Ca(NO_3)_2{\cdot}4H_2O$ was added dropwise over the $(NH_4)_2HPO_4$ solution, which was stirred vigorously at room temperature for about 1 h until a milky and somewhat gelatinous precipitate was obtained and further stirred for further 1 h to increase the reaction rate and homogenize the mixture [20]. The mixture was synthesized at 100°C for 24 h. Then, the precipitate was washed and filtered on a glass filter. After filtration, the compact and sticky cake was dried at 80°C in a furnace. The dried powder was then ground into a mortar and then calcined in an alumina crucible for 4 h [27].

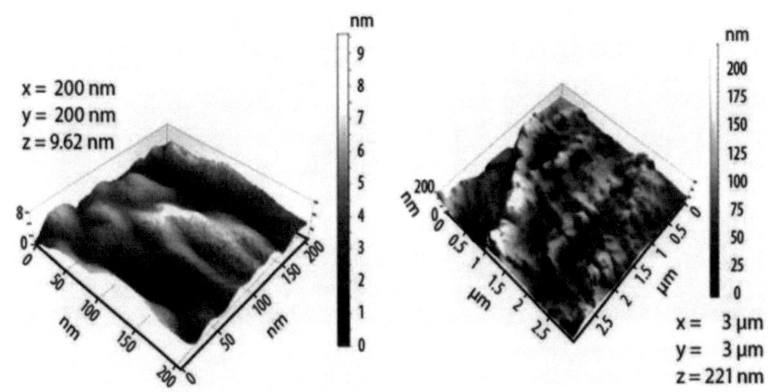

Figure 6. AFM images of not-damaged and damaged cellulose substrates.

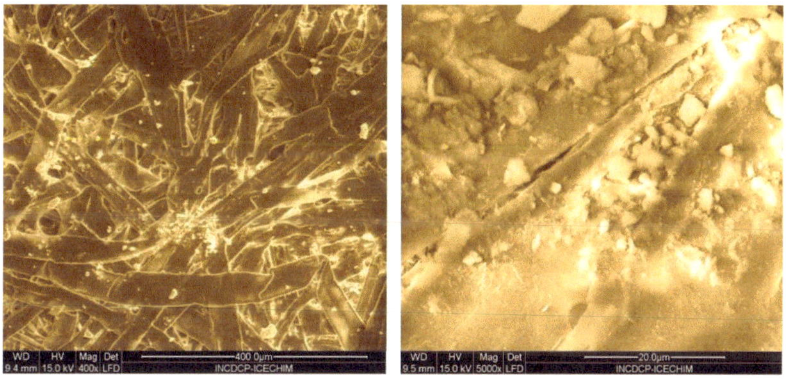

Figure 7. SEM images of not-treated and treated papers with Hap.

As a method of synthesis, a modified wet precipitation method has been chosen because this is more advantageous due to the ease with which it is achieved, the low working temperature, the relatively important percentage of the pure product and the synthesis equipment that is not expensive. It has been found that well-crystallized products with a low degree of sintering have been obtained, but relatively high calcination temperatures were required and the application of this long-term treatment, 4 h, was to obtain a finished product with the desired parameters [3]. Both X-ray diffraction and infrared spectroscopy showed the high degree of purity of the reaction products [54]. The study of SEM and AFM images was very consistent with the results obtained by other analysis techniques: thermal analysis [46]. It has been determined that the crystal size for HAp synthesized is 70–nm. It has also been concluded that a sintering temperature of the synthesis product above 850°C leads to the occurrence of a by-product reaction, namely tricalcium phosphate, and for its conversion into Hap, it is necessary to calcinate at 1200°C for 4 h.

4. Conclusions

Nanomaterials bring huge enhancements of improvement endeavors for various applications, due to the extensive scale nanomaterials for auxiliary applications. The structure-handling property acquires an imperative segment in the cultural heritage. The present paper plans to do a review of the condition of workmanship on the use of some nanomaterials to the preservation and rebuilding of the stony and paper cultural artifacts. With a smaller size, a higher penetrability, viscosity, thermal and magnetic properties, in comparison with the traditional materials, the nanomaterials can contribute to solve the problems deriving from specific phenomena that could appear during the intervention and to identify the potential newly formed products in the treated materials. In this chapter, some aspects about the nanomaterials used for conservation and restoration of stone and paper artifacts are evidenced and discussed. Distinctive sorts of nanoparticles right now used to create preservation with upgraded material properties and novel functionalities have been discussed and exemplified in this chapter ($Ca(OH)_2$, $Mg(OH)_2$, $Ba(OH)_2$, $Sr(OH)_2$, hydroxyapatite), both for their synthesis, characterization and specific applications for paper and stone surfaces.

Acknowledgements

This chapter received financial support from MCI-UEFISCDI by the projects: PNII 261/2014, PN 16.31.02.04.04, 11 BM/2016 and 31CI/2017.

Author details

Rodica-Mariana Ion[1,2]*, Sanda-Maria Doncea[1] and Daniela Ţurcanu-Caruțiu[3]

*Address all correspondence to: rodica_ion2000@yahoo.co.uk

1 ICECHIM, Research Center for Scientific Investigations and Conservation/Preservation of Industrial, Cultural and Medical Heritage (SCI-HERITAG), Bucharest, Romania

2 Materials Engineering Department, Research Center "Nanomaterials for Mechanical Microsystems", Valahia University, Targoviste, Romania

3 Center of Expertise of Artworks by Advanced Instrumental Methods (CEOAMIA), Ovidius University, Constanța, Romania

References

[1] Preservation Technologies PTLP. Preservation Technologies PTLP [Internet]. 15 January 2000 [Updated: 15 January 2000]. Available from: http://ptlp.com/base.html [Accessed: 07 October 2017]

[2] Colangiuli D, Calia A, Bianco N. Novel multifunctional coatings with photocatalytic and hydrophobic properties for the preservation of the stone building heritage. Construction and Building Materials. 2015;**93**189-196. DOI: http://dx.doi.org/10.1016/j.conbuildmat.2015.05.100

[3] Dei L, Salvadori B. Nanotechnology in cultural heritage conservation: nanometric slaked lime saves architectonic and artistic surfaces from decay. Journal of Cultural Heritage. 2006;**7**(2):110-115. DOI: http://dx.doi.org/10.1016/j.culher.2006.02.001

[4] Hensen E, Doehne E, Fidler J, Larson J, Martin B, Matteini M, Rodríguez-Navarro C, Sebastian Pardo E, Price C, de Tagle A, Teutonico J-M, Weiss NR. A review of selected inorganic consolidants and protective treatments for porous calcareous materials. Reviews in Conservation. 2003;**4**(13-25). DOI: http://dx.doi.org/10.1179/sic.2003.48.Supplement-1.13

[5] Álvarez de Buergo M, Fort R. A basic methodology for evaluating and selecting waterproofing treatments applied to carbonatic materials. Progress in Organic Coatings. 2001;**43**:258-266

[6] López-Arce P, Gomez-Villalba LS, Fernández-Valle ME, Álvarez de Buergo M, Fort R. Influence of porosity and relative humidity on consolidation of dolostone with calcium hydroxide nanoparticles: Effectiveness assessment with non-destructive techniques. Materials Characterization. 2010;**61**(2):168-184

[7] Vázquez-Calvo C, Ávarez de Buergo M, Fort R, Varas-Muriel MJ. Characterization of patinas by means of microscopic techniques. Materials Characterization. 2007;**58**(11-12): 1119-1132. DOI: http://dx.doi.org/10.1016/j.matchar.2007.04.024

[8] Daniele V, Taglieri G, Quaresima R. The nanolimes in cultural heritage conservation: Characterisation and analysis of the carbonatation process. Journal of Cultural Heritage. 2008;**9**(3):294-301. DOI: http://dx.doi.org/10.1016/j.culher.2007.10.007

[9] Ferreira Pinto AP, Delgado-Rodrigues J. Stone consolidation: The role of treatment procedures. Journal of Cultural Heritage. 2008;**9**(1):38-53. http://dx.doi.org/10.1016/j.culher.2007.06.004

[10] Naidu S, Sassoni E, Scherer GW. New treatment for corrosion-resistant coatings for marble and consolidation of limestone. In: Stefanaggi M, Vergès-Belmin V, editors. Jardins de Pierres – Conservation of Stone in Parks, Gardens and Cemeteries, Paris, 22-24 June 2011. pp. 289-294; 2011. ISBN: 2-905430-17-6

[11] Ion RM. Nano Crystalline Materials. Bucharest: FMR Ed; 2003. 189 p

[12] Baglioni P, Chelazzi D, Giorgi R. Nanotechnologies in the conservation of cultural heritage. A compendium of materials and techniques. Edit. Springer; 2015. ISBN: 978-94-017-9303-2

[13] Daniele V, Taglieri G. Nanolime suspensions applied on natural lithotypes: The influence of concentration and residual water content on carbonatation process and on treatment effectiveness. Journal of Cultural Heritage. 2010;**11**(1):102-106

[14] Ion RM, Teodorescu S, Ştirbescu RM, Dulamă ID, Şuică-Bunghez IR, Bucurică IA, Fierăscu RC, Fierascu I, Ion ML. Effects of the restoration mortar on chalk stone buildings. IOP Conference Series: Materials Science and Engineering; 2016; Jessy. 2016:012038

[15] Lopez-Arce P, Zornoza-Indart A. Carbonation acceleration of calcium hydroxide nanoparticles: induced by yeast fermentation. Applied Physics A. 2015;**120**(4):1475-1495

[16] Ion RM, Fierascu RC, Leahu M, Ion ML, Turcanu D. Nanomaterials for conservation and preservation of historical monuments. In: Proc EWCHP; Bolzano. 2013. pp. 97-104

[17] Ion RM, Bunghez RI, Pop SF, Fierascu RC, Ion ML, Leahu M. Chemical weathering of chalk stone materials from Basarabi churches. Metalurgia International. 2013;**18**(1):89-93

[18] López-Arce P, Gomez-Villalba LS, Martínez-Ramírez S, Álvarez de Buergo M, Fort R. Influence of relative humidity on the carbonation of calcium hydroxide nanoparticles and the formation of calcium carbonate polymorphs. Powder Technology. 2011;**205**(1):263-269. DOI: http://dx.doi.org/10.1016/j.powtec.2010.09.026

[19] Ion RM, Ion ML, Radu A, Şuică-Bunghez RI, Fierăscu RC, Fierăscu I, Teodorescu S. Mortar pe bază de nanomateriale pentru conservarea faţadelor construcţiilor. Romanian Journal of Materials. 2016;**46**(4):412-416

[20] Ion RM, Bunghez IR, Teodorescu S, Ion ML. Degradation of chalk stones induced by freeze–thaw action. In: Proc. XIX-th international conference "INVENTICA 2015"; June 2015; Jessy. Jessy: Performantica. 2015. pp. 141-147

[21] Kruhl JH, Nega M. The fractal shape of sutured quartz grain boundaries: application as a geothermometer. Geologische Rundschau. 1996;**85**(1):38-43

[22] Bernal SA, Mejía de Gutiérrez R, Provis JL. Engineering and durability properties of concretes based on alkali-activated granulated blast furnace slag/metakaolin blends. Construction and Building Materials. 2012;**33**:99-108

[23] Campbell A, Hamilton A, Stratford T, Modestou S, Ioannou I. Calcium hydroxide nanoparticles for limestone conservation: Imbibition and adhesion. Proceedings of Symposium, Adhesives and Consolidants for Conservation, Ottawa. 2011:1-16

[24] Drdácký M, Slížková Z, Ziegenbalg G. A nano approach to consolidation of degraded historic lime mortars. Journal of Nano Research. 2009;**8**:13-22. DOI: http://dx.doi.org/10.4028/www.scientific.net/JNanoR.8.13

[25] Ruffolo S, LaRussa MF, Aloise P, Belfiore CM, Macchia A, Pezzino A, Crisci GM. Efficacy of nanolime in restoration procedures of salt weathered limestone rock. Applied Physics A: Materials Science and Processing. 2014;**114**:753-758

[26] Slíížková Z, Drdácký M, Viani A. Consolidation of weak lime mortars by means of saturated solution of calcium hydroxide or barium hydroxide. Journal of Cultural Heritage. 2015;**16**:452-460. http://dx.doi.org/10.1016/j.culher.2014.09.003

[27] Ambrosi M, Ambrosi M, Dei L, Giorgi R, Neto C, Baglioni P. Colloidal particles of Ca(OH)2: Properties and applications to restoration of frescoes. Langmuir. 2001;**17**(14):4251-4255

[28] Kryza R, Prell M, Czechowski F, Domaradzka M. Acidic weathering of carbonate building stones: experimental assessment (preliminary results). Studia Universitatis Babeş-Bolyai. Geologia. 2009;**54**(1):33-36

[29] Sassoni E, Franzoni E. Sugaring marble in the Monumental Cemetery in Bologna (Italy): characterization of naturally and artificially weathered samples and first results of consolidation by hydroxyapatite. Applied Physics A. 2014;**117**:1893-1906. http://dx.doi.org/10.1007/s00339-014-8629-3

[30] Winkler EM. Stone: Properties, Durability in Men's Environment. Wien: Springer Verlag; 1973. p. 223

[31] Hansen E, Doehne E, Fidler J, Larson J, Martin B, Matteini M, et al. A review of selected inorganic consolidants and protective treatments for porous calcareous materials. Reviews in Conservation. 2003;**4**:13-25. http://dx.doi.org/10.1179/sic.2003.48.Supplement-1.13

[32] Pianski J, Brümmer K, Ziegenbalg G.. Nano-particles for stone conservation-state of the art, characteristics and recent developments. In: Stonecore—"Recent Progress in the Consolidation of Calcareous Materials", Litomysl, Czech Republic, 21-22 April 2010; 2010

[33] Daniele V, Taglieri G. $Ca(OH)_2$ nanoparticle characterization: microscopic investigation of their application on natural stones. Materials Characterisation. 2011;**72**:55-66

[34] Giorgi R, Ambrosi M, Toccafondi N, Baglioni P. Nanoparticles for cultural heritage conservation: Calcium and barium hydroxide nanoparticles for wall painting consolidation. Chemistry - A European Journal. 2010;**16**(31):9374-9382

[35] Delgado Rodrigues J, Ferreira Pinto AP. Laboratory and onsite study of barium hydroxide as a consolidant for high porosity limestones. Journal of Cultural Heritage. 2016;**19**:467-476. http://dx.doi.org/10.1016/j.culher.2015.10.002

[36] Ciliberto E, Condorelli GG, La Delfa S, Viscuso E. Nanoparticles of $Sr(OH)_2$: synthesis in homogeneous phase at low temperature and application for cultural heritage artefacts. Applied Physics A-Materials Science Process. 2008;**91**(1):137-141. http://dx.doi.org/10.1007/s00339-008-4464-8

[37] Sassoni E, Naidu S, Scherer GW. The use of hydroxyapatite as a new inorganic consolidant for damaged carbonate stones. Journal of Cultural Heritage. 2011;**12**:346-355. http://dx.doi.org/10.1016/j.culher.2011.02.005

[38] Ion R-M, Turcanu-Carutiu D, Fierascu R-C, Fierascu I. Chalk Stone restoration with hydroxyapatite-based nanoparticles. SBMM. 2014;**12**:9:24

[39] Ion R-M, Turcanu-Carutiu D, Fierascu R-C, Fierascu I, Bunghez I-R, Ion M-L, Teodorescu S, Vasilievici G, Raditoiu V. Caoxite-hydroxyapatite composition as consolidating material for the chalk stone from Basarabi-Murfatlar churches ensemble. Applied Surface Science. 2015;**358**:612-618. http://dx.doi.org/10.1016/j.apsusc.2015.08.196

[40] Ion RM, Fierascu RC, Fierascu I, Senin RM, Ion ML, Leahu M. Influence of Fântânița Lake (Chalk Lake) Water on the Degradation of Basarabi–Murfatlar Churches. Engineering Geology for Society and Territory. 2015;**8**:543-546

[41] Sequeira S, Casanova C, Cabrita EJ. Deacidification of paper using dispersions of $Ca(OH)_2$ nanoparticles in isopropanol. Journal Cultural Heritage. 2006;**7**:264-272

[42] Ion R-M, Doncea SM, Ion M-L, Rădiţoiu V, Amăriuţei V. Surface investigations of old book paper treated with hydroxyapatite nanoparticles. Applied Surface Science. 2013;**285**(A):27-32

[43] Taglieri G, Daniele V, Del Re G, Volpe R. A new and original method to produce $Ca(OH)_2$ nanoparticles by using an anion exchange resin. Advances in Nanoparticles. 2015;**4**:17-24. DOI: 10.4236/anp.2015.42003

[44] Roy A, Bhattacharya J. Synthesis of $Ca(OH)_2$ nanoparticles by wet chemical method. Micro & Nano letters. 2010;**3**:131-134

[45] Samanta A. Synthesis of nano calcium hydroxide in aqueous medium. Journal American Ceramic Society. 2016;**795**:787-795

[46] Doncea SM, Ion RM, Fierascu RC, Bacalum E, Bunaciu AA, Aboul-Enein HY. Spectral methods for historical paper analysis: Composition and age approximation. Instrumentation Science and Technology. 2010;**38**(1):96-106

[47] Gomez-Villalba LS, López-Arce P, de Buergo MA, Zornoza-Indart A, Fort R. Mineralogical and textural considerations in the assessment of aesthetic changes in dolostones by effect of treatments with Ca(OH)2 nanoparticles. Science and Technology for the Conservation of Cultural Heritage. 2013;235-329

[48] Doncea SM, Ion RM, Fierascu RC, Dumitriu I. Extended spectral analysis (FTIR, EDXRF, ICP-EAS) of cellulose-based artifacts. Studia Universitatis Babes Bolyai, Geologia, Special. 2009;83-86

[49] Ion RM, Ion ML, Niculescu VIR, Dumitriu I, Fierascu RC, Florea G, Bercu C, Serban S. Spectral analysis of original and restaurated ancient paper from a Romanian Gospel. Rom. Journal of Physics. 2008;**53**(5-6):781-791

[50] Mosini V, Calvini P, Mattogno G, Righini G. Derivative infrared spectroscopy and electron spectroscopy for chemical analysis of ancient paper documents. Cellulose Chemistry and Technology. 1990;**24**:263-272

[51] Doncea SM, Ion RM, Nuta A, Somoghi R, Ghiurea M. Optical methods of investigation for book papers conservation with nanoparticles. In: SPIE US, editor. Proceeding of SPIE; 2010. p. 78211 F

[52] Sierra-Fernández A, Gómez-Villalba LS, Milosevic O, Fort R, Rabanal ME. Synthesis and morpho-structural characterization of nanostructured magnesium hydroxide obtained by a hydrothermal method. Ceramics International. 2014;**40**:12285-12292. DOI: http://dx.doi.org/10.1016/j.ceramint.2014.04.073

[53] Gómez-Villalba LS, López-Arce P, Álvarez de Buergo M, Fort R. Structural stability of a colloidal solution of Ca(OH)2 nanocrystals exposed to high relative humidity conditions. Applied Physics A.2011;**104**(4):1249-1254. DOI: http://dx.doi.org/10.1007/s00339-011-6457-2

[54] Taglieri G, Mondelli C, Daniele V, Pusceddu E, Trapananti A. Synthesis and X-ray diffraction analyses of calcium hydroxide nanoparticles in aqueous suspension. Advances in Materials Physics and Chemistry. 2013;**3**:108-112. http://dx.doi.org/10.4236/ampc.2013.31A013

Chapter 11

Nanofibers and Electrospinning Method

Nabeel Zabar Abed Al-Hazeem

Additional information is available at the end of the chapter

http://dx.doi.org/10.5772/intechopen.72060

Abstract

As a result of their peculiar features like extremely high surface area to weight ratio, low density, as well as high pore volume and controllable pore size, which may not be present in other structures, nanofibers have taken centre stage in nanotechnology. The indicated properties, consequently, make non-woven nanofibers as appropriate materials for wide-spread applications. Electrospinning, because of its high productivity, simplicity, low cost, reproducibility and its potentialities of being utilised at the industrial level is regarded as one of the most potential processes in nanotechnology. This method implies the application of high voltage electric field aiming to extract very thin fibres from a polymeric fluid stream (solution or melt) potentially deliverable through a millimetre-scale needle. Electrospinning, as a technique, is reliant on various processing standards like solution properties and processing parameters. Consequently, altering these parameters could exert a considerable degree of influence on the nanofiber size, shape and morphology. Thus, by controlling those parameters well, specific fibres can be produced to benefit various applications.

Keywords: nanofibers, application, electrospinning, nanostructure

1. Introduction

Electrospinning is a simple and comprehensive process for generating an ultrafine fibre from varieties of materials which include polymer, composite and ceramic. The electrospinning setup consists of three major components namely, high voltage power supply, syringe with metal needle and a conductive collector. It is, in fact, very sophisticated, but a simple, processing mechanism of producing nanofiber. The electrospinning process can be classified into several techniques like vibration electrospinning, magneto-electrospinning, siro-electrospinning and bubble electrospinning, according to Liu et al. [1]. As the charge liquid jet moves from the syringe tip to the collector, the mode of current flow changes from ohmic to convective flow as the charge moves instead to the fibre surface.

A slow acceleration is a characteristic of the ohmic flow, since the geometry of the Taylor cone is controlled by the ratio of the surface tension to electrostatic repulsion [2]. After successfully addressing the ohmic flow, the jet travels at a rapid acceleration, which includes the transition zone from liquid to dry solid. In the end, the jet penetrates the collector [3–5]. The name 'Taylor Cone' simply represents the conical shape formed at the needle tip see (**Figure 1**).

In 1964, Sir Geoffrey Ingram Taylor described this cone [7] as a continuation of the study by Zeleny [7] on the formation of a cone-jet of glycerine exposed to high electric fields. Several others including Wilson and Taylor, Nolan and Macky [7] continued research in this area. However, it was Taylor who investigated further into the reactions between droplets and electric fields. Taylor's result is based on two assumptions: (1) that the surface of the cone is an equipotential surface and (2) that the cone exists in steady state equilibrium. Immediately after being discharged and the Taylor cone activated, the polymer jet goes through a whipping process [8] in which the solvent evaporates, precipitating a charged polymer fibre, which lays itself at random on a grounded collecting metal screen. As far as the melt is concerned, the discharged jet solidifies when it travels in the air and is collected on the grounded metal screen [7]. Fridrikh et al. [9] theorised that the terminal diameter of the 'whipping' jet (h_t) is controlled by flow rate (Q), electric current (I) and fluid surface tension (γ) as given by the equation

$$h_t = \left(\gamma\bar{\varepsilon}\frac{Q^2}{I^2}\frac{2}{\pi(2\ln x - 3)}\right)^{\frac{1}{3}} \quad (1)$$

where $\bar{\varepsilon}$ is the dielectric constant, (x) is the displacement, Eq. (1) offers a prediction that the terminal diameter of the whipping jet is controlled by the flow rate, electric current and the surface tension of the fluid, disregards the elastic effects and fluid evaporation, and also makes an assumption about the minimal jet thinning after the saturation of the whipping instability takes place.

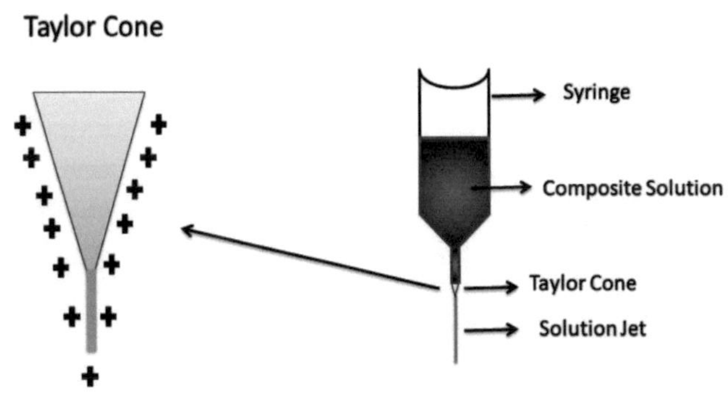

Figure 1. Illustration of Taylor Cone formations from the Syringe Needle Tip [6].

2. Electrospinning setups

2.1. Electrospinning process

The electrospinning apparatus is really a simple idea, carrying only three main components: a high voltage power supply, a polymer solution reservoir (e.g., a syringe, with a small diameter needle) with or without a flow control pump, and a metal collecting screen. A high voltage power supply with adjustable control can well provide up to 50-kV DC output and, depending on the number of electrospinning jets, the multiple outputs that function independently, are necessitated. The polymeric solution is kept in a reservoir and connected to a power supply to establish a charged polymer jet. Charging the polymer solution could be done either with a syringe with a metal needle or a capillary with a metal tip in the polymer solution. If the syringe is not placed horizontally, polymer flow can be driven by gravity. However, to remove the experimental variables, a syringe pump is engaged to control the precise flow rate. The fibre collecting screen is expected to be conductive and it can either be a stationary plate or a rotating platform or substrate. The plate can produce non-woven fibres, whereas a rotating platform can produce both nonwoven and aligned fibres.

Presently, two standard electrospinning setups are available namely the vertical and horizontal, with three new electrospinning setups with different angles for the study of the effect of the gravity. As a result of the increasing interest in this technology, many research groups have developed sophisticated mechanisms by which more complex nanofibers structures, can be fabricated in a more controlled and efficient manner [10, 11], see (**Figure 2**). For instance, motor-controlled multiple jets and fibre-collecting targets provide avenue for producing a single nanofibrous scaffold consisting of multiple layers, with each layer obtained from a different polymer type. Furthermore, this technology can be used to manufacture polymer composite scaffolds where the fibres of each layer represent a combination of various polymer types.

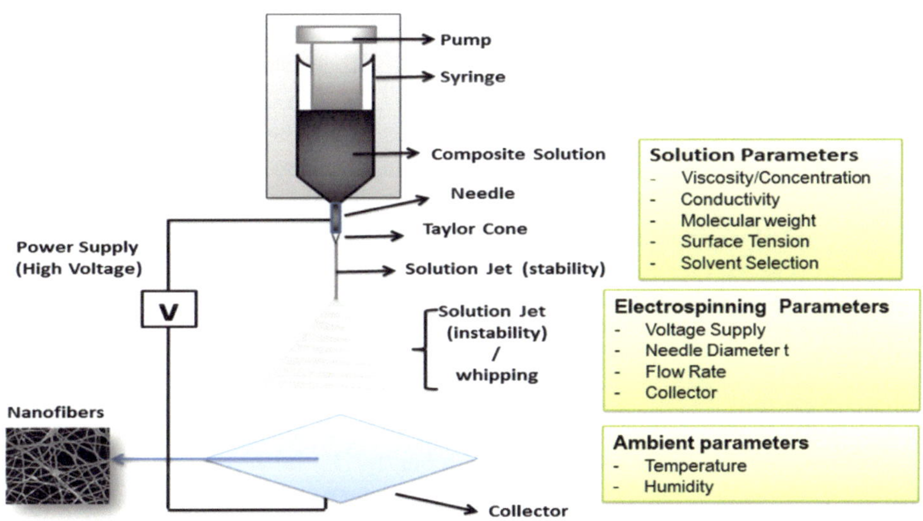

Figure 2. Electrospinning setup and controllable electrospinning process parameters [12].

3. Mechanism and technique for nanofiber formation

Electrospinning process has been studied extensively [13]. The mechanism takes place when the surface tension of the solution is overcome by an applied electric field, thereby ejecting tiny jets from the surface. Taylor [14] identified a critical voltage at which this breakdown would occur:

$$V_c^2 = 4\frac{H^2}{L^2}\left(\ln\frac{2L}{R} - \frac{3}{2}\right)(0.117\pi\gamma R) \tag{2}$$

where Vc is the critical voltage, H is the separation between the capillary and the ground, L is the length of the capillary, R is the radius of the capillary and γ is the surface tension of the liquid. A similar relationship by Carson et al. [15] had progressed for the potential electrostatic spraying from a hemispherical drop pendant from a capillary tube:

$$V = 300\sqrt{20\gamma r} \tag{3}$$

where v is electric field, γ is the surface tension of the liquid and r is the radius of the pendant drop [15]. As he investigated a small range of fluids, Taylor determined a 49.3° equilibrium angle-balanced surface tension with electrostatic forces, and he had manipulated this value in his derivation. Taylor cones are required for electrospinning because they define the onset of subtle velocity gradients in the fibre forming process. When V > Vc, a thin jet of solution will explode from the cone surface and travel towards the nearest electrode of opposite polarity, or electrical ground. As a way to describe this, electrospinning jet is a string of charged elements connected by a viscoelastic medium, with one end attached to the point of origin and the other end is let free. When a polymer solution, held in a capillary by its surface tension, is subjected to an electrical field, a charge is induced on the liquid surface [16]. Mutual charge repulsion occurs in a force opposing the surface tension forces and shear stresses are set up in the fluid. With increased intensity in the electrical field, ions in the like-polarity solution will be forced to aggregate at the surface of the drop. The length of the stable jet increases with increasing voltage. After the viscoelastic jet starts flowing away from the Taylor's cone, initially it traverses a linear path. The jet gradually starts to bend away from this linear path and complex shape changes may occur from repulsive forces generated in the charged elements inside the electrospinning jet [17]. The jet may undergo substantial reductions in cross sectional area and spiralling loops may grow from it. This phenomenon is often referred to as whipping instability. Consequently, the hemispherical surface of the solution at the tip of the capillary elongates to form a cone, called the Taylor cone (**Figure 3**). At remarkably high electrical fields (V > Vc), a charged jet of solution is ejected from the tip of the Taylor cone and will travel to an electrode of opposite polarity (or electrical ground).

A dimensionless parameter called the Berry number (B_e) [18, 19] was used by various research groups as a processing index for controlling the diameter of fibres. B_e number is defined as:

$$B_e = \eta C \tag{4}$$

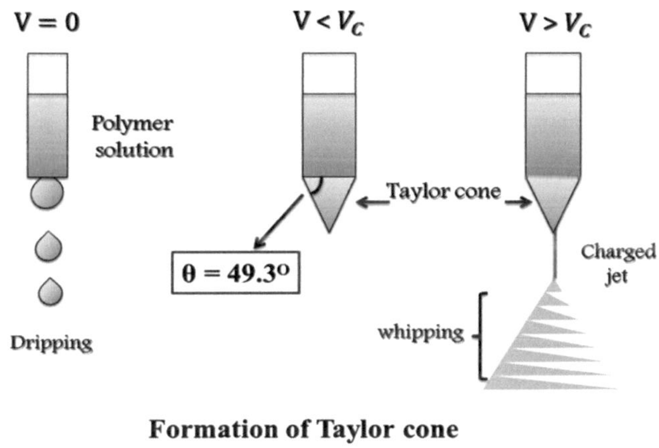

Figure 3. Schematic illustration of the effects by electric field applied to a solution in a capillary.

where [η] is the intrinsic viscosity of the polymer which is the ratio of the specific viscosity to the concentration at an infinite dilution and C is the concentration of the polymer solution. Intrinsic viscosity is dependent on polymer molecular weight. B_e also describes the level of the entanglement of polymer chains in a solution. As far as highly diluted solutions are concerned, when B_e is less than unity, the polymer molecules are sparsely distributed in the solution. There is low probability for individual molecules to be entangled with each other.

When Be is greater than one, the polymer concentration as well as the level of molecular entanglement are enhanced, leading to more favourable conditions for the fibre formation [20]. The experiment implied that as the solution viscosity increased the fibre diameter also increased (approximately proportionally) to the jet length. Baumgarten detailed the relationship between fibre diameter and solution viscosity expressed by the following equation:

$$D = \sqrt[2]{\eta} \tag{5}$$

where D is the fibre diameter and η is the solution viscosity in poise. It is also proves that fibre diameter is highly dependent on the applied electric field. An increase in the applied voltage increases the electrostatic stress, which, in turn, produces smaller diameter fibres [7, 21]. According to Huang et al. [22] further increase in the electric field, is a critical value achieved when the electrostatic repulsive force deals with the surface tension and the charged jet of the fluid is ejected from the end of the Taylor cone. In the process, the solvent evaporates, contributing to the formation of charged polymer fibre. As for the melt, the discharged jet solidifies when it moves in the air. Based on findings of the study conducted by Pham et al. [23], the shape of the base is dependent on the surface tension of the liquid and the force of the electric field; jets can be ejected from surfaces that are mostly flat if the electric field is high enough. The solvent in the polymer jet evaporates in the movement to the collecting screen, thereby increasing the surface charge on the jet. As it passes via the electric field, this increase in the surface charge induces instability in the polymer jet [24]. The polymer jet divides geometrically, first into two jets, and then into many more as the process repeats itself in order to compensate for this instability. Nanofibers are formed from the spinning force action, given by the electrostatic force on the continuous splitting of the polymer droplets. They are deposited one layer after another on the metal target plate, forming a non-woven nanofibrous mat. The mechanisms, by which nanofibers are formed, much less controlled, have not been fully elucidated yet, although the electrospraying/electrospinning technology has been used for such a long time. Not much of theoretical clarity has been achieved, although, several studies have been carried out to investigate the mechanism of fibre formation to reproducibly control scaffold design. A uniform fibrous structure is created only under optimised operating during the process electrospinning. The structural morphology of the nanofibers is affected by both extrinsic and intrinsic parameters [16]. In order to produce inform nanofibers, external parameters, like environmental humidity and temperature, in addition to intrinsic parameters, including applied voltage, working distance and conductivity and viscosity of the polymer solution need to be optimised. The intrinsic parameters are more critical in determining the nanofiber structure in general sense.

4. Operating parameters for electrospinning

Three main parameters, which are solution parameters, process parameters and ambient parameters, tend to affect the electrospinning process. These operating parameters play a big role in determining the desired quality of the electrospun fibre produced [25, 26]. The most preferred in many applications is a fibre with diameter within 10–1000 nm in scale and a smooth surface morphology. The solution properties are difficult to alter, according to Lu and Ding [25], since the relationship between one parameter will drag the other parameters; and in addition, they are very difficult to isolate as one controllable parameter. Li and Wang [27] discussed the effects of working parameters that govern the electrospinning process and the process, and discovered that these parameters could affect the fibre morphologies and diameters. In their study on the effects of the parameters on nanofiber diameter, Thompson et al. [28] found out that the jet radius could leave an impact to the production of the electrospun fibre. In their study, a number of parameters significantly affected the fibre formation compared to other parameters. For instance, the first electrospinning method by Formhals [29] had had some technical disadvantages since it was not easy to dry the fibres entirely after electrospinning as the spinning and collection zones have a very short distance; this resulted in a less aggregated web structure [29]. After half a decade, however Formhals [30], in his pioneering work, changed the distance between the nozzle and the collecting device in order to give more drying time for the electrospun fibres at a longer distance [31].

4.1. Polymer/solution parameters

4.1.1. Viscosity/concentration

The most critical factor in controlling the structural morphology of the nanofibrous structure is polymer solution viscosity, a parameter that is directly proportional to the concentration of the polymer solution. For fibre formation, polymer viscosity should be in a particular range, depending on the type of polymer and solvent used. An electrospinning method by Zeng J et al. was used to fabricate PLA nanofiber, with different concentrations or viscosities ranging from

1% to 5% by weight as shown in **Figure 4**, and below 3% for beads or beaded fibres and 3% and above for continuous nanofibers [32]. A bead that contains nanofibers structure was created below this range. Spherical beads had become longer and turned into spindle-shaped ones, with increased viscosity, and the number of beads in the structure was decreased. In the same manner, Liu et al. also talked about that a different particular range of viscosity rendered appropriate for the formation of uniform nanofibers composed of cellulose [33]. Recent studies done by Deitzel et al. [34] and Demir et al. [35], in addition, have illustrated that a more viscous polymer solution can well create larger fibres. To sum up, these studies have been able to prove that there is a polymer-specific, optimal viscosity value for electrospinning.

According to Sill and von Recum [36], polymer concentration determines the spinnability of a solution. For chain entanglements to occur, the solution must have high enough polymer concentration. However, the solution should not be either too diluted or too concentrated. Both, the viscosity and surface tension of the solution are affected by the polymer concentration.

4.1.2. Conductivity

The charge carrying capacity of polymer solutions with high conductivity is greater than solutions with low conductivity. Therefore, the fibre jet produced from a solution of high conductivity will tend to have higher tensile force when exposed to an applied voltage. Through

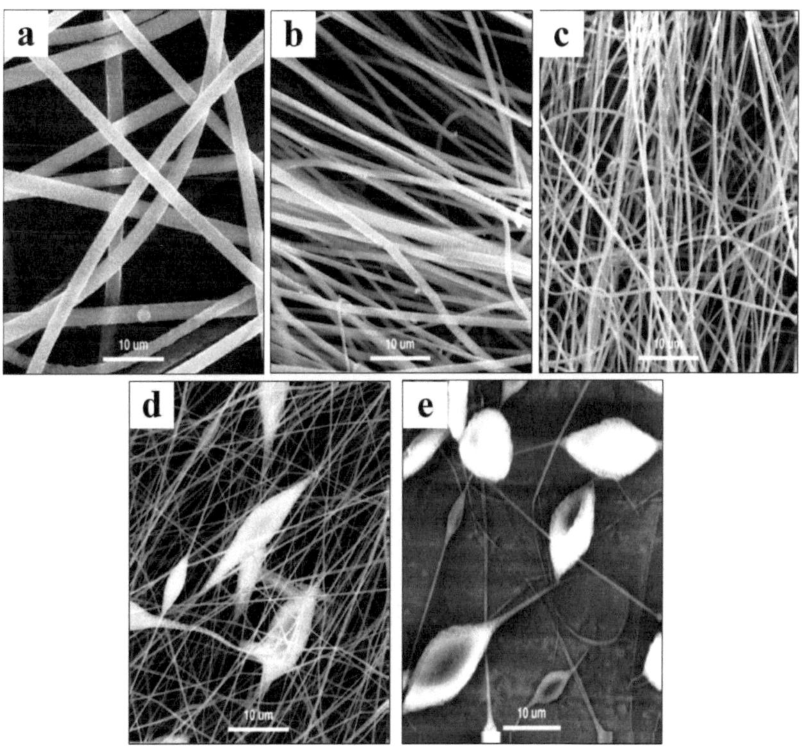

Figure 4. FESEM of electrospinning of nanofibers with different concentrations PLA. (a) 5%, (b) 4%, (c) 3%, (d) 2%, and (e) 1% [32].

observation, an increase in the solution conductivity brings about a substantial decrease in the nanofiber diameter; and also, evidently the radius of the nanofiber jet is inversely related to the cube root of the electrical conductivity of the solution [7, 37, 38].

The conductivity of a given cell has a connection with the molar conductivity following Eq. (6) [39], where k is the conductivity with units of mS/cm, c is the ion concentration with units of mol/L and therefore, the molar conductivity (Λ) has units of S cm²/mol.

$$\Lambda = \frac{k}{c} \tag{6}$$

Chitral and Shesha [40] had published the results of a comprehensive investigation of the effects of change in the conductivity of polyethylene oxide (PEO)/water solution on the electrospinning process and fibre morphology. The effects of the conductivity of PEO solution on the jet current and jet path were elaborated further, with the addition of NaCl to the solution results in the formation of protrusion on the fibre surface, as shown in **Figure 5**. The effects of the conductivity of polyethylene oxide (PEO) solution on the jet current and jet path were also considered.

Zong and colleagues [37] had done a study to see the effect of adding varying kinds of salts to poly(L-lactic acid) (PLLA) solutions in electrospinning. KH_2PO_4, NaH_2PO_4 and NaCl were studied, and each was added in separation at 1% W/V to PLLA solutions. The resulting electrospun nanofibers were smooth, bead-free and they also had smaller diameters than those of the nanofibers electrospun from solutions that did not have a salt. While $KH_2PO_4^-$ which contains solutions produced nanofibers with the largest diameter, those containing NaCl

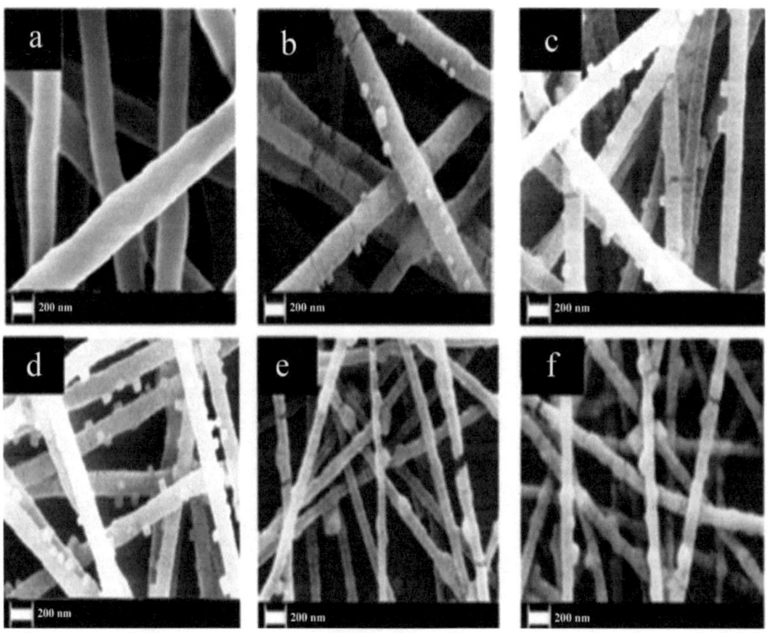

Figure 5. FESEM images of samples of electrospinning PEO/NaCl fibres for a range of conductivities. (a) 5 g/0 g, (b) 5 g/0.1 g, (c) 5 g/0.2 g, (d) 5 g/0.5 g, (e) 5 g/1.25 g, and (f) 5 g/2 g [40].

produced nanofibers with the smallest diameter. The ion size was also found to ascertain the nanofiber diameter. Ions with smaller radii had higher charge density, and thus they gave greater forces of elongation on the electrospun nanofibers [41].

4.1.3. *Molecular weight*

Molecular weight of the polymer also leaves a great effect on the morphologies of the electrospinning fibre. The entanglement of polymer chains in solutions, namely the solution viscosity, is principally a reflection of the molecular weight. Keeping the concentration fixed, and lowering the molecular weight of the polymer have the ability to form beads instead of the smooth fibre. Smooth fibre will be obtained by increasing the molecular weight. What is also worth noting is that too high molecular weight favours micro-ribbon formation even with low concentration [42, 43]. Çiğdem A et al. [44] studied the impact of the molecular weight (MW) on the fibre structure of electrospun poly(vinylalcohol) (PVA) which has molecular weights that range from 89000 to 186,000 g/mol when dissolved in water, as can be seen in **Figure 6**.

4.1.4. *Surface tension*

As the function of solvent compositions of the solution, surface tension is an important factor in electrospinning. Yang et al. [45] conducted a study on the influence of surface tensions on the morphologies of electrospun products with PVP as model with ethanol, N,N-dimethylformamide (DMF) and dichloromethane (MC) as solvents. In the process, it was discovered that different solutions may contribute different surface tensions. With the concentration fixed, and the surface tension of the solution, reduced beaded fibres can be converted into smooth fibres.

4.1.5. *Solvent selection*

For a particular polymer to solubilise and be transformed into nanofibers via the process of electrospinning, the choice of solvent is very important. The solubility of the polymer in the solvent and the boiling point of the solvent, which altogether indicate its volatility, are two major aspects worth considering when it comes to choosing a solvent. Volatile solvents are

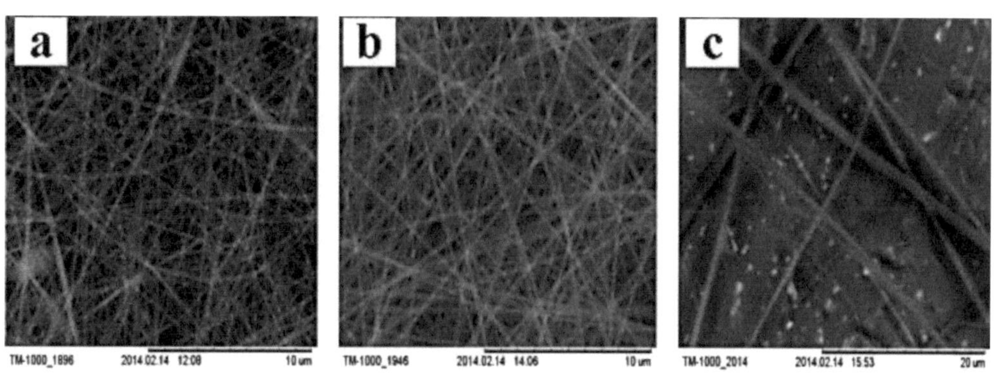

Figure 6. FESEM showing the typical structure in the electrospun PVA polymer for various molecular weights. (a) 89000–98,000 g/mol; (b) ~ 125,000 g/mol; and (c) 146,000–186,000 g/mol (solution concentration: 25 wt.%) [44].

the more favourable choice as they assist the dehydration of the nanofibers during trajectory from the capillary tip to the collector surface, because of their lower boiling point, and thus causing a rapid evaporation rate. Nonetheless, highly volatile solvents that have very low boiling points should be prevented as they may evaporate at the capillary tip and further leading to the clogging and the obstruction of the flow-rate of the polymer solution. Solvents that have high boiling points may not dehydrate entirely before reaching the collector, thus resulting in ribbon-like, flat, nanofiber morphologies or conglutination of nanofibers at the boundaries [36, 46]. The ability of the electrospinning polyvinylpyrrolidone (PVP) by Yang and Coworkers [45], was investigated with different solvents. The solvents examined were MC, ethanol and DMF, while the beaded nanofibers were formed from DCM and DMF solutions of PVP, the use of ethanol produced PVP nanofibers. Nanofibers electrospun from an integration of the ethanol and DMF had small diameters of 20 nm, while a combination of ethanol and DCM resulted in the formation of nanofibers with diameters as large as 300 nm (see **Figure 7**). It is therefore conclusive that nanofiber morphology and porosity may be regulated by the defensible use of solvents or a combination of solvents.

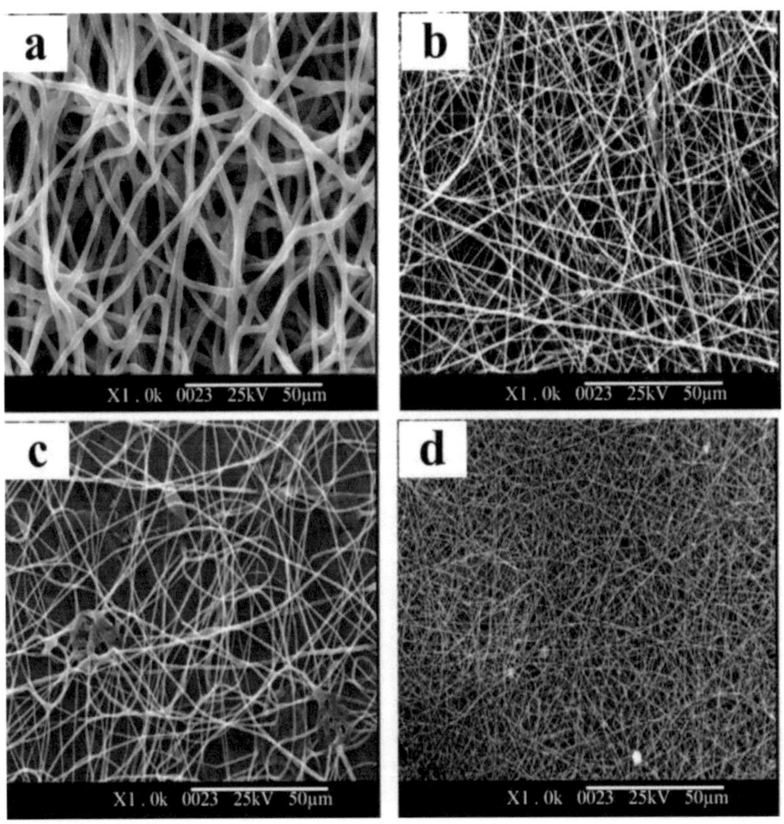

Figure 7. TEM images of PEO nanofibers electrospun different solvent. (a) chloroform (3%), (b) Ethanol (4%), (c) DMC (5%), (d) Water (7%) [47].

4.2. Electrospinning parameters

4.2.1. Voltage supply

The amount of charge per unit surface area of the polymer droplet which constitutes charge density is determined by the applied voltage, working distance and the conductivity of the polymer solution. Applied voltage is used to provide the driving force to spin fibres by imparting charge to the polymer droplet. The working distance which is the distance between the tip of syringe and the collecting plate, in addition to the applied voltage, can influence the structural morphology of nanofibers. Demir et al. [35] suggested that when higher voltages are applied, more polymer is ejected to form a larger diameter fibre. Similarly, high voltage conditions also created a rougher fibre structure. To reduce bead formation, Zong et al. [37] proposed an approach to increase charge density on the surface of the droplet by adding salt particles. However, they concluded that high-charge density produced thinner fibres, a finding not corroborated by Demir et al. [35]. The study by Pham et al. [23] shows that in the state of low voltages or field strengths, typically, a drop is suspended at the needle tip, and a jet will originate from the Taylor cone producing bead-free spinning (under the assumption that the force of the electric field is sufficient to address the surface tension). Hao Shao et al. [48] studied the effect of high voltage, and observed change in the morphology as a result of change in high voltage as shown in **Figure 8**.

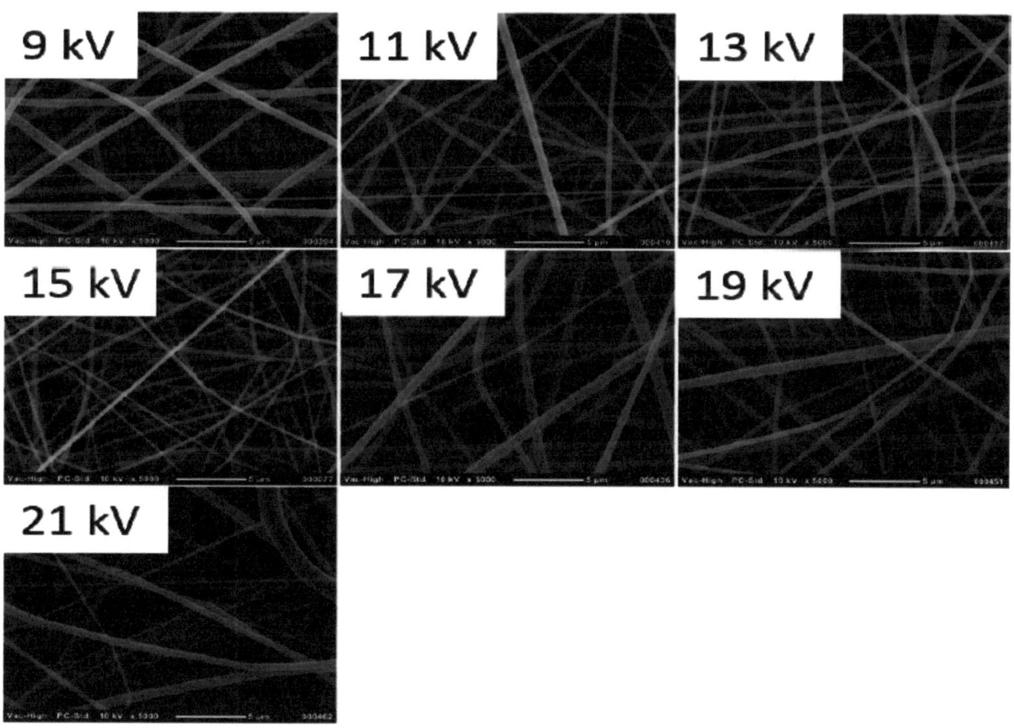

Figure 8. SEM images of the PVDF nanofibers prepared at different applied voltages [48].

4.2.2. Needle diameter (nozzle)

The size of needle has a certain effect on the nanofibers diameters. It was discovered that a reduction in the diameters of the electrospun nanofibers was caused by a decrease in the internal diameter of the needle. The nozzle (usually the syringe needle set up) determines the amount of polymer melt that comes out, which, in turn, affects the size of the drop being formed and also the pressure or the amount of force required by the pump to push the melt out. If the polymer melt is less viscous, it can easily flow out of the nozzle. The polymer melt is usually a thick highly viscous fluid. So, if the nozzle is too small, and the melt is too viscous, the melt cannot be forced out. Therefore, an appropriate nozzle should be used. Different types of nozzles or spinnerets have been used over the years [49]. The effect of needle diameters on the resulting electrospun poly(methyl methacrylate) (PMMA) average nanofiber diameters was evaluated for three different needle gauges by Javier Macossay et al. [49]. These fibres presented regular surface morphologies, with a few nanofiber bundles being evident in **Figure 9**.

4.2.3. Distance between tip and collector

The distance from the needle to the collector is very important, because by decreasing it, the electrical field increases instead; and also the stretching force does and the time at which the fibre undergoes the field is lower, causing sufficient evaporation of the solvent of the fibres. The result is that when there is reduction in the distance between the needle and the collector the fibres grow and may be subject to structural deformities like blobs. The high voltage was determined at 15 kV and the distance from the tip of the needle to the collector is in the range from 9 cm to 21 cm. The fibres' morphology was assessed from the SEM images of **Figure 10**; for the shortest distance (9 cm), the fibres came together at their intersections following the incomplete evaporation of the solvent before the jet arrived at the collector. For the other four distances used, the fibres appeared similar and the mean fibre diameter increased a little with the distance to the collector. The distance established between the tip and the collector exerted a direct influence in flight time and electric field strength. For the fibres to form, the electrospinning jet must be given ample time for most of the solvents to be evaporated. The electric field strength will increase at the same time and this will increase the acceleration of

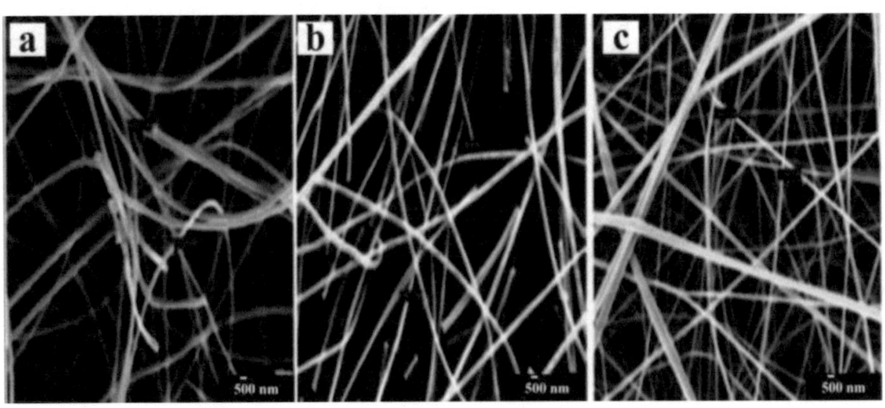

Figure 9. SEM of PMMA nanofibers, utilizing internal diameter needle. (a) 0.83 mm, (b) a 0.4 mm, (c) a 0.1 mm [49].

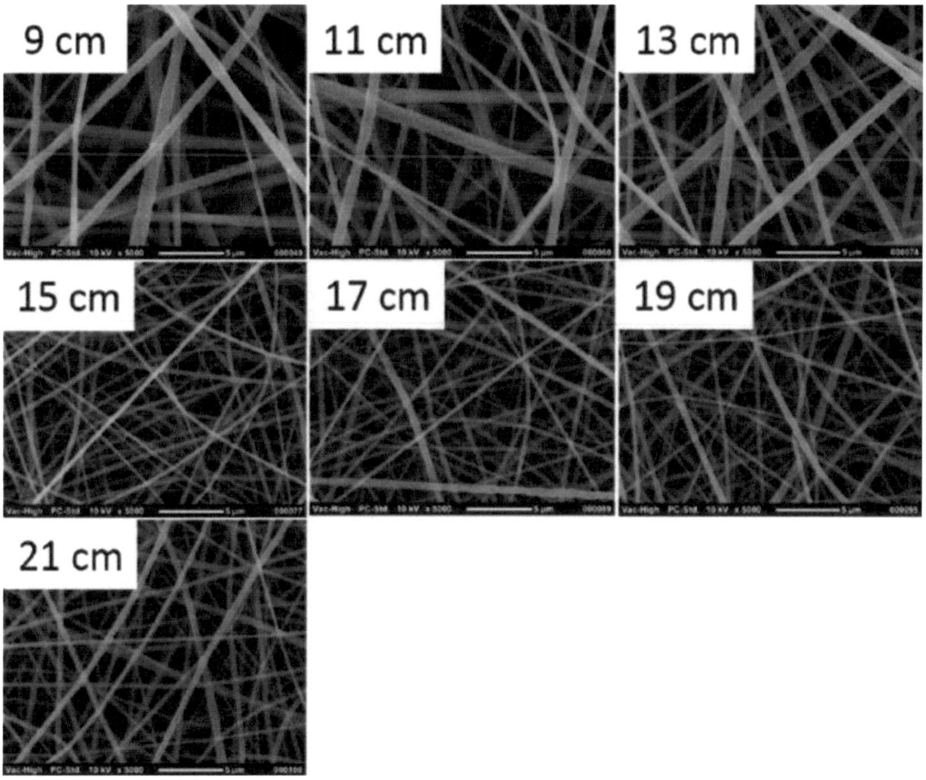

Figure 10. SEM images of the PVDF nanofibers prepared at different spinning distances [48].

the jet to the collector. As a result, there may not be enough time for solvents to evaporate when they reach the collector.

4.2.4. Flow rate

Another important parameter process is the flow rate of the polymer solution within the syringe. For the polymer solution to have enough time for polarization, lower flow rate is more preferred. If the flow rate is very high, bead fibres with thick diameter will form instead of smooth fibres with thin diameters owing to the short drying time before reaching the collector, and also due to low stretching forces. There is a corresponding rise in the fiber diameter or blobs size, as a result of greater volume of solution ejected from the needle tip, when there is increase in the feed rate. As shown in **Figure 11**, Shamim Z et al. [50] indicated that when the flow rate is decreased with other parameters kept constant; there is a decrease in the blobs size and an increase in nanofiber diameter. The inference here is that, with the decrease in flow rate, blobs size could get smaller until the non-beaded structure is obtained.

4.2.5. Collector

The formation of nanofibers can be classified into woven and non-woven nanofibers. The type of collector used plays a big role in differentiating the type or nanofiber alignments. The use

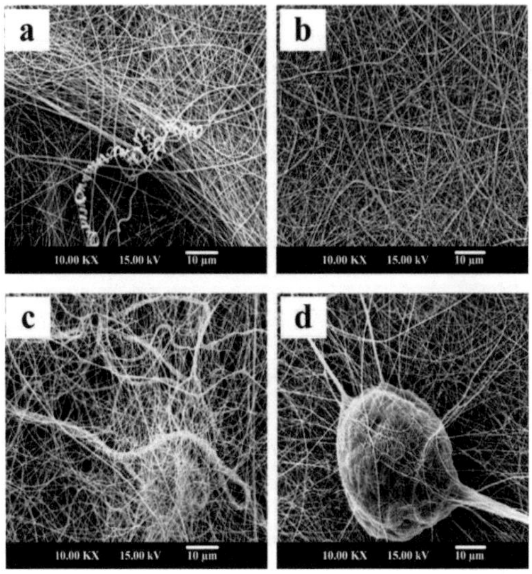

Figure 11. SEM images of PVA different flow rate. (a) 0.1 ml/h, (b) 0.5 ml/h, (c) 1 ml/h, (d) 1.5 ml/h [51].

of oriented collector, as well as static double grounded collector [2], constitutes the methods adopted in developing aligned woven nanofibers. The rotating drum collector is used for collecting the aligned arrays nanofibers, while the rotating disk is used for collecting uniaxially aligned nanofibers. The alignment fibres obtained from the rotating drum correspond to the rotational speed applied on the drum [2]. This type of electrospinning method is more complex because the speed of the rotation needs to be very properly controlled to produce nanofibers with such a good alignment. The rotating disk collector can also serve to collect continuous nanofibers, since they can very much attract the large electrical field applied at the edge of the disk [2].

As seen in **Figure 12**, SEM images of diverse collectors for many reports, have been developed including the wire mesh studied by Wang X et al. [52], pin studied by Sundaray B et al. [53], grids studied by Li D et al. [54], parallel or gridded bar and rotating rods or wheel studied by Xu CY et al. [55], and liquid bath studied by Ki CS et al. [56]. Kim et al. [57] proved that the different types of composition used in the collector affected the structure of the poly (L-lactide) (PLLA) and poly (lactide-co-glycolide) ($PLA_{50}GA_{50}$) fibres.

4.3. Ambient parameters

Fibre diameters and morphologies such as humidity and temperature could also be affected by ambient parameters. Increasing temperature, as noted by Mituppatham et al. [58] for instance, favours the thinner fibre diameter with polyamide-6 fibres for the inverse relationship between the solution viscosity and the temperature, as shown in **Figure 13**. With regards to humidity, low humidity could dry the solvent totally and increase the velocity of the solvent evaporation. On the contrary, high humidity will lead to thick

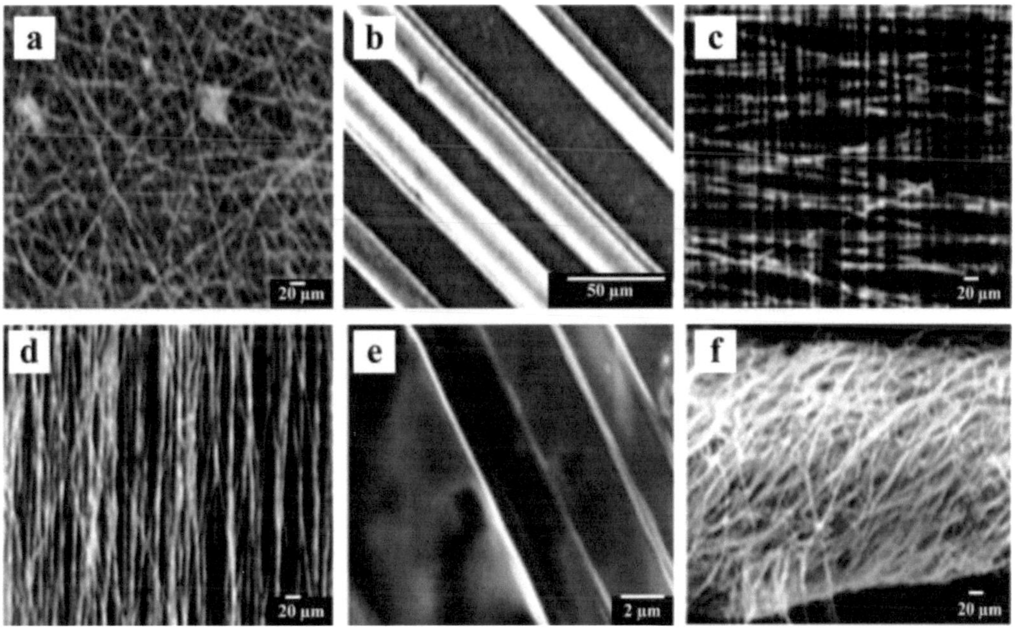

Figure 12. SEM images of the different electrospun products with various types of collectors. (a) Wire Screen, (b) Pin, (c) Gridded Bar, (d) Parallel Bar, (e) Rotating Wheel, and (f) Liquid Bath [52–56].

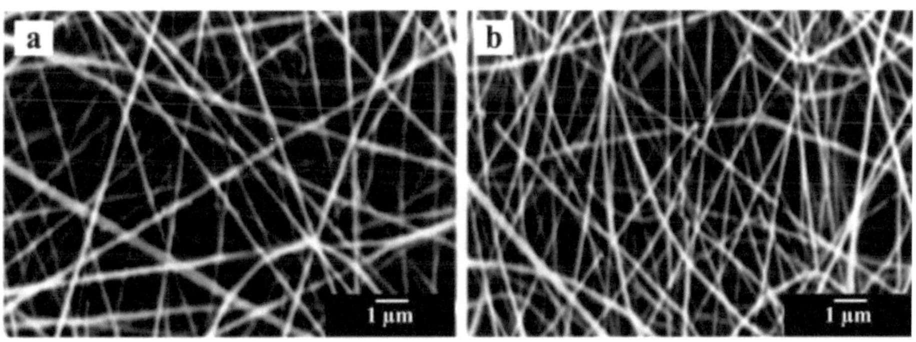

Figure 13. SEM images of the electrospun PA-6-32 fibers under different temperatures. (a) 30° and (b) 60° [58].

fibre diameter because the charges on the jet can be neutralised and the stretching forces become small.

The variety of humidity can also affect the surface morphologies of electrospun PS fibres, as recently show by Casper et al. [59]. Nezarati et al. [60] observed that low humidity (5% RH) resulted in beads connected by thin fibres, but increasing the RH (20–75% RH) resulted in smooth, uniform fibres for poly(ethylene glycol) (PEG). In addition as relative humidity was increased from 50–75%, fibre density decreased, see (**Figure 14**).

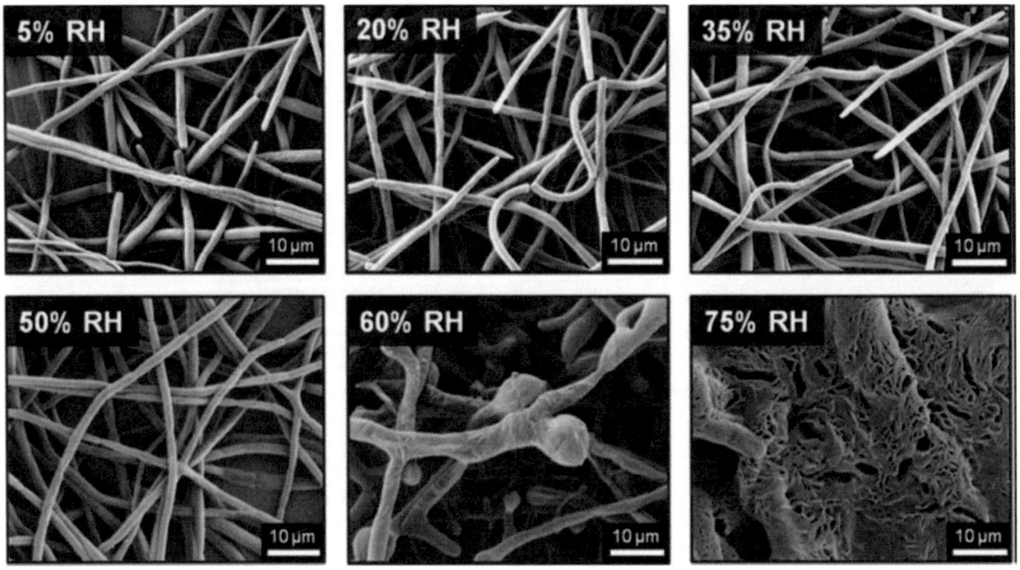

Figure 14. SEM images of poly(ethylene glycol) (PEG) electrospun at relative humidity (RH) ranging from 5 to 75% [60].

5. Conclusions

In this chapter, nanofibers are successfully grown using the electrospinning method at different parameters. These nanofibers, fabricated on several substrates, were investigated using FESEM, observations revealed the formation of nanofibers. Effect of the parameters on the morphology and the diameter was shown. The optimal condition was selected to study the effects of parameters on surface morphology and diameter through the study of characterisation of semiconductor/polymer NFs through the above observations and structural. The semiconductor/polymer nanofibers thin films have offered interesting due to having multi-applications. The sensor, solar cell, supercapacitors, drug delivered, filtering, and more of that its success through been fabricated nanofibers using electrospinning. The electrospinning will not use for fabricating nanofibers only in future work, however, will expand to produce other nanostructures such as nanorods, due to the electronic process has the ability to control the parameters.

Author details

Nabeel Zabar Abed Al-Hazeem[1,2*]

*Address all correspondence to: nabeelnano333@gmail.com

1 Gifted School in Anbar, Gifted Guardianship Committee, Ministry of Education, Iraq

2 Institute of Nano-Optoelectronics Research and Technology Laboratory (iNOR), USM – School of Physics, Penang, Malaysia

References

[1] Liu Y et al. Controlling numbers and sizes of beads in electrospun nanofibers. Polymer International. 2008;**57**(4):632-636

[2] Baji A et al. Electrospinning of polymer nanofibers: effects on oriented morphology, structures and tensile properties. Composites Science and Technology. 2010;**70**(5):703-718

[3] Cavaliere S et al. Electrospinning: Designed architectures for energy conversion and storage devices. Energy & Environmental Science. 2011;**4**(12):4761-4785

[4] Chronakis IS. Novel nanocomposites and nanoceramics based on polymer nanofibers using electrospinning process—A review. Journal of Materials Processing Technology. 2005;**167**(2):283-293

[5] Bognitzki M et al. Nanostructured fibers via electrospinning. Advanced Materials. 2001;**13**(1):70-72

[6] Haslauer CM et al. Collagen–PCL sheath–core bicomponent electrospun scaffolds increase osteogenic differentiation and calcium accretion of human adipose-derived stem cells. Journal of Biomaterials Science, Polymer Edition. 2011;**22**(13):1695-1712

[7] Baumgarten PK. Electrostatic spinning of acrylic microfibers. Journal of Colloid and Interface Science. 1971;**36**(1):71-79

[8] Larrondo L, St John Manley R. Electrostatic fiber spinning from polymer melts. I. Experimental observations on fiber formation and properties. Journal of Polymer Science, Polymer Physics Edition. 1981;**19**(6):909-920

[9] Fridrikh SV et al. Controlling the fiber diameter during electrospinning. Physical Review Letters. 2003;**90**(14):144502

[10] Kidoaki S, Kwon IK, Matsuda T. Mesoscopic spatial designs of nano-and microfiber meshes for tissue-engineering matrix and scaffold based on newly devised multilayering and mixing electrospinning techniques. Biomaterials. 2005;**26**(1):37-46

[11] Stankus JJ et al. Microintegrating smooth muscle cells into a biodegradable, elastomeric fiber matrix. Biomaterials. 2006;**27**(5):735-744

[12] Ke P et al. From macro to micro: structural biomimetic materials by electrospinning. RSC Advances. 2014;**4**(75):39704-39724

[13] Buchko CJ et al. Processing and microstructural characterization of porous biocompatible protein polymer thin films. Polymer. 1999;**40**(26):7397-7407

[14] Taylor G. Electrically driven jets. In: Proceedings of the Royal Society of London, Ser. A; 1969;**313**(1515):453-475

[15] Carson R et al. Photomicrography of electrically sprayed heavy particles. AIAA Journal. 1964;**2**(4):733-737

[16] Doshi J, Reneker DH. Electrospinning process and applications of electrospun fibers. In: Industry Applications Society Annual Meeting, 1993. Conference Record of the 1993 IEEE. IEEE; 1993

[17] Reneker DH et al. Bending instability of electrically charged liquid jets of polymer solutions in electrospinning. Journal of Applied Physics. 2000;**87**(9):4531-4547

[18] Berry G. Thermodynamic and conformational properties of polystyrene. II. Intrinsic viscosity studies on dilute solutions of linear polystyrenes. The Journal of Chemical Physics. 1967;**46**(4):1338-1352

[19] Hager B, Berry G. Moderately concentrated solutions of polystyrene. I. Viscosity as a function of concentration, temperature, and molecular weight. Journal of Polymer Science, Polymer Physics Edition. 1982;**20**(5):911-928

[20] Ko F et al. Electrostatically generated nanofibres for wearable electronics. In: Wearable Electronics and Photonics. Woodhead Publishing in Textiles; 2005. p. 1340

[21] Bronstein LM, Nalwa HS. Encyclopedia of nanoscience and nanotechnology. American Scientific Publishers. 2004;**7**:193-206

[22] Huang Z-M et al. A review on polymer nanofibers by electrospinning and their applications in nanocomposites. Composites Science and Technology. 2003;**63**(15): 2223-2253

[23] Pham QP, Sharma U, Mikos AG. Electrospinning of polymeric nanofibers for tissue engineering applications: a review. Tissue Engineering. 2006;**12**(5):1197-1211

[24] Taylor G. Electrically driven jets. In: Proceedings of the Royal Society of London A: Mathematical, Physical and Engineering Sciences. The Royal Society; 1969

[25] Lu P, Ding B. Applications of electrospun fibers. Recent Patents on Nanotechnology. 2008;**2**(3):169-182

[26] Maleki M, Latifi M, Amani-Tehran M. Optimizing electrospinning parameters for finest diameter of nano fibers. World Academy of Science, Engineering and Technology. 2010;**64**:389-392

[27] Li Z, Ce Wang. One-dimensional nanostructures: Electrospinning technique and unique nanofibers. Heidelberg, New York, Dordrecht, London. 2013

[28] Thompson C et al. Effects of parameters on nanofiber diameter determined from electrospinning model. Polymer. 2007;**48**(23):6913-6922

[29] Subbiah T et al. Electrospinning of nanofibers. Journal of Applied Polymer Science. 2005;**96**(2):557-569

[30] Formhals A. Method and Apparatus for Spinning. 1939. US Patent No. 2160962

[31] Pelipenko J et al. The impact of relative humidity during electrospinning on the morphology and mechanical properties of nanofibers. International Journal of Pharmaceutics. 2013;**456**(1):125-134

[32] Zeng J et al. Poly-L-lactide nanofibers by electrospinning–influence of solution viscosity and electrical conductivity on fiber diameter and fiber morphology. e-Polymers; 2003:**3**(1)

[33] Liu H, Hsieh YL. Ultrafine fibrous cellulose membranes from electrospinning of cellulose acetate. Journal of Polymer Science Part B: Polymer Physics. 2002;**40**(18):2119-2129

[34] Deitzel J et al. The effect of processing variables on the morphology of electrospun nanofibers and textiles. Polymer. 2001;**42**(1):261-272

[35] Demir MM et al. Electrospinning of polyurethane fibers. Polymer. 2002;**43**(11):3303-3309

[36] Sill TJ, von Recum HA. Electrospinning: applications in drug delivery and tissue engineering. Biomaterials. 2008;**29**(13):1989-2006

[37] Zong X et al. Structure and process relationship of electrospun bioabsorbable nanofiber membranes. Polymer. 2002;**43**(16):4403-4412

[38] Huang L et al. Engineered collagen–PEO nanofibers and fabrics. Journal of Biomaterials Science, Polymer Edition. 2001;**12**(9):979-993

[39] Stanger JJ. Charge transfer mechanisms in electrospinning. University of Canterbury. 2008

[40] Angammana CJ, Jayaram SH. Analysis of the effects of solution conductivity on electrospinning process and fiber morphology. IEEE Transactions on Industry Applications. 2011;**47**(3):1109-1117

[41] Garg K, Bowlin GL. Electrospinning jets and nanofibrous structures. Biomicrofluidics. 2011;**5**(1):013403

[42] Koski A, Yim K, Shivkumar S. Effect of molecular weight on fibrous PVA produced by electrospinning. Materials Letters. 2004;**58**(3):493-497

[43] Zhao Y et al. Study on correlation of morphology of electrospun products of polyacrylamide with ultrahigh molecular weight. Journal of Polymer Science Part B: Polymer Physics. 2005;**43**(16):2190-2195

[44] Akduman Ç, Kumabasar EPA, Çay A. Effect of molecular weight on the morphology of electrospun poly (vinyl alcohol) nanofibers. 2014

[45] Yang Q et al. Influence of solvents on the formation of ultrathin uniform poly (vinyl pyrrolidone) nanofibers with electrospinning. Journal of Polymer Science Part B: Polymer Physics. 2004;**42**(20):3721-3726

[46] Lannutti J et al. Electrospinning for tissue engineering scaffolds. Materials Science and Engineering: C. 2007;**27**(3):504-509

[47] Son WK et al. The effects of solution properties and polyelectrolyte on electrospinning of ultrafine poly (ethylene oxide) fibers. Polymer. 2004;**45**(9):2959-2966

[48] Shao H et al. Effect of electrospinning parameters and polymer concentrations on mechanical-to-electrical energy conversion of randomly-oriented electrospun poly (vinylidene fluoride) nanofiber mats. RSC Advances. 2015;**5**(19):14345-14350

[49] Luo C et al. Electrospinning versus fibre production methods: from specifics to technological convergence. Chemical Society Reviews. 2012;**41**(13):4708-4735

[50] Zargham S et al. The effect of flow rate on morphology and deposition area of electrospun nylon 6 nanofiber. Journal of Engineered Fabrics & Fibers (JEFF). 2012;7(4)

[51] Rodoplu D, Mutlu M. Effects of electrospinning setup and process parameters on nanofiber morphology intended for the modification of quartz crystal microbalance surfaces. Journal of Engineered Fibers and Fabrics. 2012;**7**(2):118-123

[52] Wang X et al. Formation of water-resistant hyaluronic acid nanofibers by blowing-assisted electro-spinning and non-toxic post treatments. Polymer. 2005;**46**(13):4853-4867

[53] Sundaray B et al. Electrospinning of continuous aligned polymer fibers. Applied Physics Letters. 2004;**84**(7):1222-1224

[54] Li D, Wang Y, Xia Y. Electrospinning nanofibers as uniaxially aligned arrays and layer-by-layer stacked films. Advanced Materials. 2004;**16**(4):361-366

[55] Xu C et al. Aligned biodegradable nanofibrous structure: a potential scaffold for blood vessel engineering. Biomaterials. 2004;**25**(5):877-886

[56] Ki CS et al. Electrospun three-dimensional silk fibroin nanofibrous scaffold. Journal of Applied Polymer Science. 2007;**106**(6):3922-3928

[57] Kim HS et al. Morphological characterization of electrospun nano-fibrous membranes of biodegradable poly (L-lactide) and poly (lactide-co-glycolide). In: Macromolecular Symposia; 2005. Wiley Online Library

[58] Mituppatham C, Nithitanakul M, Supaphol P. Ultrafine electrospun polyamide-6 fibers: Effect of solution conditions on morphology and average fiber diameter. Macromolecular Chemistry and Physics. 2004;**205**(17):2327-2338

[59] Casper CL et al. Controlling surface morphology of electrospun polystyrene fibers: Effect of humidity and molecular weight in the electrospinning process. Macromolecules. 2004;**37**(2):573-578

[60] Nezarati RM, Eifert MB, Cosgriff-Hernandez E. Effects of humidity and solution viscosity on electrospun fiber morphology. Tissue Engineering, Part C: Methods. 2013;**19**(10):810-819

Chapter 12

Palladium (II) Oxide Nanostructures as Promising Materials for Gas Sensors

Alexander M. Samoylov, Stanislav V. Ryabtsev,
Vasily N. Popov and Petre Badica

Additional information is available at the end of the chapter

http://dx.doi.org/10.5772/intechopen.72323

Abstract

One of the most important environment monitoring problems is the detection of oxidizing gases in the ambient air. Negative impact of noxious oxidizing gases (ozone and nitrogen oxides) on human health, sensitive vegetation, and ecosystems is very serious. For this reason, palladium (II) oxide nanostructures have been employed for oxidizing gas detection. Thin and ultrathin films of palladium (II) oxide were prepared by thermal oxidation at dry oxygen of previously formed pure palladium layers on polished poly-Al_2O_3, SiO_2/Si (100), optical quality quartz, and amorphous carbon/KCl substrates. At ozone and nitrogen dioxide detection, PdO films prepared by oxidation at $T = 870$ K have demonstrated good values of sensitivity, signal stability, operation speed, and reproducibility of sensor response. In comparison with other materials, palladium (II) oxide thin and ultrathin films have some advantages at gas sensor fabrication. Firstly, for oxidizing gas detection, PdO films with *p*-type conductivity are more perspective than the material with *n*-type conductivity. Secondly, at ambient conditions, palladium (II) oxide is insoluble in water and does not react with it. These facts are favorable for the fabrication of gas detectors because they make possible to minimize the air humidity influence on PdO sensor response values. Thirdly, the synthesis procedure of PdO films is rather simple and is compatible with planar processes of microelectronic industry.

Keywords: palladium (II) oxide, nanostructure, gas sensor, ozone, nitrogen dioxide

1. Introduction

Nowadays, the detection of oxidizing gases in the ambient air is one of the most important environment monitoring problems for industrialized countries. During the last 25 years, the steady increase in concentration of nitrogen dioxide and tropospheric (low level) ozone is observed. As it is known, three out of six common air pollutants (also called "criteria pollutants") are oxidizing gases: sulfur dioxide, nitrogen oxides, and tropospheric ozone [1, 2]. One part of ecologists is sure that increase in the content of low-level ozone in atmospheric air is caused mainly by an intensification of industrial production, motor and air transport. Undoubtedly, ozone gas is applied in many fields such as food, pharmaceutical, textile, and chemical industries, water treatment, and purification of gases. However, there is an opinion that emergence of tropospheric ozone in ambient air is a consequence of the "greenhouse" effect [3].

Under sunlight, the interaction of ozone, nitrogen oxides, and volatile hydrocarbons can produce many toxic organic compounds (**Figure 1**). By the action of sunlight, oxygen atoms freed from nitrogen dioxide attack oxygen molecules to make ozone. Nitrogen oxide can combine with ozone to reform nitrogen dioxide, and the cycle repeats.

Moreover, at interaction with ozone, the ultraviolet component of sunlight leads to the formation of excess quantity of the reactive oxygen species (ROS): oxygen ions, free radicals, and peroxides. In living bodies, even the trace amounts of ROC can provoke an oxidative stress. For human, the oxidative stress is a reason for atherosclerosis, hypertension, Alzheimer's disease, diabetes, and geromorphism [4–9]. The negative impact on human health of an aspiration of noxious oxidizing gases (ozone and nitrogen oxides) is more serious, particularly for children, the elderly, and people who suffer from lung diseases [1, 2]. Nitrogen oxides and tropospheric ozone can also have harmful effects on sensitive vegetation and ecosystems [10–13].

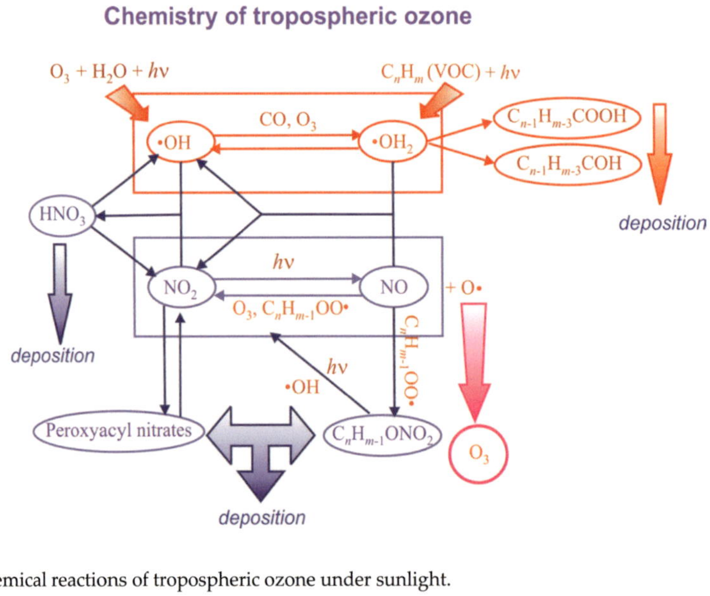

Figure 1. The chemical reactions of tropospheric ozone under sunlight.

For these reasons, various types of the binary, ternary and quaternary metal-oxide semiconductors have been widely applied for oxidizing gas detection. In most cases, for this purpose, the *n*-type semiconductors such as SnO_2 [14–20], ZnO [21–26], WO_3 [27–32], In_2O_3 [33, 34], and TiO_2 [35] are used traditionally. In recent years, the search of the materials, which would be capable to lower the detection limit of oxidizing gases, became more active.

The study of palladium (II) oxide nanostructures as materials for gas sensor fabrication was started only since 2014. The assumption to use palladium (II) oxide, which is a *p*-type semiconductor with the energy band gap $\Delta E_g = 2.2–2.7$ eV [36–38], as the material for the detection of toxic and highly inflammable gas in ambient air has not been accidental for some reasons [39, 40]. Firstly, for a long time, palladium and its compounds in (+2) oxidation state were exploited as very effective catalysts for oxidation reactions of hydrocarbons, including automobile catalytic converters and the catalytic combustion of methane in advanced gas turbines. In catalytic converters, the key processes are the complete oxidation of any hydrocarbon in the exhaust gas stream, the simultaneous oxidation of carbon monoxide, and reduction of nitrogen oxides. Secondly, due to the extremely high catalytic activity, palladium and palladium (II) oxide were applied as additives to improve gas-sensing performance of tin dioxide SnO_2 to a wide range of gases [41–43]. Thirdly, the opinion that long recovery process and high stability could be referred to the main disadvantages of the oxidizing gas sensors based on tin dioxide has been expressed earlier [44, 45]. Fourthly, the metal oxide semiconductors with *p*-type conductivity are more perspective for oxidizing gas detection than the materials with *n*-type conductivity. In this case, the chemical adsorption of oxidizing gas molecule on *p*-type semiconductors surface leads to decrease in the sensor resistance that has simplified the detection process [35, 36]. Thus, at oxidizing gas detection, palladium (II) oxide nanostructures should demonstrate the increase in sensor response value in comparison with the traditional *n*-type conductivity materials using for the same purpose.

2. Fabrication of palladium (II) oxide nanostructures

Initially, the sensing properties to oxidizing gases of palladium (II) oxide nanostructures were tested on ultrathin and thin films at detection of ozone and nitrogen dioxide [46, 47]. The procedure of PdO thin and ultrathin films synthesis was realized by two stages. First, the initial palladium films (thickness 5–30 nm) were formed by thermal sublimation of palladium foil (purity is 99.99%) in high vacuum chamber evacuated to 5×10^{-7} Torr using a turbo molecular pump. In vacuum chamber, the condensation of Pd metal vapors was performed on different substrates: SiO_2/Si (100), Si (100), optical quality quartz, and KCl (100) with buffer layer of amorphous carbon (**Figure 2**). The values of tungsten heater temperature in order to fabricate initial palladium films with average rate within interval 0.01–0.016 nm per second were determined as a result of Pd films cross-sections by high-resolution scanning and transmission electron microscopy (HR STEM) study.

The substructure of initial palladium layers was studied by an X-ray analysis and the HEED method. As it is shown in **Figure 3a** and **3b**, the initial Pd films were polycrystalline and

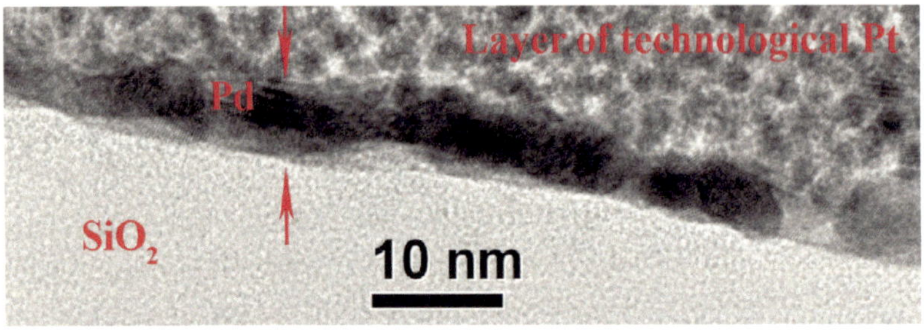

Figure 2. High-resolution TEM image of $Pd/SiO_2/Si$ (100) heterostructure cross-section prepared by focused ion beam (FIB) technique.

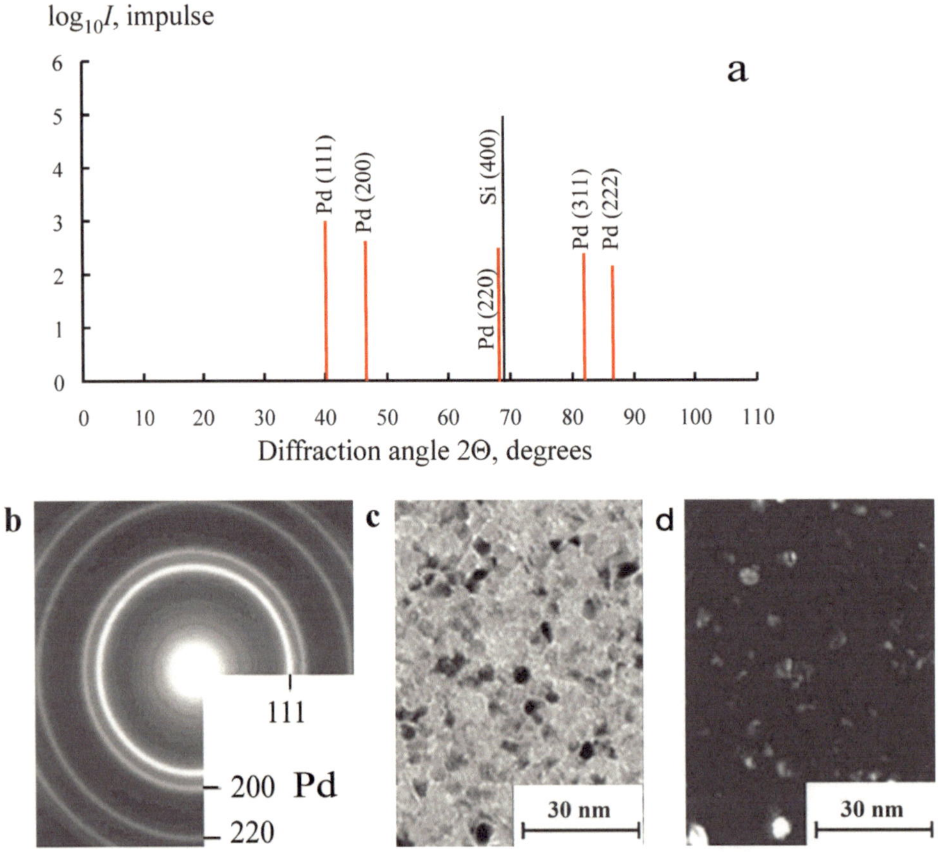

Figure 3. Experimental results of initial Pd films (thickness ~10 nm) crystal structure study: (a) XRD patterns of Pd film deposited on Si (100) substrate; (b) HEED patterns of Pd film deposited on amorphous carbon/KCl substrate; (c) bright-field TEM image of Pd film deposited on amorphous carbon/KCl substrate; (d) dark-field TEM image of Pd film deposited on amorphous carbon/KCl substrate.

http://dx.doi.org/10.5772/intechopen.72323

highly dispersive with random orientation of grains irrespective of the substrate nature (SiO_2/Si (100), optical quality quartz, and amorphous carbon/KCl). The analysis of *bright-field* (**Figure 3c**) *and dark-field* (**Figure 3d**) TEM images proves that palladium crystalline grains form a continuous coating without an axial texture with very low density of micropores. On bright-field image, the light contrast (**Figure 3c**) testifies to the decrease of film thickness at grain borders [48].

Prepared Pd nanostructures on different substrates were annealed at dry oxygen atmosphere for 1 h for layers with thickness 5–15 nm and for 2 h for layers with thickness 30 ± 5 nm at temperatures T_{ox} = 510, 570, 670, 770, 870, and 1070 K. The dehumidification of oxygen at pressure 120–130 kPa (1.2–1.3 Bar) was carried out by gas flow passage through gas bubbler with concentrated sulfuric acid and further through silica tube full of ground zeolite [48].

3. Phase composition and crystal structure of palladium (II) oxide nanostructures

X-ray diffraction (XRD) patterns of samples prepared by oxidation of Pd films on SiO_2/Si (100) wafers at dry oxygen atmosphere at T_{ox} = 510, 570, 770, 870, and 970 K are shown in **Figure 4**.

It is necessary to note that in **Figure 4**, the values of XRD reflex intensities are presented in a logarithmic scale because the intensity of Si (400) peak practically exceeds the intensity of palladium and palladium (II) oxide peaks by two orders of magnitude owing to a small thickness of the prepared films. The comparison of the as-grown Pd films XRD patterns with XRD patterns of Pd film after the annealing at T_{ox} = 510 K (**Figure 4a**) did not reveal any quality changes. The increase in intensities of palladium reflexes was found only.

Thus, it has been established that the annealing of Pd layers at T_{ox} < 570 K (**Figure 4a**) did not result in the change of their phase constitution. The annealing at T_{ox} = 570 K resulted in the formation of two phase films (**Figure 4b**). XRD patterns have shown the presence of Pd with PdO simultaneously.

According to XRD results, the rise of the oxidation temperature up to T_{ox} = 770 K and T_{ox} = 870 K led to the formation of the homogenous polycrystalline PdO films. It has been determined (**Figure 4c** and **4d**) that palladium (II) oxide films were characterized by tetragonal crystal lattice (space group $P4_2/mmc$ and PtS structure type). XRD patterns show (**Figure 4c–4e**) that the peaks became sharper and higher with the oxidizing temperature increasing from T_{ox} = 770 K up to T_{ox} = 970 K. Moreover, the peaks of palladium (II) oxide prepared by oxidation at T_{ox} = 970 K are much sharper and higher than those for films oxidized at T_{ox} = 870 K. This fact can be interpreted as one of the evidences of the crystalline perfection enhancement of palladium (II) oxide films and the grain size enlargement with the increase in the oxidation temperature.

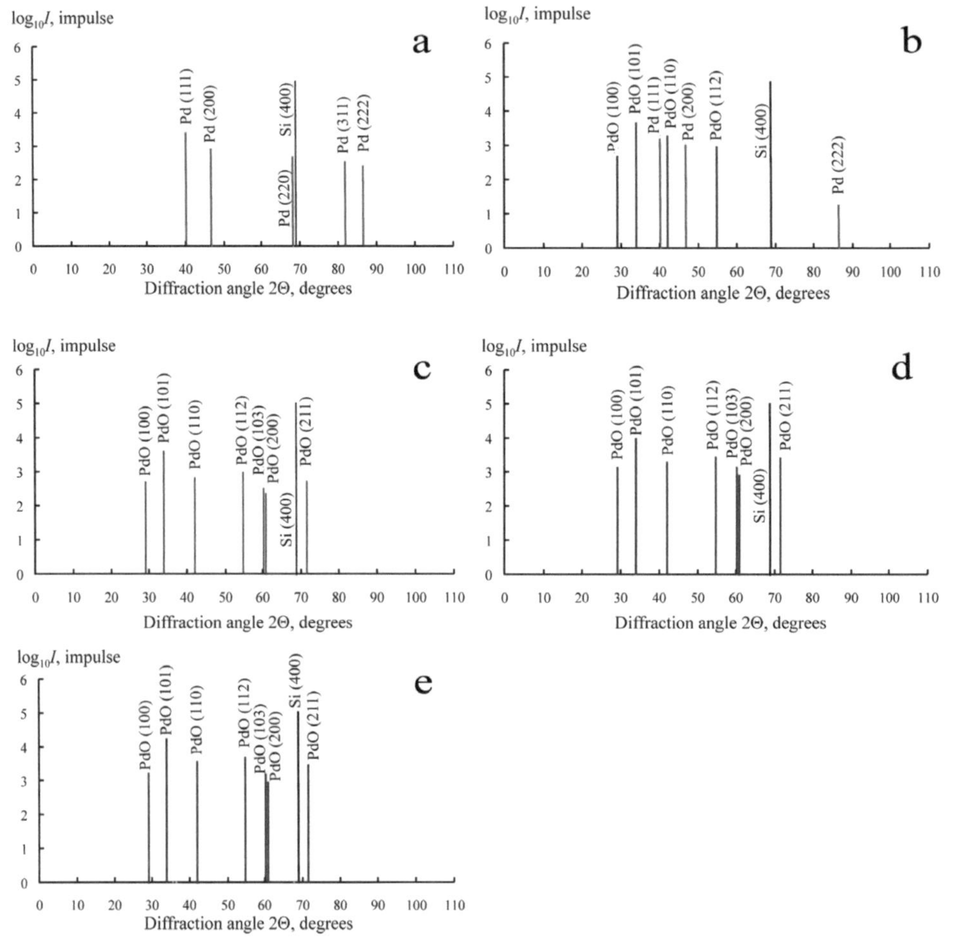

Figure 4. X-ray diffraction patterns of palladium film deposited on SiO_2/Si (100) substrate after oxidation in dry oxygen at different temperatures: *a*—T_{ox} = 510 K; *b*—T_{ox} = 570 K; *c*—T_{ox} = 770 K; *d*—T_{ox} = 870 K; and *e*—T_{ox} = 970 K.

High energy electron diffraction (HEED) technique was used as an alternative method to study PdO film phase composition (**Figure 5**).

Table 1 compares the results of X-ray analysis (layers on SiO_2/Si substrates), the HEED method (layers on optical quartz and Al_2O_3 substrates), and TEM micro diffraction (layers on amorphous carbon/KCl). An examination of the data presented in **Table 1** shows that the X-ray analysis, HEED method, and TEM micro diffraction gave the identical results for the films oxidized at temperatures T_{ox} = 510, 570, 770, 870, and 970 K. The results of these two methods confirm that: (1) the annealing of Pd films at T_{ox} = 510 K does not induce the changes in their phase composition; (2) after annealing of Pd films at the T_{ox} = 570 K, the partial oxidation takes place and gives two phase samples – a mixture of Pd and PdO; and (3) after the annealing of palladium layers at T_{ox} = 770–970 K, the total oxidation gives homogeneous PdO films [48].

http://dx.doi.org/10.5772/intechopen.72323

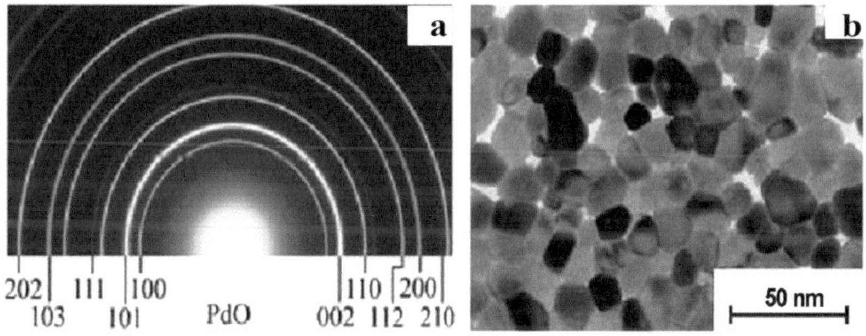

Figure 5. HEED patterns (*a*) and bright-field TEM image (*b*) of PdO film after oxidation at T_{ox} = 870 K.

Oxidation temperature T_{ox}, K	Phase composition		
	X-ray analysis	HEED	TEM microdiffraction
510	Pd	Pd	Pd
570	Pd + PdO	Pd + PdO	Pd + PdO
670	—	PdO	—
770	PdO	PdO	PdO
870	PdO	PdO	PdO
970	PdO	PdO	PdO

Table 1. Results of phase composition study of Pd films after oxidation at T = 500–970 K.

4. Electrical properties of palladium (II) oxide nanostructures

The type of conductivity of PdO films synthesized at T_{ox} = 770–970 K was determined by the Seebeck effect study and by calculation of the electromotive force E_{emf} values:

$$E_{emf} = -S \nabla T, \tag{1}$$

where S is thermo-power (Seebeck coefficient), and ∇T is the temperature gradient. Copper-constantan thermocouples have been used to measure a temperature difference.

Experimental values of E_{emf} have proved the p-type conductivity for all homogeneous palladium (II) oxide films (**Figure 6**). The fact of p-type conductivity of PdO bulk samples was reported in the previous publications [36, 37]. The values of thermo power (Seebeck coefficient) S have been calculated using the Eq. (1). Depending on thickness of palladium oxide films and oxidation temperature, the Seebeck coefficient values changed within the limits from +120 to +220 μV/K. The relative error at thermo-power measurement did not exceed 7%.

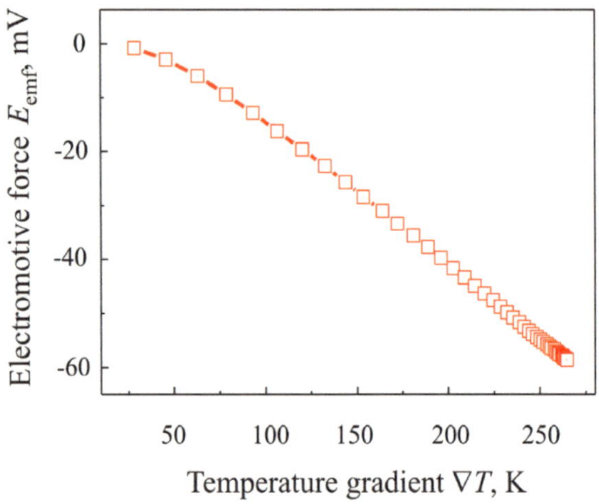

Figure 6. Electromotive force E_{emf} dependence upon the temperature gradient for PdO film prepared by oxidation at T_{ox} = 870 K (thickness ~ 30 nm).

In view of *p*-type conductivity, palladium (II) oxide films are characterized with the cation deficiency regarding the stoichiometric 1:1 ratio. Thus, for PdO, the Kröger-Vink defect reactions can be written as follows:

a. with the cations in deficiency on the lattice sites:

$$Pd_{Pd}^{\times} + O_{O}^{\times} + \frac{1}{2}O_{2}^{(gas)} \rightleftarrows Pd_{Pd}^{\times} + V_{Pd}'' + 2\,O_{O}^{\times} + 2\,h^{\cdot\cdot} \quad (2)$$

b. with the anions in excess on the interstitial sites:

$$Pd_{Pd}^{\times} + O_{O}^{\times} + \frac{1}{2}O_{2}^{(gas)} \rightleftarrows Pd_{Pd}^{\times} + O_{O}^{\times} + O_{i}'' + 2\,h^{\cdot\cdot} \quad (3)$$

The results obtained in the present work correlate with the capacitance voltage characteristics of PdO films on silicon [49]. Previously it was found that within the band gap of PdO films, one single energy state is realized only [49]. Therefore, only one type of point defects, which have generated holes, dominates in palladium (II) oxide films. The experimental study of the point defects nature will be the subject of further investigations.

5. Gas sensor properties of palladium (II) oxide nanostructures

Ozone and nitrogen dioxide sensitivity has been measured using the specially fabricated test samples of gas sensors based on thin and ultrathin PdO films oxidized at T_{ox} = 870 K. During sensor response measurements of PdO, ultrathin and thin films prepared by oxidation at T_{ox} = 873 K, synthetic air, calibrated gas mixtures with fixed nitrogen dioxide concentration, and standardized ozone generator produced by Optec were used. The ozone gas was generated by oxidizing oxygen molecules of synthetic air (SA) by a pen-ray ultraviolet (UV) lamp calibrated to give the O_3 concentration range between 0.03 ppb and 800 ppb. The synthetic air containing ozone was blown directly on the sensor placed on the top of holder within the test chamber. The operating temperature T_d of the sensor ranging from room temperature to 670 K was controlled by chromel-alumel thermocouple. The measurement started after the sample resistance achieved a steady value [50, 51].

Sensor response S was determined as the ratio of the sensor resistance in synthetic air R_0 to the sensor resistance in gas R:

$$S = \frac{R_0}{R} \tag{4}$$

The measurements of NO_2 and O_3 concentration were performed in flow path conditions with the rates of 300 cm^3 per minute and 2.4 dm^3 per minute, respectively. The gas flow rate was measured by controllers produced by Bronkhorst.

As it possible to see in **Figures 7** and **8**, at rather low operation temperature T_d, the sensors based on thin (T_d = 490 K) and ultrathin (T_d = 448 K) PdO films show good sensitivity to rather low concentrations of ozone. **Figures 7** and **8** show that at process of ozone quantitative detection at SA atmosphere within concentration interval 100–250 ppb, palladium (II) oxide films have demonstrated high values of sensor response, signal stability, and reproducibility of sensor response also. This fact was proved by the results of multiple measurement cycles with the same O_3 concentrations. PdO films with thickness of about 35 nm are characterized by higher values of sensor response (on the average in 7–8 times) in comparison with ultrathin films at the same ozone concentrations (**Figure 9**). It is possible to explain

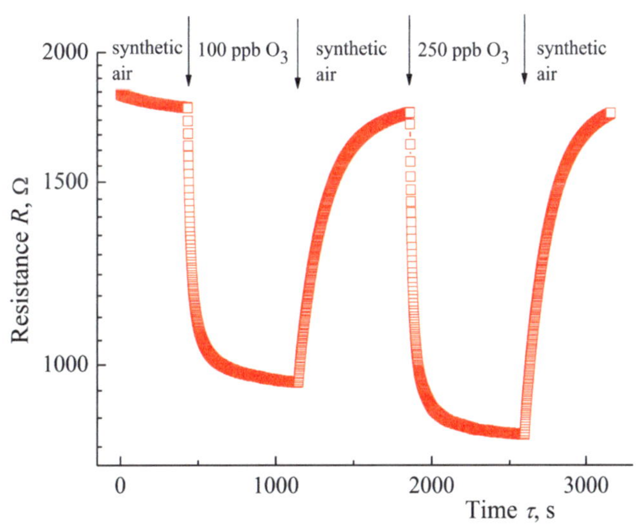

Figure 7. Time dependence of PdO ultrathin film (thickness ~ 10 nm) sensor resistance R at ozone different concentrations (operation temperature T_d = 448 K).

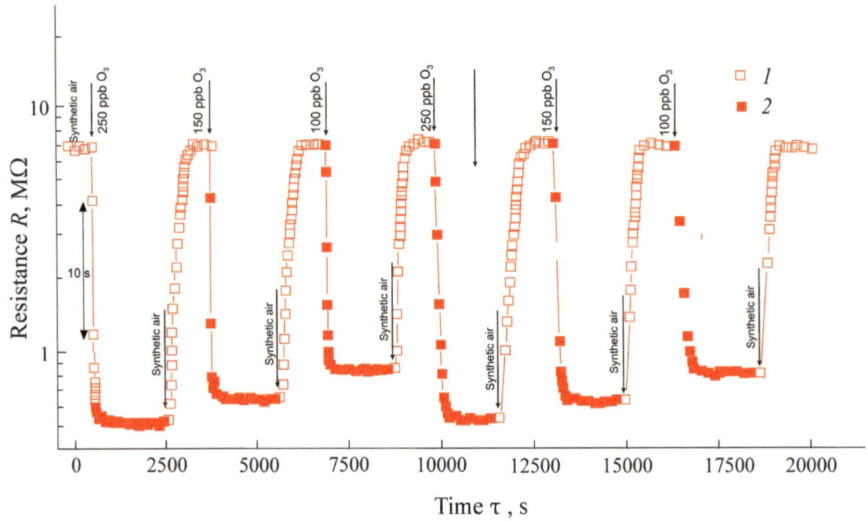

Figure 8. Time dependence of PdO thin film (thickness ~ 35 nm) sensor resistance R at ozone different concentrations (operation temperature T_d = 490 K).

this fact that the contribution in integrated conductivity of near-surface layers with high defects density is essentially higher for ultrathin PdO films than for films with thickness of about 35 nm (**Figure 9**). It is necessary to emphasis that the established feature demands a detailed study.

It has been established that PdO thin and ultrathin film sensors gave the stable signal, and the resistance values reliably returned to the baseline at SA atmosphere [50, 51]. It is necessary to note that the recovery period is quite long (600–700 s). It is necessary to note that the similar sensor behavior is typical for other materials used oxidizing gas detection. Usually in this case, the long recovery period is explained by the absence of oxidizing gas immediate interaction with oxygen molecules adsorbed on sensor material surface. At reducing gas detection, the direct interaction with oxygen molecules takes place; therefore, the recovery time is quite short. Moreover, the recovery time depends significantly on the operating temperature.

The sensitivity of palladium (II) oxide ultrathin films to nitrogen dioxide (another toxic oxidizing gas) has also been tested (**Figure 10**). As it can be seen in **Figure 10**, at the process of NO_2 quantitative detection within concentration interval 500 ppb–200 ppm, PdO ultrathin films have demonstrated good values of sensor response, signal stability, and reproducibility of sensor response [51]. It is necessary to note that the recovery period at NO_2 detection is longer than that at O_3 detection (**Figures 7, 8,** and **10**).

During the determination of ozone (concentration $\varphi = 100$ ppb) and nitrogen dioxide (concentration $\varphi = 10$ ppm), the temperature dependences of PdO ultrathin film sensor response S are presented in **Figure 11**.

It is found that within interval of operation temperature $323 < T_d < 623$ K, the maximum values of response S have been observed at $T_d = 448$ K (NO_2 detection) and at $T_d = 490$ K (O_3 detection).

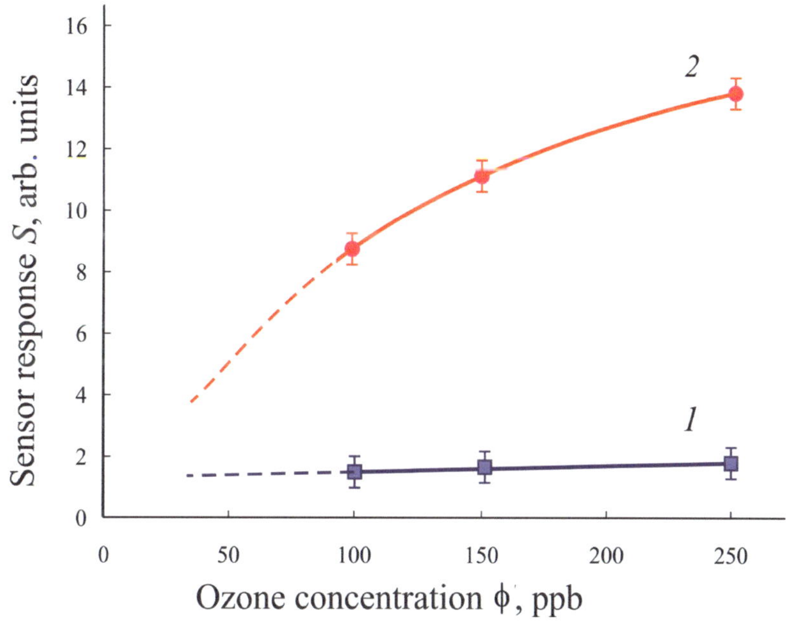

Figure 9. Dependence of PdO ultrathin and thin film sensor response *S* at ozone different concentrations: 1—Ultrathin film (thickness ~ 10 nm, operation temperature T_d = 448 K) and 2—Thin film (thickness ~ 35 nm, operation temperature T_d = 490 K).

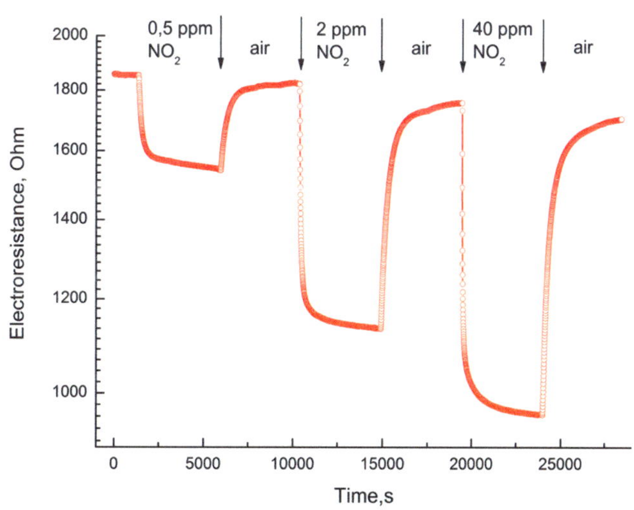

Figure 10. Time dependence of PdO ultrathin film (thickness ~ 10 nm) sensor resistance *R* at nitrogen dioxide different concentrations (operation temperature T_d = 448 K).

As it can be seen in **Figure 11**, approximately equal values of sensor response *S* of palladium (II) oxide films are realized at different concentration of oxidizing gases: $\phi(O_3)$ = 0.1 ppm and $\phi(NO_2)$ = 10 ppm.

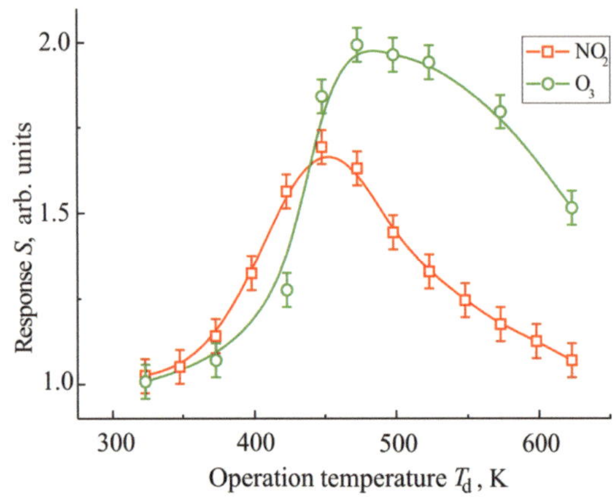

Figure 11. Dependence of PdO ultrathin film (thickness ~ 10 nm) sensor response *S* upon the operation temperature T_d at detection of ozone (O_3 concentration 0.1 ppm) and nitrogen dioxide (NO_2 concentration 10 ppm).

6. Discussion

Data presented in **Table 2** show that physical properties (molecular mass, electric dipole moment) and chemical properties (structure of molecule, high oxidative activity) of detected gases are very similar. According to experimental evidence from microwave spectroscopy, ozone and nitrogen dioxide are bent molecules with C_{2v} symmetry (**Table 2**). O_3 and NO_2 are the polar molecules with a dipole moment of 0.66 and 0.39 D, respectively.

Data in **Table 2** show that ozone and nitrogen dioxide molecules essentially differ with magnetic properties only. Ozone is diamagnetic, which means that its electrons are all paired. Unlike ozone, the ground electronic state of nitrogen dioxide is a doublet state. Owing to since nitrogen atom has one unpaired electron NO_2 molecule is paramagnetic.

Nevertheless, at detection of ozone and nitrogen dioxide, the temperature that has matched the maximum values of sensor response differs only 25°. Thus, there is prerequisite for the increase in selectivity of palladium (II) oxide sensors at O_3 and NO_2 detection after studying in detail that oxidation procedure conditions influence on microstructure and stoichiometry deviation.

Moreover, at ambient conditions, palladium (II) oxide is insoluble in water and does not react with it. As the bottom sediment, the palladium (II) hydroxide is formed only at interaction of soluble palladium (II) salt and alkali [52]. These facts are favorable for fabrication of gas detectors because they make possible to minimize the air humidity influence on PdO sensor response values.

To estimate such perspective for palladium (II) oxide nanostructures, we have allowed the speculative extrapolation of experimental data to the point that corresponds to zero ozone concentration (**Figure 12**). At ozone concentration $\phi = 10$ ppb, which corresponds to 0.1 × PEL (permissible exposure limit), sensor response *S* would be about 2 (*S* ~ 2 is an open circle in **Figure 12**). The extrapolated sensitivity value at concentration $\phi(O_3) = 0.1 \times$ PEL arouses hope that palladium (II) oxide films will be used in fabrication of ozone sensors.

At interaction with PdO surface ozone molecules are more active than nitrogen dioxide ones. This interaction is accompanied by more essential increase in the hole density of palladium (II) oxide ultrathin films. In general case, the surface interaction of PdO nanostructures can be written within the framework of Kröger-Vink notation:

$$Pd_{Pd}^{\times} + O_{O}^{\times} + O_{3}^{(gas)} \rightleftarrows Pd_{Pd}^{\times} + O_{O}^{\times} + O_{i}'' + 2h^{\cdot\cdot} + O_{2}^{(gas)} \quad (5)$$

According to Eq. (5), oxygen atom is integrated with palladium (II) oxide structure and O_2 molecule is desorbed from the surface. As result of this reaction (5), two holes are formed.

From this point of view, it is possible to explain high efficiency of palladium (II) oxide films at ozone detection. The attempt to distinguish the real reason of PdO nanostructures' higher

Molecule	Molar mass M, g × mol^{-1}	Space group symmetry	Magnetic Properties	Magnetic susceptibility $\chi \times 10^6$, cm^3/mol	Dipole moment $\mu \times 10^{30}$, C·m	PEL, ppm
O_3	48.00	C_{2v}	Diamagnetic	+6.7	2.2	0.1
					0.66D	(0.2 mg/m3)
NO_2	46.0055	C_{2v}	Paramagnetic	+150.0	1.3	5
					0.39D	(9 mg/m3)

Table 2. Physicochemical properties and permissible exposure limit (PEL) of ozone and nitrogen dioxide [52–56].

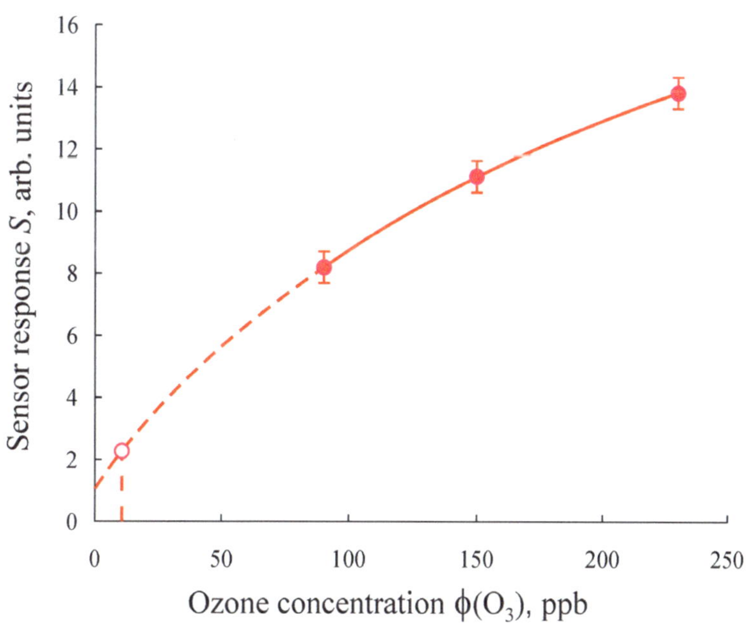

Figure 12. Dependence of PdO sensor response S upon the ozone concentration in synthetic air (operating temperature T_d = 490 K (220°C).

sensitivity to ozone, it should be looked for ozone's extremely high oxidizing ability. As it can be seen in **Table 2**, the PEL value of ozone is smaller than the similar characteristic of nitrogen dioxide by 50 times practically. This fact is the indirect evidence of ozone-exclusive oxidizing activity. The difference in sensitivity of palladium (II) oxide nanostructures at ozone and nitrogen dioxide detection will be a subject of the subsequent experiments and discussions. Now, it is possible to designate the direction of these future researches only. It is reasonable to assume that under ozone molecules impact, the metastable nanoclusters are formed on the surface of PdO, in which the oxidation states of palladium are higher than (II), for example, (III) or (IV).

7. Conclusion

The results of X-ray analysis, HEED, and HR TEM have demonstrated the possibility of the synthesis of homogeneous nanocrystalline thin and ultrathin films of palladium (II) oxide on different substrates. The very first examinations of sensitivity to different nitrogen dioxide and ozone concentration at rather low operating temperature have shown the high values of sensor response, signal stability, operation speed, and reproducibility of PdO films sensor response. The possibility of work at quite low temperatures will allow decreasing in the energy consumption of the analytical instruments. The detection of O_3 and NO_2 by palladium (II) oxide sensors can be applied in the fields of the human health and environment protection. Because the synthesis procedure is rather simple and compatible with planar processes of the microelectronic industry PdO nanostructures have a good perspective to be one of the main materials for commercial fabrication of oxidizing gases (ozone, nitrogen dioxide, chlorine etc.) sensors.

Author details

Alexander M. Samoylov[1]*, Stanislav V. Ryabtsev[1], Vasily N. Popov[1] and Petre Badica[2]

*Address all correspondence to: samoylov@chem.vsu.ru

1 Voronezh State University, Universitetskaya, Voronezh, Russian Federation

2 National Institute of Materials Physics, Atomistilor, Magurele, Ilfov, Romania

References

[1] Estimating Mortality Risk Reduction and Economic Benefits from Controlling Ozone Air Pollution. Report of the US National Academies of Sciences. Washington: The National Academies Press; 2008. 247 p

[2] Health Aspects of Air Pollution with Particulate Matter, Ozone and Nitrogen Dioxide. Report on a WHO Working Group, 13-15-01-2003. Bonn, Germany: ©World Health Organization. 2003. 94 p

[3] Amos P. K. Tai, Maria Val Martin, Colette L. Heald. Threat to future global food security from climate change and ozone air pollution. Nature Climate Change. Letters. Published online: 27-07-2014. DOI:10.1038/nclimate2317

[4] Chandra Kala, Syed Salman Ali, Abid Mohd, Sweety Rajpoot, Najam Ali Khan. Protection against fca induced oxidative stress induced dna damage as a model of arthritis and in vitro anti-arthritic potential of *Costus speciosus* rhizome extract. International Journal of Pharmacognosy and Phytochemical Research. 2015;7(2):383-389

[5] Parellada M, Moreno C, Mac-Dowell K, Leza JC, Giraldez M, Bailón C, Castro C, Miranda-Azpiazu P, Fraguas D, Arango C. Plasma antioxidant capacity is reduced in Asperger syndrome. Journal of Psychiatric Research. 2012;**46**:394-401

[6] Joseph N, Zhang-James Y, Perl A, Faraone SV. Oxidative stress and ADHD: A meta-analysis. Journal of Attention Disorders. 2015;**19**:915-924

[7] Hwang O. Role of oxidative stress in Parkinson's disease. Experimental Neurology. 2013;**22**:11-17

[8] Giacco F, Brownlee M. Oxidative stress and diabetic complications. Circulation Research. 2010;**107**(9):1058-1070

[9] Romá-Mateo C, Aguado C, García-Giménez JL, Ibáñez-Cabellos JS, Seco-Cervera M, Pallardó FV, Knecht E, Sanz P. Increased oxidative stress and impaired antioxidant response in lafora disease. Molecular Neurobiology. 2015;**51**:932-946

[10] Popov VN. Possible role of free oxidation processes in the regulation of reactive oxygen species production in plant mitochondria. Biochemical Society Transactions. 2003;**31**, part 6:1316-1317

[11] Shiri Avnery, Denise L. Mauzerall, Junfeng Liu, Larry W. Horowitz. Global crop yield reductions due to surface ozone exposure: 1. Year 2000 crop production losses and economic damage. Atmospheric Environment. 2011;**45**:2284-2296

[12] Avnery S, Mauzerall DL, Liu J, Horowitz LW. Global crop yield reductions due to surface ozone exposure: 2. Year 2030 potential crop production losses and economic damage under two scenarios of O_3 pollution. Atmospheric Environment. 2011;**45**:2297-2309

[13] Popov VN. Feedback loop of non-coupled respiration and reactive oxygen species production in plant mitochondria. In: Kapuganti Jagadis Gupta and Abir U. Igamberdiev. Reactive Oxygen and Nitrogen Species Signaling and Communication in Plants. Vol. 23., ed. Switzerland: ©Springer International Publishing; 2015. 316 p

[14] Krivetskiy V, Ponzoni A, Comini E, Badalyan S, Rumyantseva M, Gaskov A. Selectivity modification of SnO_2-based materials for gas sensor arrays. Electroanalysis. 2010;**22**:1-8

[15] Badalyan SM, Rumyantseva MN, Nikolaev SA, Marikutsa AV, Smirnov VV, Alikhanian AS, Gaskov AM. Effect of au and NiO catalysts on the NO_2 sensing properties of nanocrystalline SnO_2. Inorganic Materials. 2010;**46**:232-236

[16] Jung S-H, Choi S-W, Kim SS. Fabrication and properties of trench-structured networked SnO_2 nanowire gas sensors. Sensors and Actuators B. 2012;**171-172**:672-678

[17] Oros C, Horprathumb M, Wisitsoraat A, Srichaiyaperk T, Samransuksamer B, Limwichean S, Eiamchai P, Phokharatkul D, Nuntawong N, Chananonnawathorn C, Patthanasettakul V, Klamchuen A, Kaewkhaoe J, Tuantranont A, Chindaudom P. Ultra-sensitive NO_2 sensor based on vertically aligned SnO_2 nanorods deposited by DC reactive magnetron sputtering with glancing angle deposition technique. Sensors and Actuators B. 2016;**223**:936-945

[18] Adelina Stanoiu, Simona Somacescu, Jose Maria Calderon-Moreno,Valentin Serban Teodorescu, Ovidiu Gabriel Florea, André Sackmann, Cristian Eugen Simion. Low level NO_2 detection under humid background and associated sensing mechanism for meso-porous $SnO_{2,}$ Sensors and Actuators B. 2016;**231**:166-174

[19] Zhao X, Shi W, Mu H, Xie H, Liu F. Templated bicontinuous tin oxide thin film fabrication and the NO_2 gas sensing. Journal of Alloys and Compounds. 2016;**659**:60-65

[20] Yuling Wei, Changlong Chen, Guangzheng Yuan, Shuai Gao. SnO_2 nanocrystals with abundant oxygen vacancies: Preparation and room temperature NO_2 sensing. Journal of Alloys and Compounds. 2016;**681**:43-49

[21] Saboor FH, Ueda T, Kamada K, Hyodo T, Mortazavi Y, Khodadadi AA, Shimizu Y. Enhanced NO_2 gas sensing performance of bare and Pd-loaded SnO_2 thick film sensors under UV-light irradiation at room temperature. Sensors and Actuators B. 2016;**223**:429-439

[22] Jiao M, Chien NV, Duy NV, Hoa ND, Van Hieu N, Hjort K, Nguyen H. On-chip hydrothermal growth of ZnO nanorods at low temperature for highly selective NO_2 gas sensor. Materials Letters. 2016;**169**:231-235

[23] Katoch A, Sun G-J, Choi S-W, Byun J-H, Kim SS. Competitive influence of grain size and crystallinity on gas sensing performances of ZnO nanofibers. Sensors and Actuators B. 2013;**185**:411-416

[24] Lontio Fomekong R, Lahem D, Debliquy M, Yunus S, Lambi Ngolui J, Delcorte A. $Ni_{0.9}Zn_{0.1}O$/ZnO nanocomposite prepared by malonate coprecipitation route for gas sensing. Sensors and Actuators B. 2016;**231**:520-528

[25] Lingmin Y, Guo F, Liu S, Yang B, Jiang Y, Qi L, Fan X. Both oxygen vacancies defects and porosity facilitated NO_2 gas sensing response in 2D ZnO nanowalls at room temperature. Journal of Alloys and Compounds. 2016;**682**:352-356

[26] Ganbavle VV, Inamdar SI, Agawane GL, Kim JH, Rajpure KY. Synthesis of fast response, highly sensitive and selective Ni:ZnO based NO_2 sensor. Chemical Engineering Journal. 2016;**286**:36-47

[27] Jaina R, Leib Y, Maric R. Ultra-low NO_2 detection by gamma WO_3 synthesized by reactivespray deposition technology. Sensors and Actuators B. 2016;**236**:163-172

[28] Van PTH, Do DD, Duy NV, Hoa ND, Hieu NV. Ultrasensitive NO_2 gas sensors using tungsten oxide nanowires with multiple junctions self-assembled on discrete catalyst islands via on-chip fabrication. Sensors and Actuators B. 2016;**227**:198-203

[29] Hi Gyu Moon, Soo Deok Han, Min-Gyu Kang, Woo-Suk Jung, Beomjin Kwon,Chulki Kim, Taikjin Lee, Seok Lee, Seoung-Hyub Baek, Jin-Sang Kim, Hyung-Ho Park, Chong-Yun Kang. Glancing angle deposited WO_3 nanostructures for enhanced sensitivity and selectivity to NO_2 in gas mixture. Sensors and Actuators B. 2016;**229**:92-99

[30] Kim J-S, Yoon J-W, Hong YJ, Kang YC, Abdel-Hady F, Wazzan AA, Lee J-H. Highly sensitive and selective detection of ppb-level NO_2 using multi-shelled WO_3 yolk–shell spheres. Sensors and Actuators B. 2016;**229**:561-569

[31] Giancaterini L, Emamjomeh SM, De Marcellis A, Palange E, Resmini A, Anselmi-Tamburini U, Cantalini C. The influence of thermal and visible light activation modes on the NO_2 response of WO_3 nanofibers prepared by electrospinning. Sensors and Actuators B. 2016;**229**:387-395

[32] Zhang W, Ming H, Liu X, Wei Y, Na L, Qin Y. Synthesis of the cactus-like silicon nanowires/tungsten oxide nanowires composite for room-temperature NO_2 gas sensor. Journal of Alloys and Compounds. 2016;**679**:391-399

[33] Ilin A, Martyshov M, Forsh E, Forsh P, Rumyantseva M, Abakumov A, Gaskov A, Kashkarov P. UV effect on NO_2 sensing properties of nanocrystalline In_2O_3. Sensors and Actuators B. 2016;**231**:491-496

[34] Xiao B, Wang F, Zhai C, Wang P, Xiao C, Zhang M. Facile synthesis of In_2O_3 nanoparticles for sensing properties at low detection temperature. Sensors and Actuators B. 2016;**235**:251-257

[35] Navale ST, Tehare KK, Shaikh SF, Patil VB, Pawar BN, Naushad M, Stadler FJ, Mane RS. Hexamethylenetetramine-mediated TiO_2 films: Facile chemical synthesis strategy and their use in nitrogen dioxide detection. Materials Letters. 2016;**173**:9-12

[36] Rey E, Kamal MR, Miles RB, Royce BSH. The semiconductivity and stability of palladium oxide. Journal of Materials Science. 1978;**13**:812-816

[37] Nilsson PO, Shivaraman MS. Optical properties of PdO in the range of 0.5-5.4 eV. Journal of Physics C: Solid State Physics. 1979;**12**:1423-1427

[38] Ryabtsev SV, Shaposhnik AV, Samoylov AM, Sinelnikov AA, Soldatenko SA, Kuschev SB, Ievlev VM. Thin films of palladium oxide for gas sensors. Doklady Physical Chemistry. 2016;**470**:158-161

[39] Kneer J, Wöllenstein J, Palzer S. Manipulating the gas–surface interaction between copper (II) oxide and mono-nitrogen oxides using temperature. Sensors and Actuators B. 2016;**229**:57-62

[40] Kim J-H, Katoch A, Choi S-W, Kim SS. Growth and sensing properties of networked p-CuO nanowires. Sensors and Actuators B. 2015;**212**:190-195

[41] Yamazoe N, Kurokawa Y, Seiyama T. Effects of additives on semiconductor gas sensors. Sensors and Actuators B. 1983;**4**:283-289

[42] Ievlev VM, Kushchev SB, Sinel'nikov AA, Soldatenko SA, Ryabtsev SV, Bosykh MA, Samoilov AM. Structure of heterosystems formed by a SnO_2 film and island metal (Ag, au, or Pd) condensate. Inorganic Materials. 2016;**52**:700-707. DOI: 10.1134/S0020168516070062

[43] Marikutsa AV, Rumyantseva MN, Gaskov AM, Samoylov AM. Nanocrystalline tin dioxide: Basics in relation with gas sensing phenomena. Part II. Active Centers and Sensor Behavior, Inorganic Materials. 2016;**52**:1327-1354

[44] Korotcenkov G, Cho BK. Ozone measuring: What can limit application of SnO_2-based conductometric gas sensors? Sensors and Actuators B. 2012;**161**:28-44. DOI: 10.1016/j.snb.2011.12.003

[45] Marikutsa AV, Rumyantseva MN, Gaskov AM, Samoylov AM. Nanocrystalline tin dioxide: Basics in relation with gas sensing phenomena. Part I. Physical and chemical properties and sensor signal formation. Inorganic Materials. 2015;**51**:1329-1347. DOI: 10.1134/S002016851513004X

[46] Ievlev VM, Ryabtsev SV, Shaposhnik AV, Samoylov AM, Kuschev SB, Sinelnikov AA. Ultrathin films of palladium oxide for oxidizing gases detecting. Procedia Engineering. 2016;**168**:1106-1109

[47] Samoylov A, Ryabtsev S, Shaposhnik A, Kuschev S, Soldatenko S, Ievlev V. Palladium oxide thin film for oxidizing gases detecting. The 16-th International Meeting on Chemical Sensors IMCS 2016. Final Program & Absrtacts Book: Jeju, Jeju Island, Korea, July 10-13; 2016. 96 p

[48] Ryabtsev SV, lyevlev VM, Samoylov AM, Kuschev SB, Soldatenko SA. Real microstructure and electrical properties of palladium oxide thin films for oxidizing gases detecting. Science and Application of Thin Films, Conference & Exhibition (SATF-2016). Çeşme, Izmir, Turkey, September 19-23: Book of Abstracts; 2016. 44 p

[49] Sobolev VV, Mordas DO, Sobolev VV. Optical properties and electronic structure of PdO. Inorganic Materials. 2004;**40**:166-170

[50] Ryabtsev SV, Ievlev VM, Samoylov AM, Kuschev SB, Soldatenko SA. Microstructure and electrical properties of palladium oxide thin films for oxidizing gases detection. Thin Solid Films. 2017;**636**:751-759

[51] Ievlev VM, Ryabtsev SV, Samoylov AM, Shaposhnik AV, Kuschev SB, Sinelnikov AA. Thin and ultrathin films of palladium oxide for oxidizing gases detection. Sensors & Actuators: B. Chemical Part 2. 2018;**255**:1335-1342

[52] Greenwood NN, Earnshaw A. Chemistry of the Elements. 2nd ed. Oxford: Butterworth-Heinemann; 1997. 1340 p

[53] Hodgeson JA, Sibert EE, Curl RF Jr. Dipole moment of nitrogen dioxide. Journal of Physical Chemistry. 1963;**67**(12):2833-2835. DOI: 10.1021/j100806a079

[54] Boggs JE. The dipole moments and polarizabilities of nitrogen dioxide and nitrogen tetroxide. Journal of Physical Chemistry. 1964;**68**(8):2379-2381. DOI: 10.1021/j100790a513

[55] The National Institute for Occupational Safety and Health (NIOSH). NIOSH Pocket Guide to Chemical Hazards. https://www.cdc.gov/niosh/npg/npgd0454.html

[56] General Chemistry Is A Free Introductory Textbook On Chemistry. http://www.vias.org/genchem/index.html

Chapter 13

Functionalized Carbon Nanomaterials in Drug Delivery: Emergent Perspectives from Application

Nabanita Saikia

Additional information is available at the end of the chapter

http://dx.doi.org/10.5772/intechopen.71889

Abstract

Carbon nanotubes (CNTs) have attracted substantial research interest in biomedical sciences and bionanotechnology, rendered from its unique structure, electronic, mechanical, and optical properties. Despite the diverse potential applications, the integration of CNTs in biomedical research is one of the most challenging areas where nanotubes fall under much scrutiny. Pristine nanotubes are highly hydrophobic, and non-dispersible in most of the common aqueous and organic solvents and to render nanotubes biocompatible, functionalization is one of the key prerequisites. In this regard, covalent and noncovalent functionalization are the two widely adopted approaches for co-tethering biologically active molecules on the CNTs. Likewise, the hollow cavity of the nanotube facilitates in the endohedral encapsulation of biomolecules, peptides, DNA oligonucleotides, and proteins, thereby retaining the physiological attributes of the biological molecules. The chapter focuses on the emerging approaches to the functionalization of single-wall CNTs (SWCNTs) and the potential application of functionalized SWCNTs in tuberculosis and cancer chemotherapy using state-of-the-art density functional theory, molecular docking and molecular dynamics simulation methods.

Keywords: carbon nanotubes, drug delivery, molecular dynamics, density functional theory

1. Introduction

1.1. Carbon: The fundamental building block of life

Carbon is the most versatile element in the periodic table that forms the basis of all kinds of life on earth. Elemental carbon displays a complex allotropy depending on the nature of hybridization; diamond (*sp*^{3} hybridized), graphite, graphene, fullerenes, and carbon nanotubes (*sp*^{2} hybridized). Graphite is the most common allotrope of carbon and the word graphite in Greek means *'to write'*. *Graphene* an acronym for the 2D layered graphite, is the mother of all carbon materials [1], as a graphene sheet can be wrapped to form 0D fullerenes, rolled to form 1D nanotubes, or stacked to form 3D graphite as depicted in **Figure 1**. The unearthing of *"ground-breaking experiments regarding the two-dimensional material graphene"* by Geim and Novoselov in 2010, heralded graphene as the next generation carbon material [2].

CNTs are hexagonally arranged, honey-combed lattice of carbon atoms formed by the rolling of graphene into seamless cylindrical structures (see **Figure 2a**). Nanotubes like graphene have a high diameter to length ratio (aspect ratio) [3] and demonstrate high electrical, mechanical, and thermal conductivity along with structural stability [4–7]. CNTs are broadly classified as single-wall CNTs (SWCNTs) and multi-wall CNTs (MWCNTs). The SWCNT comprise of a single graphene sheet, with diameter ~ 0.5–1.5 nm and length of ~100 μm [8], while MWCNT is formed from the co-axial stacking of SWCNTs, with diameter ~1.4–100 nm, length between 1 nm-μm, and internuclear distance of 0.3–0.4 nm between the co-axial tubes. The representation of a zigzag ($m = 0$), armchair ($n = m$), and chiral ($n \neq m$) nanotube is depicted in **Figure 2b-d**. The (n, m) indices render remarkable electronic properties to the CNTs [9] and the sp^2 hybridization along the tubular axis makes it chemically inert by nature.

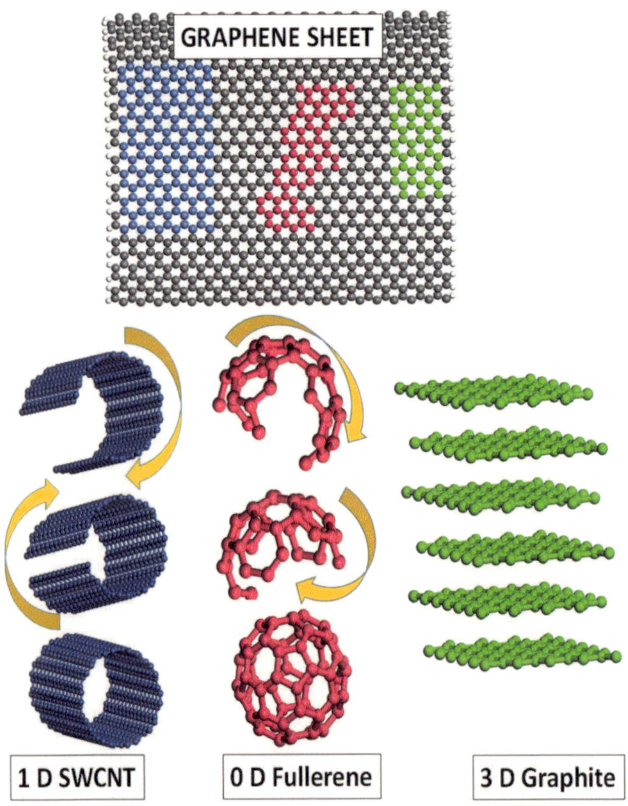

Figure 1. Formation of SWCNT, fullerene and graphite from a single graphene monolayer.

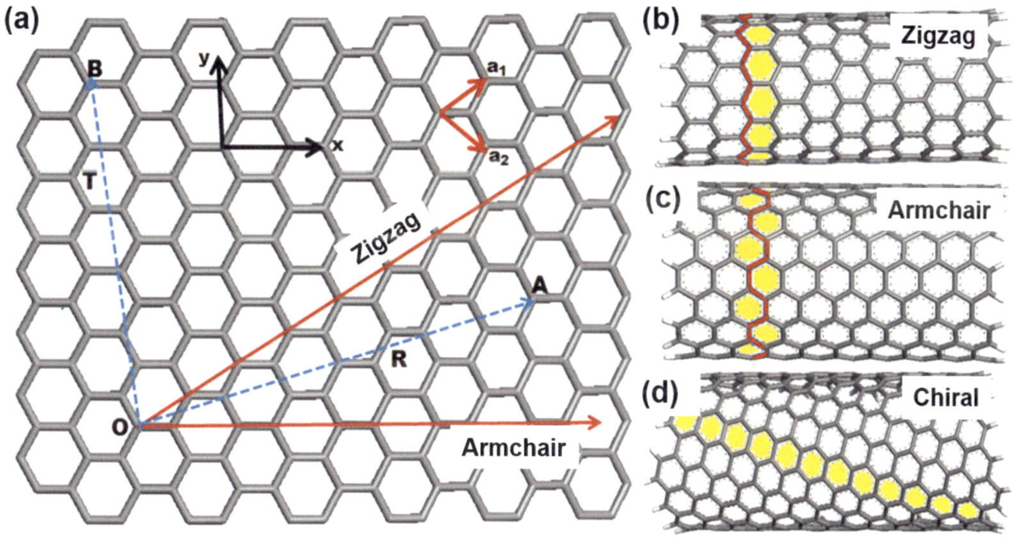

Figure 2. (a) A graphene sheet depicting the (b) zigzag and (c) armchair and (d) chiral CNT based on rolling of carbon atoms along chiral vectors through the circumference (OA) of nanotube.

The unique electronic properties exhibited by CNT are governed by the quantum confinement of electrons where the periodic boundary conditions come into interplay. Because of the quantum confinement, electrons can propagate along the tube axis: forward and backward, along with the conservation of energy and momentum. Unlike metals which have a smooth density of states (DOS), CNTs are characterized by many van Hove singularities [10], and the DOS depends on diameter and chirality of the nanotube [11]. The conducting properties of CNT is an inverse function of its diameter, that is, with increase in diameter, band gap between the valence and conduction bands decreases and at a certain point both the bands overlap to give rise to metallic nanotubes. Semiconducting nanotubes on the other hand (with similar diameter as metallic nanotubes) possess similar van Hove singularities near the Fermi level [12].

These unprecedented properties have largely contributed to the extensive biomedical research, especially as nanocapsules for therapeutic drugs, proteins, and gene delivery [13]. CNTs find application in bionanotechnology and pharmaceutical sciences, and current drug delivery modules have been incorporating CNTs for improved target specific detection and treatment of diseases. Although the potential application of carbon nanomaterials, particularly CNTs are farfetched, the concerns raised over the effect of large-scale synthesis of CNTs to the environment, biocompatibility, toxicity, biodegradation and remediation cannot be undermined and needs to be addressed thoroughly. Hence, a detailed *in vivo* and *in vitro* toxicity analyses is mandatory in understanding CNT-based therapeutic regimes for sustained drug delivery applications.

Some of the questions that underlie the importance of the study are (i) understanding the mechanism of CNT uptake followed by the subsequent release of therapeutic molecules, (ii) *in vivo* biocompatibility, and (iii) long-term practical implication to direct exposure to the physiological environment. Although theoretical and/or experimental studies have attempted to address the main questions like mechanism of drug-nanotube interaction, the preferable binding sites of drugs onto nanotubes, drug activity under confinement, and change in redox properties of drugs under the physiological conditions, these studies are rather limited in predicting the likelihood of using CNT as carrier vehicles for the long-term storage and release of therapeutic and biologically active molecules *in vivo*.

The chapter in a very comprehensive yet succinct way addresses the potential applications of SWCNTs in drug delivery, managing to draw a fine line between the scopes of application and practical viability of integrating carbon nanomaterials in biomedical research. Herein, we report the theoretical aspects of modeling novel SWCNT-based drug delivery systems using the covalent and noncovalent functionalization schemes. Nanotubes of varying chirality and length are considered for functionalization, drug loading, and targeting onto the active binding sites of receptor proteins. With the successful incorporation of CNTs in cancer therapy, we propose a novel approach toward integrating CNTs in Tuberculosis (TB) therapy. To the best of our knowledge, theoretical studies on the potential application of CNTs in pharmaceutical sciences pertaining to TB and other bacterial diseases have not been discussed extensively. We address the recent theoretical advancements using the state-of-the-art density functional theory (DFT), molecular docking, and molecular dynamics (MD) simulation methods. Molecular docking serves as an instrumental tool in computer-aided drug design for predicting the preferred binding mode of a ligand to a receptor (protein). Docking studies help characterize the protein binding cavity, understand the orientation of ligand with respect to the receptor protein, and the nature of interaction between the protein with functionalized nanomaterial, which can aid in the structure-based design of novel drug delivery systems for future experimental studies.

2. Functionalization of CNTs

Traditional approaches to drug delivery function over a broad spectrum, resulting which, specificity toward drug administration and delivery are rarely accomplished. Development of polymer-based nanocomposite materials has enabled the successful engineering of drug delivery modules via the incorporation of nanomaterials and nanoparticles as nanocapsules for sustained release of therapeutics in a dosage-dependent manner. Nanotechnology, on the other hand, has revolutionized the pharmaceutical sector with the assimilation of functionalized nanomaterials like CNTs in drug, gene delivery, and tissue engineering [14, 15]. CNTs play dual role by rendering directionality in targeting the tumor (malignant) cells and facilitating the controlled mediated release of therapeutic molecules. The application of CNTs as carrier payloads for anticancer drugs cisplatin [16], carboplatin [17], doxorubicin (DOX) [18–20], mechlorethamine [21], paclitaxel [22] and antitubercular drugs like isoniazid (INH) [23], rifampicin, pyrazinamide (PZA) [24] have been reported.

With the inherent limitations in application of pristine, unmodified nanotubes, functionalization is the collective approach toward tailoring nanotubes electronic properties. Pristine

CNTs are generally hydrophobic with low solubility in most of the common aqueous and organic solvents and the hydrophobicity is accounted to the size, structure, and bundling effect which restricts the uptake and assimilation within the biological environment [25]. Functionalization assists in reducing the bundling effect which arises due to the van der Waals (vdW) attractive forces between adjacent nanotube surfaces and is efficient in increasing the biocompatibility thereby facilitating cellular internalization and trafficking. It has been reported that the functionalized nanotubes (*f*CNT) exhibit better biocompatibility with reduced *in vivo* and *in vitro* toxicity [26–28]. The extent of functionalization depends on the nature and reactivity of sidewall (curvature), number of functional groups that can be co-tethered along the sidewall, and steric hindrance between functional groups and nanotube sidewall. The subsequent sections discuss some of the adopted approaches in the functionalization of SWCNTs at the level of experiment and theory.

2.1. Solubilization of CNT through covalent functionalization

Some of the alternative schemes to functionalization of CNT is through covalent method using 1,3-dipolar cycloaddition [29], [2 + 1] cycloaddition of dichlorocarbene, silylene, germylene [30], hydroboration [31], arylation, hydrogenation by Birch reduction [32], carboxylic acid groups [33], Diels-Alder reaction, esterification of carboxylated nanotubes [34] and fluorination reactions [35]. Experimental and theoretical studies have shown that the extent of covalent functionalization depends on the curvature of nanotube as an increase in curvature decreases the reactivity toward sidewall functionalization [36].

2.1.1. 1,3-dipolar Cycloaddition (DC)

The solubility of CNT can be enhanced by the covalent functionalization using 1,3-DC reactions. Azomethine ylide (CH_2NHCH_2), ozone (O_3), nitrone ($CH_2N(H)O$), nitrile ylide ($CHNCH_2$), nitrile imine (CHNNH) are the commonly used functional groups for 1,3-DC reaction. The intrinsic physical properties of CNTs such as photoluminescence and Raman scattering decreases upon covalent functionalization, due to chemical bond formation between the functional group and carbon atoms. Theoretical studies by Lu et al. [30] reported the reaction energies (E_r), barrier heights (E_a) and retro barrier height values of a series of 1,3-dipolar molecules on (5, 5) SWCNT using two-layered ONIOM (B3LYP/6-31G*:AM1) approach. Likewise, experimental studies by Prato and co-workers [37] substantiated the theoretical findings on the 1,3-DC functionalization of CNTs. 1,3-DC functionalization was also achieved through the addition of ozone wherein ozone adds onto the end caps and kink regions rather than nanotube sidewall due to increased strain and loss in conjugation. Lu et al. [38] reported the 1,3-DC reaction of ozone onto nanotube sidewalls using two-layered ONIOM (B3LYP/6-31G*:AM1) method.

Prato and co-workers [39] investigated the 1,3-DC reaction of azomethine ylide on both SWCNT and MWCNT. The water-soluble amine functionalized CNT was highly suitable for immobilization of biomolecules, and purification of pristine nanotubes during syntheses. Attachment of peptide molecules onto covalently functionalized SWCNT was reported by Prato and co-workers [40]. The C-terminal group of peptide chain was attached to N-terminal side group to form a supramolecular complex of peptide wrapped nanotubes. Gallo et al. [23], incorporated *f*SWCNT and fullerenes as nanovectors for the functionalization of INH drug. Armchair (5, 5) SWCNT was functionalized via 1,3-DC reaction of azomethine ylide with the polyethylene glycol (PEG) oligomer tailored to the INH drug. Increasing the number of functionalized units leads to an increase in HOMO-LUMO energy gap and global hardness, and decrease in binding (−3.52 to −6.65 eV) and solvation energy (−31.60 to −49.99 eV) values. An increase in global hardness with increase in functionalization suggests a net stabilization of the complex. It is noteworthy to mention that the optimum length of PEG oligomer used as a linker for the 1,3-DC functionalization is essential as longer PEG chains can interfere with drug administration, block the interaction between the nanotube and cell lines of the body, cellular uptake of drug, and degrade the therapeutic activity of drug molecules [41]. The PEG units with superior hydrophilicity, biocompatibility, and low immunogenicity can resist the opsonisation and increase the retention time of the nanotube-drug conjugate system *in vivo* [42, 43].

The structure, electronic properties, and reactivity of a series of 1,3-DC functionalized armchair (*n, m*) and zigzag (*n, 0*) SWCNTs with antitubercular drugs 2-methyl heptyl isonicotinate (MHI) and PZA via. PEG linker was investigated using first-principles DFT calculations [44–46]. With increase in sidewall functionalization, the global hardness and HOMO-LUMO energy gap decreases suggesting an overall decrease in stability of the complex, which is indicative of the localized induced deformation in the nanotube at the site of covalent attachment. On the other hand, the solubility of bare INH and PZA drugs was enhanced in presence of nanotube support. We showed that the optimum length and chirality of the nanotube is central to understand the electronic properties of functionalized nanotubes, particularly from a drug delivery perspective.

2.1.2. Functionalization using organic acids

Covalent functionalization of CNTs using carboxylic (−COOH) group was realized through oxidation with strong organic acids like H_2SO_4/HNO_3 [47], phosphates, and sulfur-containing units. Acid functionalized CNTs are highly soluble in water under a wide range of pH and exhibit a significant reduction in the aggregation of nanotube bundles. The dispersibility facilitates in the sidewall, endohedral and end tip functionalization of CNT with organic acids and different functional groups. The sidewall functionalization of CNTs via cycloaddition reaction with azide, ozone, transition metal oxides, and carbenes [48] is illustrated in **Figure 3**.

Lu et al. [31] using two-layered ONIOM (B3LYP/6-31G*:AM1) approach reported the reaction pathway and site selectivity for [2 + 1] cycloaddition of dichlorocarbene, silylene, germylene, and oxycarbonylnitrene onto (5, 5) SWCNT. Dichlorocarbene addition occurs preferentially at the 1,2-pair site. The silylene addition at 1,2-pair site was predicted to be exothermic (−20.7 kcal mol^{-1}) and follows a barrier less reaction pathway. Germylene addition was exothermic by 8.5 kcal mol^{-1}, lower than dichlorocarbene and silylene and proceeds in absence of a transition state pathway. Oxycarbonylnitrene addition onto 1,2-pair site of SWCNT was exothermic by 66.2 kcal mol^{-1}, higher than the other three cycloaddition groups. The transition state had an activation barrier of 7.2 kcal mol^{-1}, which suggested that the cycloaddition reaction was facile in nature.

http://dx.doi.org/10.5772/intechopen.71889

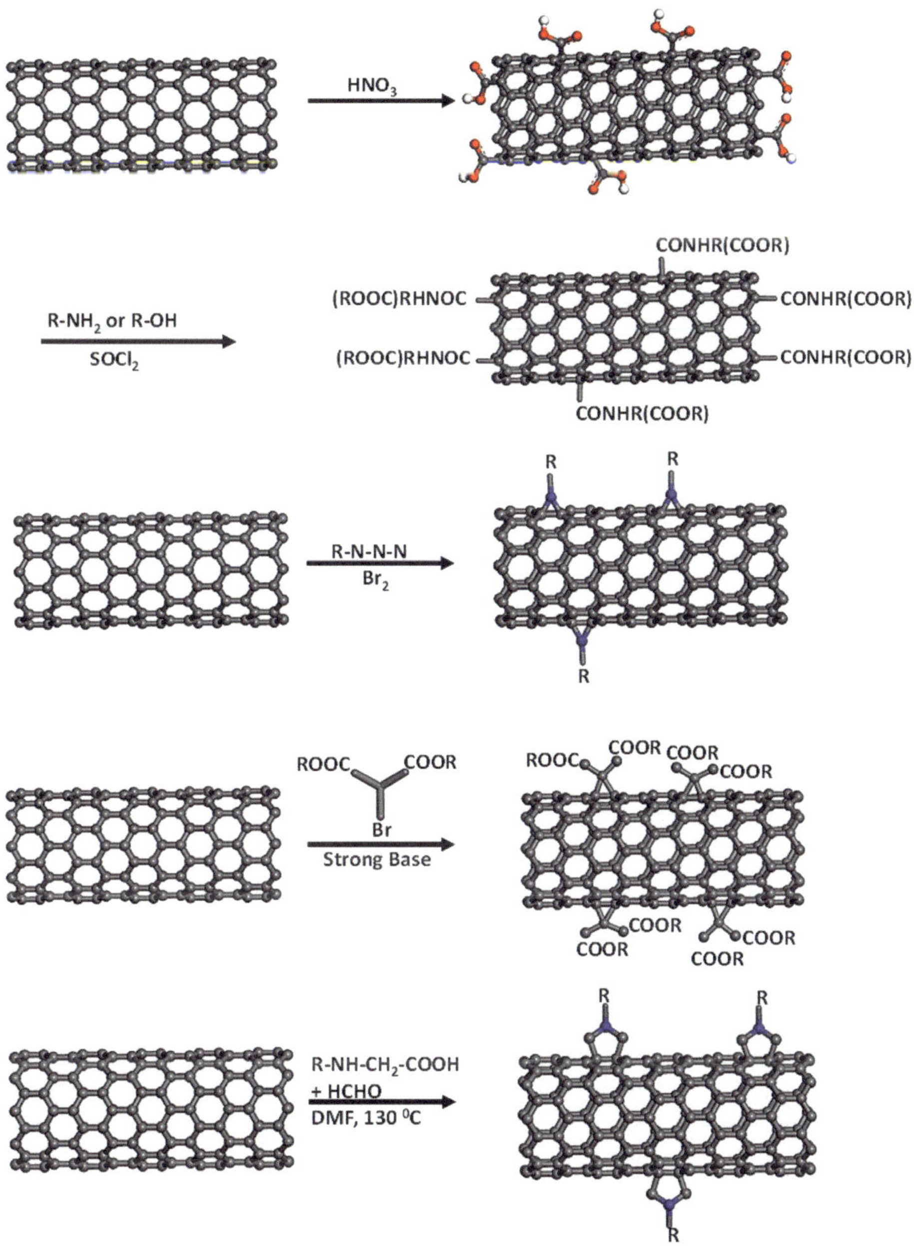

Figure 3. Schemes for sidewall functionalization of SWCNT using covalent bonds. Adopted from Ref. [48].

2.2. Noncovalent functionalization of CNTs

Although covalent functionalization improves the solubility of CNT, it modifies the intrinsic electronic properties by deforming the C-C bond length, perturbing the π-delocalization, and shortening the length of the nanotube. Noncovalent functionalization provides the alternative approach to improving the solubility of nanotubes without deforming the π-delocalization. For example, exohedral wrapping with polymeric molecules like PEG [49], polymers [50], ss-DNA, and endohedral filling can help in increasing the solubility. Likewise, the polymer molecules can form a surface coating via π-stacking interactions, mediated by weak vdW forces, and hydrophobic interactions. The following subsection discusses some of the widely adopted approaches to the noncovalent functionalization of CNTs.

2.2.1. Functionalization via π-π stacking

The π-π stacking of organic molecules namely pyrene, anthracene, and porphyrin increases the solubility of pristine nanotubes and facilitates in the binding of proteins, polysaccharides, and peptides. Dai and co-workers [51] investigated the noncovalent functionalization of CNT, wherein succinimidyl ester group was co-tethered onto pyrene rings via butanoic acid side chains, facilitating the immobilization of proteins. The amide group of the protein replaces the N-hydroxysuccinimide group that propagates the transportation of biomolecules. Falvo and co-workers [52] reported the functionalization of MWCNT with streptavidin protein, wherein the MWCNT pre-functionalized with 1-pyrene butanoic acid succidymidyl ester (1-pbase) was co-tethered on the nanotube sidewall. The pyrenes formed π-π bonds with the MWCNT sidewall under the influence of which MWCNT undergoes a phase transfer with 1-base acting as a phase transfer catalyst.

The noncovalent functionalization of pyrene on SWCNT was investigated by Cosnier and co-workers [53] for application as modified electrodes in biosensing devices. Calomel electrode was taken as the reference and Pt electrode (5 mm diameter) modified by casting 20 μl THF solution of pristine SWCNT and B-doped SWCNT polished with 20 μm diamond paste as the counter electrode. The SWCNT/pyrene-biotin and B-SWCNT/pyrene-biotin was incubated in 20 μl avidin solutions for 20 min, and the response time for glucose sensing was measured using amperometric response technique. The enzyme-modified SWCNT electrodes were incorporated as electrodes for glucose sensing. In-situ polymerization of MWCNT with polyimide (PI) results in the π-stacking interactions between the imide and aromatic benzene rings of CNT with subsequent wrapping of PI along the nanotube circumferential axis [54]. Polymer wrapping improves the thermal stability and renders the conjugated complex suitable for nanoelectronics devices with improved electronic, thermal and optical properties.

Noncovalent functionalization of porphyrin molecules with SWCNTs have been extensively studied as high yield light-harvesting systems with tunable electronic properties for biological and optoelectronic applications [55, 56]. Roquelet et al. [57], reported an efficient method for the synthesis of porphyrin/SWCNT complex utilizing a micelle-swelling technique in presence of organic solvent. The organic solvent leads to swelling of the micelle facilitating the interaction of porphyrin molecules to the micelle core and SWCNT. Dispersion corrected DFT calculations on the structure, electronic and optical properties of SWCNT functionalized tetraphenylporphyrin (TPP) molecule showed that diameter rather than chirality of the nanotube stabilizes the π-π stacking of TPP molecule [58]. The optical absorption of TPP was not affected by the diameter or chirality of CNT and the optical spectra showed the absorption of π-stacked TPP at almost the same position as the isolated TPP, indicating that the TPP absorption properties were preserved in the complex.

2.2.2. *Functionalization using biomolecules and nucleobases*

Functionalization of CNTs with biomolecules is useful in the development of nanobio composite devices. Immobilization of DNA in DNA-based biosensors is possible with the incorporation of CNTs in nucleic acid sensing, gene therapy, and biosensor fabrication [59–61]. DNA because of the base pairing sequence facilitates in the alignment of nanotube assembly [62]. Rodger and co-workers [63] investigated the interaction of CNT with DNA using linear dichroism (LD) method. DNA/CNT hybrid exhibited higher LD signals, higher than the sum of the LD spectrum of individual DNA and SWCNT. Jung et al. [64] developed methods for covalent linking of DNA oligonucleotides onto SWCNT films which were later immobilized onto solid surfaces. The carboxylated/aminated DNA oligonucleotides were covalently attached to functionalized SWCNT, the length of which was controlled via oxidation with strong organic acids, leading to the formation of carboxylated SWCNT.

Li et al. [65] investigated the self-assembly of CNT and gold nanoparticles into multicomponent structures using DNA oligonucleotides. The CNT pre-functionalized with -COOH facilitated in the grafting of ssDNA strands and multiple assemblies of nanotubes were thus possible using this technique. In another combined theoretical and experimental study by Sood and co-workers [66], interaction of DNA nucleobases namely adenine (A), guanine (G), cytosine (C), thymine (T) with (5, 5) SWCNT was reported. The *ab initio* studies showed that binding energies of nucleobases onto SWCNT was governed mainly by vdW forces and follow the order: C > G > A > T, respectively. Likewise, the binding energies of A, G, C, T, U nucleobases on (7, 0) SWCNT was predicted to follow the order: G > C > A > T > U, and bears a monotonic dependence on nanotube diameter; that is, nanotubes with small diameter due to low curvature exhibits low interaction energy whereas for nanotubes with high diameter the interaction energy tend to be on the higher side [67].

2.2.3. *Noncovalent functionalization using polymers*

Polymer wrapping of CNTs mediated via noncovalent functionalization toward the synthesis of highly dispersed, stable and reinforced functional dispersants in aqueous and organic solvents was reviewed by Fujigaya and Nakashima [68]. The polymer wrapping forms a thermodynamically stable coating and any unbound polymer could be removed via dialysis, ultra-centrifugation or chromatographic separation techniques. Similarly, noncovalent functionalization of CNTs using polyethylene glycol (PEG) PEGylated-phospholipid chains forms an effective means of high loading of the drug and biomolecules at the free end of PEG chain and onto nanotube sidewall. PEGylated-phospholipid chain facilitates the high loading (about ~400%) of drug molecules onto CNT and characterizes as a potent carrier vehicle in drug delivery applications. PEG tailored SWCNT exhibit no toxicity for over several months, which was further substantiated from time-dependent assays performed onto mice. Drugs which normally remained insoluble within the biological systems, upon conjugation with PEG modified CNT revealed high solubility as well as retention time within the body. The noncovalent functionalization retained the physical properties of nanotubes without drastically perturbing the overall electronic properties.

First-principles studies on the interaction of conjugated polymers with (8, 0) SWCNT and (10 × 10) graphene sheet was investigated by Jilili et al. [69], to confirm the experimental observation that polymers are suitable for noncovalent functionalization. The GGA approximation was predicted to be inadequate in describing the physisorbed systems, whereas LDA and vdW corrected GGA yield conclusive results. The electronic structure of SWCNT/graphene was maintained around the Fermi energy with negligible charge transfer between the conjugates. The interaction of polymer-SWCNT/graphene was of weak vdW type with minimal effects on the physical and electronic properties of SWCNT/graphene, important for an effective noncovalent functionalization.

3. Functionalized carbon Nanomaterials in drug delivery

Drug delivery is a process of administering drugs in a controlled, sustained manner to achieve maximal therapeutic efficacy upon transdermal administration. The foremost objectives in developing novel drug delivery systems are to improve the therapeutic competence by [1] increasing bioavailability, [2] preventing toxicity, harmful side-effects by increasing the persistence of a drug, [3] reducing drug exposure toward non-target cells, and [4] minimizing drug degradation and loss [70–73]. Drug delivery systems are designed to improve the pharmacological and therapeutic profile of drug molecules with an ability to cross the cell membrane upon administration [74, 75]. The most important characteristic of SWCNT as a drug delivery system is its ability to penetrate the cell membrane [76], and facilitate the intracellular internalization and trafficking within the cell cytoplasm [77]. A major breakthrough in nanoscience was the advent of CNTs as one of the most sought-after materials for designing novel drug delivery carrier modules to comply with the biotechnological and pharmaceutical objectives. CNT due to its needle-like cylindrical structure can easily penetrate the cell membrane and enter the cell nuclei, while the cell does not recognize it as an intruder.

Functionalized CNTs can act as carriers for antimicrobial agents like amphotericin B [78, 79] and transport it within the mammalian cells. This reduces the antifungal toxicity as compared to the toxicity of free drug (40% of the cells being killed by CNTs-free formulation compared to no cell death by CNTs formulation). The surface-engineered CNTs can capture the pathogenic bacteria in liquid media [80, 81]. In addition, SWCNTs exhibit unique optical properties such as near-infrared region (NIR) fluorescence, and Raman scattering. The fluorescence range spans the entire biological tissue transparent window and is, therefore, promising for drug detection and biological imaging [82–84].

3.1. SWCNTs in tuberculosis therapy

The science of bacteriology is credited to the contributions of Louis Pasteur and Robert Koch. It was the discovery of *Mycobacterium tuberculosis* by Koch that revolutionized medical history [85]. TB is a chronic disease caused by the infection of *Mycobacterium tuberculosis* [86] and is a leading cause of mortality worldwide. The World Health Organization (WHO) 2017 annual report prompted to 10.4 million new TB cases, of which, India and Indonesia alone accounted for a third of the world's TB-burden [87]. In 2016, a total of 9287 new TB cases were reported in the United States [88]. The drastic widespread of TB is mainly accounted to poverty, homelessness, synergy with HIV/AIDS pandemic, multi-drug resistant (MDR), and extensively drug resistant (XDR) stains of *M. tuberculosis* [89].

Streptomycin was the first antitubercular drug discovered in 1943 [90] and since then several therapeutic drugs like para-amino salicylic acid (1946), isoniazid (INH) (1952), pyrazinamide (PZA) (1952), cycloserine (1952), ethionamide (1956), rifampicin (1957) and ethambutol (1962) have been discovered. The TB therapy involves a combination of four first-line drugs namely; INH, PZA, rifampicin, and ethambutol administered for a period of 2 months followed by minimum 4 months' treatment regimen of INH and rifampicin [91]. PZA (pyrazine-2-carboxamide) is one of the first-line drugs used in TB treatment recommended by the WHO. PZA is metabolized into its active form (pyrazoic acid) by the amidase activity of *M. tuberculosis* nicotinamidase/pyrazinamidase (MtPncA) encoded by the pncA gene [92]. The administration of PZA in high dosage can cause minor to detrimental health problems and the antibiotic resistance of bacteria under prolonged exposure triggers the need for better drug delivery methods to directly bind with the TB bacteria.

We performed DFT, molecular docking and MD studies on the SWCNT-mediated PZA delivery onto the active site of *M. Tuberculosis* pncA enzyme [93]. The DFT calculations predict that the covalent functionalization was thermodynamically favored with negative binding energy values. The decrease in binding energy of PZA/SWCNT with increase in nanotube diameter illustrates that the curvature of nanotube plays an important role in determining the reactivity, and nanotubes with narrow diameter are thermodynamically favorable compared to tubes with larger diameter. The molecular docking studies supported the DFT results thereby establishing that, incorporation of SWCNT facilitates in target specific delivery of PZA within the binding site of pncA as shown in **Figure 4**. The narrow diameter nanotubes were better docked compared to the larger diameter nanotubes and length of PEG chain was predicted to be reasonably adequate for the delivery of PZA within the binding site of pncA. The presence of nanotube did not result in any structural deformation in pncA, rather the incorporation of SWCNT facilitated in the stabilization of PZA conjugated complex.

Noncovalent functionalization of SWCNT and boron nitride nanotubes (BNNTs) with PZA was investigated using DFT and MD simulation (see **Figure 5**) to comprehend the role of nanotube chirality on the electronic properties of the complexes [94]. BNNTs are structural analogs of CNTs with a wide band gap of ~5.5 eV, high chemical, and thermal stability. The potential application of BNNTs is rather limited in terms of its high chemical stability and poor dispersibility. The theoretical results predict the modification in electronic structure of both SWCNTs and BNNTs with the enhancement of electronic states, significant lowering in HOMO-LUMO energy gap and the presence of new dispersionless states within the band gap. Depending on the nanotube chirality, PZA exhibits a preferential selectivity for adsorption, which is further confirmed from the band structure, DOS, total projected DOS, and frontier orbital analysis.

The functionalized nanotube facilitates in the loading and delivery of a PZA onto the active entering pathway of pncA without the nanotube affecting the structural conformation of

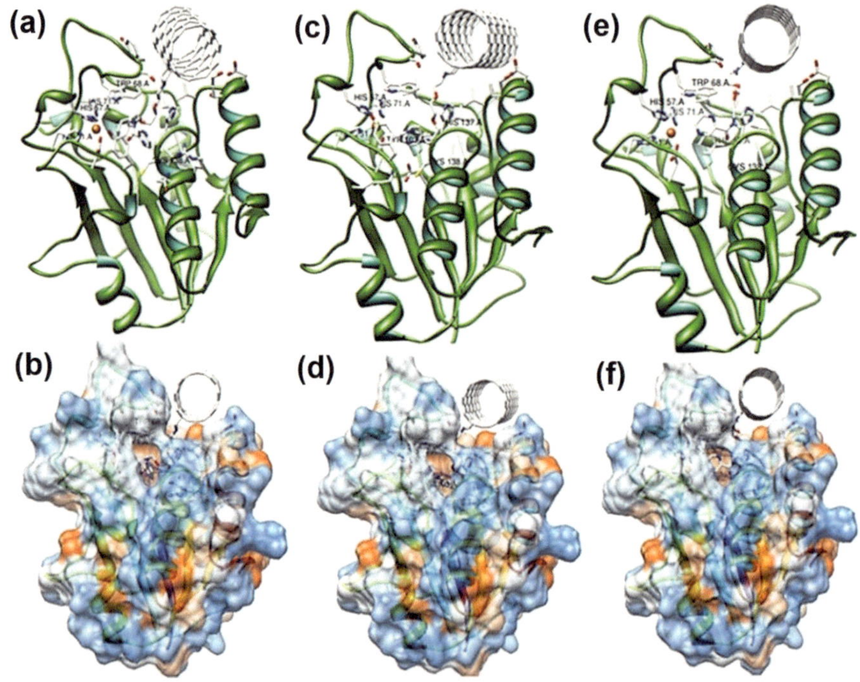

Figure 4. The docked conformation and hydrophobic surface of sidewall functionalized PZA/SWCNT within the active binding site of pncA protein, (a–b) PZA/(9,0) SWCNT (3 unit cells), (c–d) PZA/(9,0) SWCNT (5 unit cells), (e–f) edge functionalized PZA/(9,0) SWCNT (5 unit cells). Reprinted with permission from Ref. [93]. Copyright 2017, Elsevier.

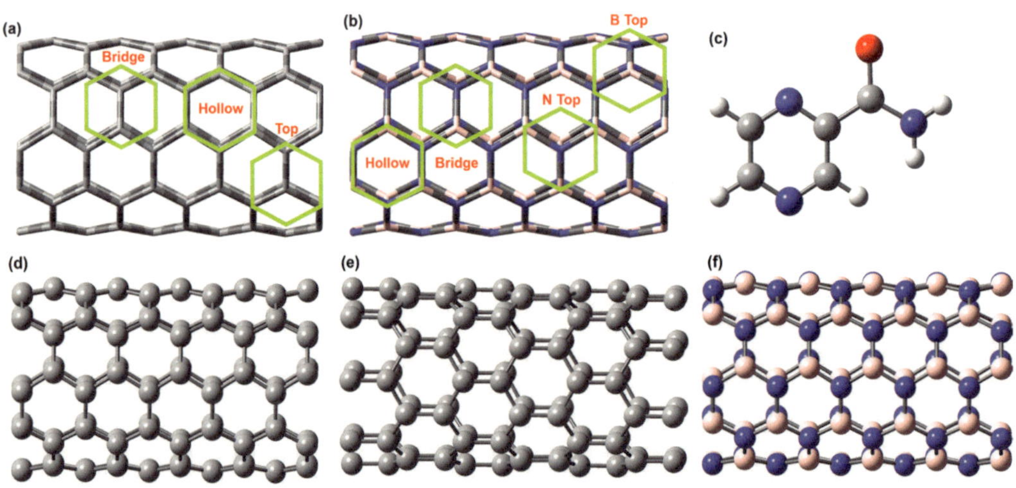

Figure 5. (a) Adsorption sites in the model SWCNT, (b) Adsorption sites in a model (5, 5) BNNT; optimized geometries of (c) PZA, (d) (5, 5) CNT, (e) (8, 0) CNT and (f) (5, 5) BNNT. Adopted from Ref. [94].

pncA as shown in **Figure 6**. The incorporation of nanotube yields better docking scores for PZA then the drug being administered in bare form. Although covalent functionalization aids in achieving target specific delivery of PZA within the active site of pncA, noncovalent functionalization was predicted to be effective for engineering nanotube structure and electronic properties for successful drug delivery applications.

Zanella et al. [95] performed theoretical studies on the interaction of non-steroid anti-inflammatory drug nimesulide with pristine and Si-doped capped SWCNT. The adsorption of nimesulide on Si-doped capped SWCNT exhibit a higher binding energy of 1.8 eV compared to pristine capped SWCNT (0.32 eV) which was due to the high reactive bonding sites on Si atom. The strong interaction of nimesulide with Si-doped SWCNT served as better drug delivery carriers in comparison to pristine capped SWCNT.

Wang and co-workers [96] performed MD studies to investigate the mechanism of encapsulation of nifedipine drug within (10, 10) SWCNT. Their studies showed that the internal

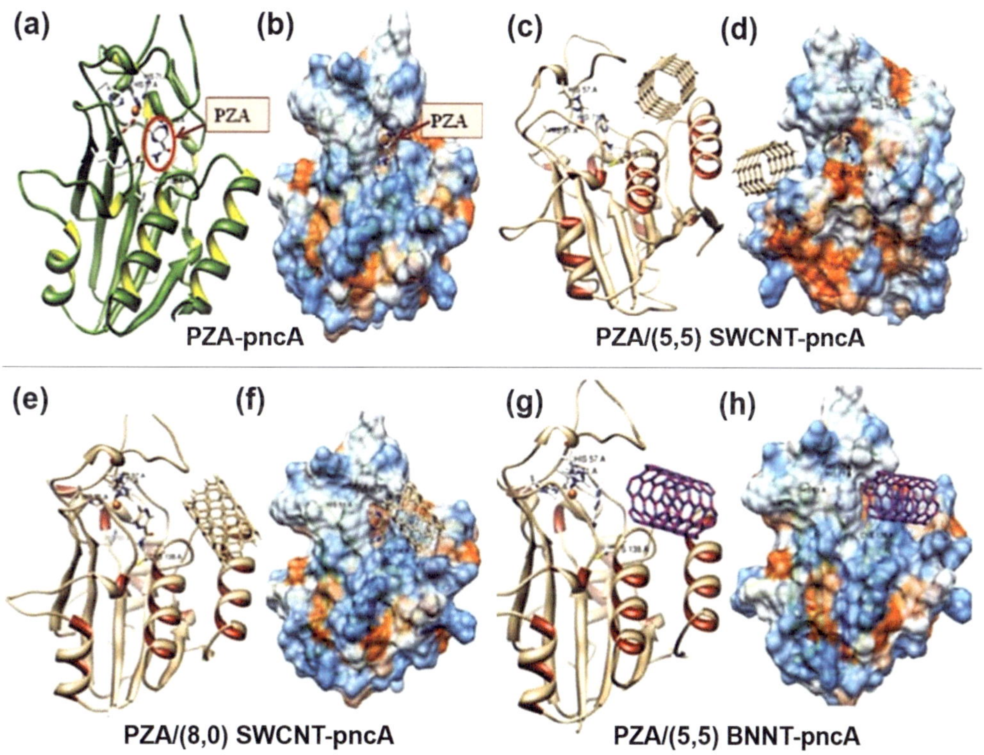

Figure 6. (a) Docked PZA/pncA, (b) electrostatic surface (c) PZA/(5, 5) SWCNT docked onto pncA, (d) electrostatic surface, (e) PZA/(8, 0) SWCNT docked onto pncA, (f) electrostatic surface, (g) docked PZA/(5, 5) BNNT with pncA, (h) electrostatic surface. Adopted from Ref. [94].

adsorption of nifedipine was more stable than external adsorption by 5.3 to 7.8 kcal/mol. In solvent phase, the encapsulation of nifedipine was impeded due to competitive vdW and hydrophobic interactions in SWCNT-nifedipine-water complex. Encapsulation of nifedipine orients the distribution of water molecules inside SWCNT accompanied by the H-bond formation between water molecules and oxygen atom of nifedipine. During the encapsulation process, SWNT undergoes weak fluctuations due to the oscillatory behavior of nifedipine encapsulated within the CNT.

4. SWCNTs in cancer therapy

Platinum-based Phase II and Phase III anticancer drugs hold promise in the treatment of cancer with new drugs being discovered, some of which are still under clinical trials. The two main limitations in use of Pt-based anticancer drugs are [1] the anticancer drugs undergo poor circulation in tissue cells and its activity is reduced with time due to the complex formation with plasma and tissue cells, and [2] tumor cells demonstrate resistance toward Pt-based drugs under prolonged exposure, rendering them ineffective as potent anti-tumor agents. Lippard and co-workers [16, 17] incorporated capped *f*CNT as longboat delivery vehicles for cisplatin anticancer drug through clathrin-dependent endocytosis and measured the changes in redox potential before and after release of the drug. The substituted c,c,t $[Pt(NH_3)_2Cl_2(OEt)(O_2CCH_2CH_2COOH)]$ pro-drug was attached to SWCNT functionalized with phospholipid tethered amine with PEG to solubilize the nanotube. Burger et al. [97] investigated the encapsulation of cisplatin in a phospholipid formulation. The lipid-coated cisplatin nanocapsules exhibit drug-lipid ratio and *in vitro* cytotoxicity 1000 times higher than free cisplatin. This method thus formed an effective approach in drug delivery and the means of producing lipid-based nanocapsules for encapsulating different bio- and therapeutic molecules. Hilter and Hill [98] suggested three preferred orientations of cisplatin toward the entry into CNT and probable interactions using mechanical principles and mathematical modeling. The atomic interaction between nanotubes and cisplatin was calculated using hybrid-discrete-continuum approximation. In this approximation, cisplatin was taken as a collection of discrete atoms and the CNT was treated as a continuum body of repeating carbon atoms. Non-bonded interaction, suction, and acceptance energies were calculated using the Lennard-Jones (LJ) potential. For nanotube radius of 5.3 Å, cisplatin exhibited maximum suction energy, depending on the orientation of nanotubes as a function of radii.

We performed density functional studies on the noncovalent functionalization of non-Pt-based anticancer drug camptothecin (CPT) on graphene-based nanomaterials and its prototypes, including graphene oxide (GO) [99]. The noncovalent adsorption of CPT induces a significant strain within the nanosheets and the interaction was thermodynamically favored from energetics perspective. In case of GO, surface incorporation of functional groups resulted in significant crumpling along the basal plane and the interaction was mediated by H-bonding rather than π-π stacking. The molecular docking studies of CPT onto Top1 (**Figure 7a**) showed CPT to be stacked between the Watson Crick AT and GC base pairs and the interaction was mediated via π-π stacking (**Figure 7b**). For the binding of CPT functionalized graphene and GO

http://dx.doi.org/10.5772/intechopen.71889

with topoisomerase I (top 1) CPT interacts through π stacking with AT and GC base pairs of DNA. The optimum interacting distance of CPT from AT and GC bases was calculated at 3.87 and 3.38 Å, from the central aromatic rings (**Figure 7b**). The re-rank score of bare CPT drug was calculated as −89.01 a.u. with an H-bond score of −2.53 a.u. as shown in **Table 1**.

Likewise, for the docking of CPT/8 × 8 graphene with Top1 (**Figure 8a**) CPT gets docked between the AT and GC base pairs. However, graphene gets docked along the phosphate backbone of the ds-DNA helix as shown from **Figure 8b** indicating a strong interaction between the polar phosphate groups of the DNA helix. Compared to the docking of bare CPT drug, presence of graphene stabilizes the intercalation of CPT between the AT and GC base pairs, as observed from the increase in re-rank score values.

The docking of CPT/GO with Top1 as illustrated in **Figure 9a**, depicts CPT to get docked between AT and GC base pairs of DNA, mediated by π-π stacking interaction similar to that observed for bare CPT and CPT/8 × 8 graphene. However, in the presence of GO, GO undergoes strong interactions with DNA bases and gets docked between the DNA helix and the interaction is stabilized by intermolecular H-bond between polar functional units on the basal plane of GO and DNA nucleobases (**Figure 9b**). The molecular docking studies on bare CPT and CPT functionalized graphene and GO systems showed that the interaction of CPT with Top1 is mediated by π-stacking interaction between the aromatic rings of CPT and the A and C bases of DNA. In presence of graphene and GO, CPT undergoes a similar trend in adsorption while the graphene and GO nanomaterial gets docked along the phosphate backbone indicating a strong preferential interaction with DNA.

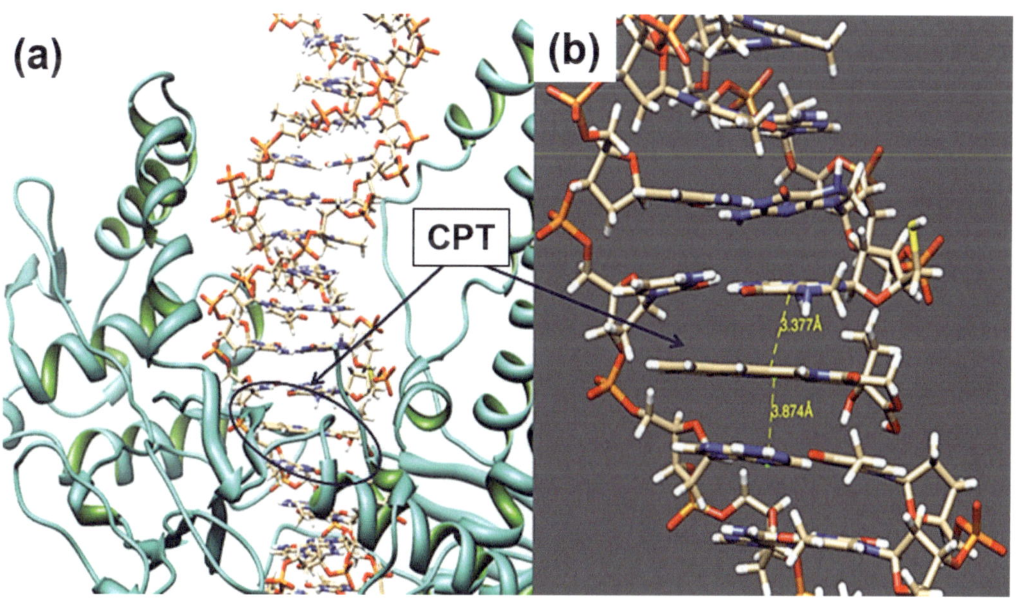

Figure 7. (a) Secondary structure of Top1 protein with the CPT drug docked within the DNA, (b) interacting distance between CPT and the DNA base pairs of top 1. Reprinted with permission from Ref. [99]. Copyright 2017, Elsevier.

System	Re-rank score CPT	Re-rank score nanosheet	H-bond score CPT	H-bond score nanosheet
CPT_Top1	−89.01	—	−2.53	—
8 × 8 graphene/CPT docked onto Top1	−89.10	95.87	−2.57	0.00
8 × 8 GO/CPT docked onto Top1	−90.21	126.21	−2.29	−4.44

Table 1. The re-rank scores and H-bond scores for the best docked conformations of CPT and CPT/8 × 8 graphene, and CPT/8 × 8 GO sheets, respectively.

Boucetta et al. [20] investigated the supramolecular MWCNT-DOX-copolymer complex for anticancer activities. Since DOX, a clinically acclaimed anticancer drug belonging to the family of anthracyclines exhibit fluorescence properties, its uptake and interaction with nanotubes upon administration can be monitored using fluorescent spectrophotometry. The copolymer coated MWCNT formed supramolecular complexes with DOX via π-stacking and revealed enhanced cytotoxicity leading to highly efficient cell killing efficiency. Likewise, Liu et al. reported the use of DOX loaded PEG functionalized CNT for targeted delivery of anticancer drugs in tumor

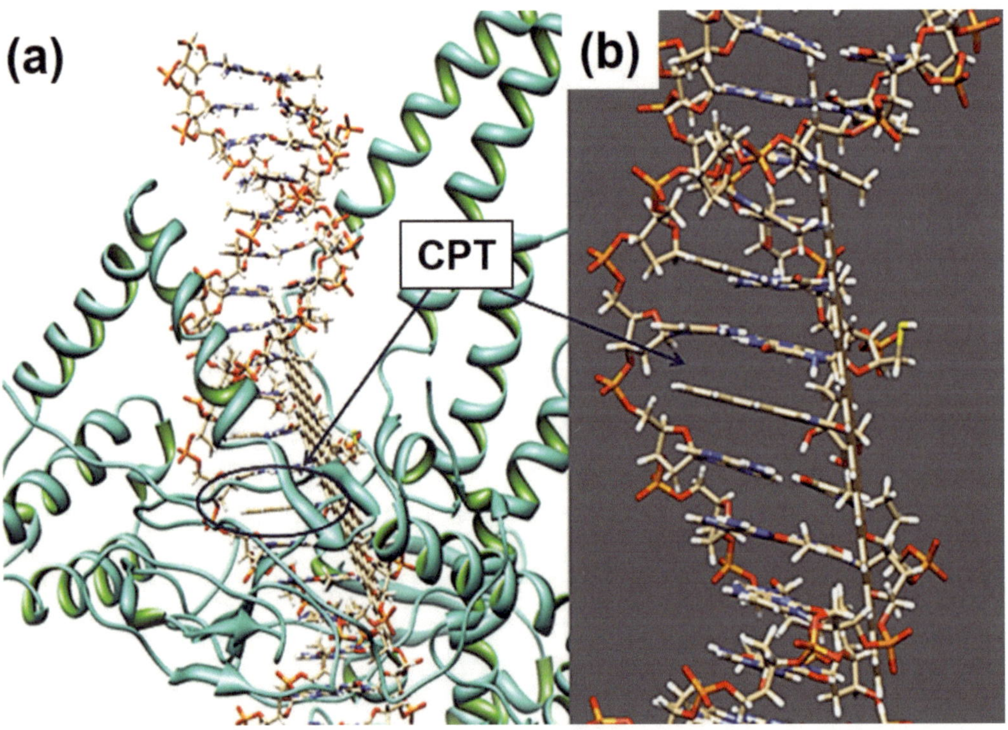

Figure 8. (a) Secondary structure of Top1 with the CPT/8 × 8 graphene sheet docked within the DNA, (b) binding of CPT/8 × 8 graphene sheet with DNA base pairs of top 1. Reprinted with permission from Ref. [99]. Copyright 2017, Elsevier.

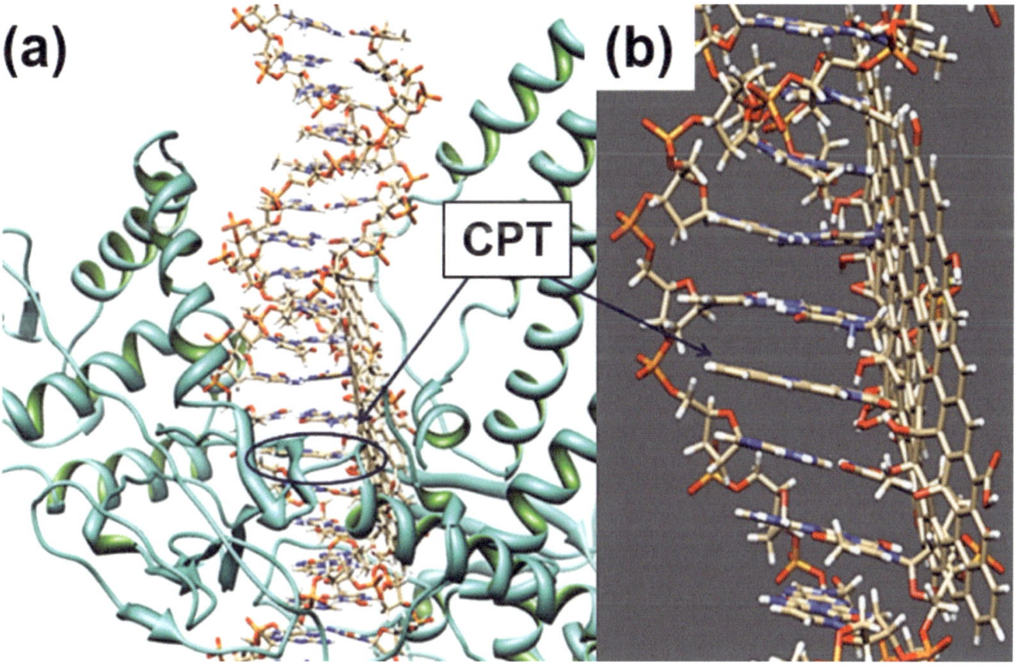

Figure 9. (a) Secondary structure of Top1 protein with the CPT/8 × 8 GO sheet docked within the DNA, (b) the binding of CPT/8 × 8 GO sheet with DNA base pairs of top 1. Reprinted with permission from Ref. [99]. Copyright 2017, Elsevier.

cells [22]. The SWCNT was pre-functionalized with PEG and DOX and a fluorescence probe (fluorescein) was loaded onto the nanotube via π stacking. The loading and subsequent release of DOX were found to be pH dependent; decrease in pH from 9 to 5 showed a decrease in DOX loading. Under acidic conditions (pH 5.5), DOX exhibited increased hydrophilicity and solubility with lysosomes and endosomes, facilitating the release of drug molecules from the nanotube. On decreasing the pH, surface loading of DOX onto nanotube surface lowered and at lower acidic pH, amine group of DOX underwent protonation resulting in increased solubility of DOX molecules.

5. Summary

Carbon nanotubes have proffered as one of the novel functional materials of the 21st century, broadening the theoretical and experimental perspectives in research to explore its novel and intriguing properties. Due to the conjugated π-electron backbone and curvature (properties very similar to fullerene and graphene) they are highly reactive and depending on the size, length and (n, m) indices, the electronic properties can be tuned to fit the desired functionality. Since the synthesis of CNT yields a mixture of both metallic and semiconducting tubes of varying diameter and chirality, separation and purification of nanotubes pose a major problem which restricts the applicability. Secondly, nanotubes are highly hydrophobic and non-dispersible in most of the common aqueous and organic solvents and tend to aggregate in bundles. To improve nanotube dispersibility, surface modification via functionalization is thus a sought-after approach and covalent and noncovalent functionalization methods can reduce the bundling effect and hydrophobicity. Covalent functionalization although renders high stability to the nanotubes, it tends to distort the structural and inherent electronic properties. Noncovalent functionalization, on the other hand, retains the intrinsic properties of the nanotube, as it forms a surface coating on the nanotube sidewall, and facilitates the uptake of drugs, biomolecules, peptides, proteins, DNA, RNA, and genes within the biological systems.

Although CNTs demonstrate practical applicability in all facets of science, be it biology, physics, medicine, nanotechnology, catalysis, or materials science, its long-term implications need to be assessed from the perspective of human health to environmental risks. The long-term fate of CNTs released into the environment depends on the structural, morphological and synthetic treatments [100]. Methods of reducing toxicity *in vivo* and *in vitro* can be envisaged via functionalization of CNT. Proper assessment and in-depth study are essential to render nanotubes useful for diverse and environmentally benign applications.

We investigated the potential application of SWCNTs, graphene-based nanomaterials and its prototypes in TB and cancer chemotherapy using conventional DFT methods supported by molecular docking and MD simulation on the nature of interaction of therapeutic drug functionalized SWCNTs/graphene with the binding site of the protein. The functionalization of SWCNTs with therapeutic drugs using covalent and noncovalent schemes were adopted to investigate the drug binding with the nanotube and the stability of the conjugated complexes. DFT results supported by molecular docking and MD simulation helps in contemplating the feasibility of SWCNT-based novel drug delivery in cancer and TB therapy.

Author details

Nabanita Saikia

Address all correspondence to: nabanita16@gmail.com

Department of Physics, Michigan Technological University, Houghton, Michigan, United States

References

[1] Geim AK, Novoselov KS. The rise of graphene. Nature Materials. 2007;**6**:183-191

[2] Novoselov KS, Geim AK, Morozov SV, Jiang D, Zhang Y, Dubonos SV, Grigorieva IV, Firsov AA. Electric field effect in atomically thin carbon films. Science. 2004;**306**:666-669

[3] Liu L, Guo GY, Jayanthi CS, Wudate SY. Colossal paramagnetic moments in metallic carbon nanotori. Physical Review Letters. 2002;**88**:217206

[4] Dresselhaus MS, Dresselhaus G, Jorio A. Unusual properties and structures of carbon nanotubes. Annual Review of Materials Research. 2004;**34**:247

[5] Hersam MC. Progress towards monodisperse single-walled carbon nanotubes. Nature Nanotechnology. 2008;**3**:387

[6] Meyyappan M. Carbon Nanotubes Science and Applications. Boca Raton: CRC Press, LLC; 2005. p. 104

[7] Ando T. The electronic properties of graphene and carbon nanotubes. NPG Asia Materials. 2009;**1**:17-21

[8] Tran PA, Zhang L, Webster T. Carbon nanofibers and carbon nanotubes in regenerative medicine. Advanced Drug Delivery Reviews. 2009;**61**:1097-1114

[9] Reich S et al. Carbon Nanotubes Basic Concepts and Physical Properties. Weinheim: WILEY-VCH, Verleg GmbH & Co. KGaA; 2004

[10] Kim P, Odom TW, Huang JL, Lieber CM. Electronic density of states of atomically resolved single-walled carbon nanotubes: Van Hove singularities and end states. Physical Review Letters. 1999;**82**:1225

[11] Charlier JC, Lambin P. Electronic structure of carbon nanotubes with chiral symmetry. Physical Review B. 1998;**57**:R15037-R15039

[12] White CT, Mintmire JW. Density of states reflects diameter in nanotubes. Nature. 1998;**394**:29-30

[13] Kostarelos K, Lacerda L, Pastorin G, Wu W, Wieckowski S, Luangsivilay J, Godefroy S, Pantarotto D, Briand JP, Muller S, Prato M, Bianco A. Cellular uptake of functionalized carbon nanotubes is independent of functional group and cell type. Nature Nanotechnology. 2007;**2**:108-113

[14] Veetil JV, Ye K. Tailored carbon nanotubes for tissue engineering applications. Biotechnology Progress. 2009;**25**:709-721

[15] Harrison BS, Atala A. Carbon nanotube applications for tissue engineering. Biomaterials. 2007;**28**:344-353

[16] Feazell RP, Nakayama-Ratchford N, Dai H, Lippard SJ. Soluble single-walled carbon nanotubes as longboat delivery systems for platinum(IV) anticancer drug design. Journal of the American Chemical Society. 2007;**129**:8438-8439

[17] Dhar S, Liu Z, Thomale J, Dai H, Lippard SJ. Targeted Single-Wall carbon nanotube-mediated Pt(IV) Prodrug delivery using Folate as a homing device. Journal of the American Chemical Society. 2008;**130**:11467-11476

[18] Pastorin G, Wu W, Wieckowski S, Briand J-P, Kostarelos K, Prato M, Bianco A. Double functionalisation of carbon nanotubes for multimodal drug delivery. Chemical Communications. 2006;(11):1182-1184

[19] Liu Z, Sun X, Nakayama RN, Dai H. Supramolecular chemistry on water-soluble carbon nanotubes for drug loading and delivery. ACS Nano. 2007;**1**:50-56

[20] Ali-Boucetta H, Al-Jamal KT, McCarthy D, Prato M, Bianco A, Kostarelos K. Multiwalled carbon nanotube-doxorubicin supramolecular complexes for cancer therapeutics. Chemical Communications. 2008;(4):459-461

[21] Shukla PK, Mishra PC, Suhai S. Reactions of DNA bases with the anti-cancer nitrogen mustard mechlorethamine: A quantum chemical study. Chemical Physics Letters. 2007;**449**:323-328

[22] Liu Z, Chen K, Davis C, Sherlock S, Cao Q, Chen X, Dai H. Drug delivery with carbon nanotubes for in vivo cancer treatment. Cancer Research. 2008;**68**:6652-6660

[23] Gallo M, Favila A, Mitnik DG. DFT studies of functionalized carbon nanotubes and fullerenes as nanovectors for drug delivery of Antitubercular compounds. Chemical Physics Letters. 2007;**447**:105-109

[24] Saikia N, Deka RC. Theoretical study on pyrazinamide adsorption onto covalently functionalized (5,5) metallic single-walled carbon nanotube. Chemical Physics Letters. 2010;**500**:65-70

[25] Upadhyayula VKK, Gadhamshetty V. Appreciating the role of carbon nanotube composites in preventing biofouling and promoting biofilms on material surfaces in environmental engineering: A review. Biotechnology Advances. 2010;**28**:802-816

[26] Schipper ML, Nakayama-Ratchford N, Davis CR, Kam NWS, Chu P, Liu Z, Sun X, Dai H, Gambhir SS. A pilot toxicology study of single-walled carbon nanotubes in a small sample of mice. Nature Nanotechnology. 2008;**3**:216-221

[27] Cherukuri P, Bachilo SM, Litovsky SH, Weisman RB. Near-infrared fluorescence microscopy of single-walled carbon nanotubes in phagocytic cells. Journal of the American Chemical Society. 2004;**126**:15638-15639

[28] Yang ST, Guo W, Lin Y, Deng XY, Wang HF, Sun HF, Liu YF, Wang X, Wang W, Chen M, Huang YP, Sun YP. Biodistribution of pristine single-walled carbon nanotubes in vivo. Journal of Physical Chemistry C. 2007;**111**:17761-17764

[29] Lu X, Tian F, Xu X, Wang N, Zhang Q. A theoretical exploration of the 1,3-dipolar Cycloadditions onto the sidewalls of (n,n) Armchair single-wall carbon nanotubes. Journal of the American Chemical Society. 2003;**125**:10459-10464

[30] Lu X, Tian F, Zhang Q. The [2+1] cycloadditions of dichlorocarbene, silylene, germylene, and oxycarbonylnitrene onto the sidewall of armchair (5,5) single-wall carbon nanotube. The Journal of Physical Chemistry. B. 2003;**107**:8388-8391

[31] Long L, Lu X, Tian F, Zhang Q. Hydroboration of C(100) surface, fullerene, and the sidewalls of single-wall carbon nanotubes with borane. The Journal of Organic Chemistry. 2003;**68**:4495-4498

[32] Pekker S, Salvetat PP, Jakab E, Bonard JM, Forro L. Hydrogenation of carbon nanotubes and graphite in liquid ammonia. The Journal of Physical Chemistry. B. 2001;**105**:7938-7943

[33] Zhao J, Park H, Han J, LU JP. Electronic properties of carbon nanotubes with covalent sidewall functionalization. The Journal of Physical Chemistry. B. 2004;**108**:4227-4230

[34] Qin YJ, Shi JH, Wu W, Li XG, Guo ZX, Zhu DB. Concise route to functionalized carbon nanotubes. The Journal of Physical Chemistry. B. 2003;**107**:12899

[35] Hamwi A, Alvergnat H, Bonnamy S, Béguin F. Fluorination of carbon nanotubes. Carbon. 1997;**35**:723-728

[36] Niyogi S, Hamon MA, Hu H, Zhao B, Bhowmik P, Sen R, Itkis ME, Haddon RC. Chemistry of single-walled carbon nanotubes. Accounts of Chemical Research. 2002;**35**:1105-1113

[37] Bianco A, Kostarelos K, Partidos CD, Prato M. Biomedical applications of functionalised carbon nanotubes. Chemical Communications. 2005;(5):571-577

[38] Lu X, Zhang L, Xu X, Wang N, Zhang Q. Can the sidewalls of single-wall carbon nanotubes be ozonized? The Journal of Physical Chemistry. B. 2002;**106**:2136-2139

[39] Georgakilas V, Kordatos K, Prato M, Guldi DM, Holzinger M, Hirsch A. Organic functionalization of carbon nanotubes. Journal of the American Chemical Society. 2002; **124**:760-761

[40] Pantarotto D, Partidos CD, Graff R, Hoebeke J, Briand JP, Prato M, Bianco A. Synthesis, structural characterization, and immunological properties of carbon nanotubes functionalized with peptides. Journal of the American Chemical Society. 2003;**125**:6160-6164

[41] Liu Z, Winters M, Holodniy M, Dai H. siRNA delivery into human T cells and primary cells with carbon-nanotube transporters. Angewandte Chemie, International Edition. 2007;**46**:2023-2027

[42] Zerda ADL, Zavaleta C, Keren S, Vaithilingam S, Bodapati S, Liu Z, Levi J, Smith BR, Ma TJ, Oralkan O, Cheng Z, Chen X, Dai H, K-Yakub BT, Gambhir SS. Carbon nanotubes as photoacoustic molecular imaging agents in living mice. Nature Nanotechnology. 2008; **3**:557-562

[43] Liu Z, Cai W, He L, Nakayama N, Chen K, Sun X, Chen X, Dai H. In vivo biodistribution and highly efficient tumour targeting of carbon nanotubes in mice. Nature Nanotechnology. 2007;**2**:47-52

[44] Saikia N, Deka RC. Theoretical study on pyrazinamide adsorption onto covalently functionalized (5,5) metallic single-walled carbon nanotube. Chemical Physics Letters. 2010;**500**:65-70

[45] Saikia N, Deka RC. Density functional calculations on adsorption of 2-methyl heptyl isonicotinate antitubercular drug onto functionalized carbon nanotube. Computational & Theoretical Chemistry. 2011;**964**:257-261

[46] Saikia N, Deka RC. A comparison of the effect of nanotube chirality and electronic properties on the π-π interaction of single-wall carbon nanotubes with pyrazinamide antitubercular drug. International Journal of Quantum Chemistry. 2013;**113**:1272-1284

[47] Rosca ID, Watari F, Uo M, Akaska T. Oxidation of multiwalled carbon nanotubes by nitric acid. Carbon. 2005;**43**:3124-3131

[48] Karousis N, Tagmatarchis N, Tasis D. Current progress on the chemical modification of carbon nanotubes. Chemical Reviews. 2010;**110**:5366-5397

[49] Huang WJ, Fernando S, Lin Y, Zhou B, Allard LF, Sun YP. Preferential solubilization of smaller single-walled carbon nanotubes in sequential functionalization reactions. Langmuir. 2003;**19**:7084-7088

[50] Carrillo A, Swartz JA, Gamba JM, Kane RS. Noncovalent functionalization of graphite and carbon nanotubes with polymer multilayers and gold nanoparticles. Nano Letters. 2003;**3**:1437-1440

[51] Chen RJ, Zhang Y, Wang D, Dai H. Noncovalent sidewall functionalization of single-walled carbon nanotubes for protein immobilization. Journal of the American Chemical Society. 2001;**123**:3838-3839

[52] Prakash R, Superfine R, Washburn S, Falvo MR. Functionalization of carbon nanotubes with proteins and quantum dots in aqueous buffer solutions. Applied Physics Letters. 2006;**88**:063102-063101

[53] Haddad R, Holzinger M, Maaref A, Cosnier S. Pyrene functionalized single-walled carbon nanotubes as precursors for high performance biosensors. Electrochimica Acta. 2010;**55**:7800-7803

[54] Yang Z, Chen X, Chen C, Li W, Zhang H, Xu L, Yi B. Noncovalent-wrapped sidewall functionalization of multiwalled carbon nanotubes with polyimide. Polymer Composites. 2007;**28**:36-41

[55] Nel AE, Mdler L, Velegol D, Xia T, Hoek EMV, Somasundaran P, Klaessig F, Castranova V, Thompson M. Understanding biophysicochemical interactions at the nano-bio interface. Nature Materials. 2009;**8**:543-547

[56] Ehli C, Oelsner C, Guldi DM, Mateo-Alonso A, Prato M, Schmidt C, Backes C, Hauke F, Hirsch A. Manipulating single-wall carbon nanotubes by chemical doping and charge transfer with perylene dyes. Nature Chemistry. 2009;**1**:243-249

[57] Roquelet C, Lauret JS, Alain-Rizzo V, Voisin C, Fleurier R, Delarue M, Garrot D, Loiseau A, Roussignol P, Delaire JA, Deleporte E. Π-stacking functionalization of carbon nanotubes through micelle swelling. ChemPhysChem. 2010;**11**:1667-1672

[58] Orellana W, Correa JD. Noncovalent functionalization of carbon nanotubes and graphene with tetraphenylporphyrins: Stability and optical properties from ab-initio calculations. Journal of Materials Science. 2015;**50**:898-905

[59] Wei F, Lillehoj PB, Ho C-M. Self-assembled monolayer based electrochemical nucleic acid sensor for vibrio cholerae detection. Pediatric Research. 2010;**67**:458-468

[60] Abu-Salah KM, Zourob MM, Mouffouk F, Alrokayan SA, Alaamery MA, Ansari AA. DNA-based nanobiosensors as an emerging platform for detection of disease. Sensors. 2015;**15**:14539-14568

[61] Balasubramanian K, Burghard M. Biosensors based on carbon nanotubes. Analytical and Bioanalytical Chemistry. 2006;**385**:452-468

[62] Douglas SM, Chou JJ, Shi WM. DNA-nanotube-induced alignment of membrane proteins for NMR structure determination. Proceedings of the National Academy of Sciences of the United States of America. 2007;**104**:6644-6688

[63] Rajendra J, Baxendale M, Rap LGD, Rodger A. Flow linear Dichroism to probe binding of aromatic molecules and DNA to single-walled carbon nanotubes. Journal of the American Chemical Society. 2004;**126**:11182

[64] Jung DH, Kim BH, Ko YK, Jung MS, Jung S, Lee SY, Jung HT. Covalent attachment and hybridization of DNA oligonucleotides on patterned single-walled carbon nanotube films. Langmuir. 2004;**20**:8886-8891

[65] Li S, He P, Dong J, Guo Z, Dai L. DNA-directed self-assembling of carbon nanotubes. Journal of the American Chemical Society. 2005;**127**:14-15

[66] Das A, Sood AK, Maiti PK, Das M, Varadarajan R, Rao CNR. Binding of nucleobases with single-walled carbon nanotubes: Theory and experiment. Chemical Physics Letters. 2008;**453**:266

[67] Shukla MK, Dubey M, Zakar E, Namburu R, Czyznikowska Z, Leszcznski J. Interaction of nucleic acid bases with single-walled carbon nanotube. Chemical Physics Letters. 2009;**480**:269

[68] Fujigaya T, Nakashima N. Non-covalent polymer wrapping of carbon nanotubes and the role of wrapped polymers as functional dispersants. Science and Technology of Advanced Materials. 2015;**16**:024802

[69] Jilili J, Abdurahman A, Gülseren O, Schwingenschlögl U. Non-covalent functionalization of single wall carbon nanotubes and graphene by a conjugated polymer. Applied Physics Letters. 2014;**105**:013103-013105

[70] Prasad PV, Kumar A, Pal PC, Sharma RS, Rao DN, Shrivastav TG, Ge R-S. Bird's eye view on the recent advances in drug delivery systems. Journal of Biomaterials and Nanobiotechnology. 2011;**2**:544-556

[71] Torchilin VP. Lipid-Core micelles for targeted drug delivery. Current Drug Delivery. 2005;**2**:319-327

[72] Torchilin VP. Micellar Nanocarriers: Pharmaceutical perspectives. Pharmaceutical Research. 2007;**24**:1-16

[73] Sawant RR, Torchilin VP. Polymeric micelles: Polyethylene glycol-Phosphatidylethanolamine (PEG-PE) – Based micelles as an example. Methods in Molecular Biology. 2010;**624**:131-149

[74] Kostarelos K. Rational design and engineering of delivery systems for therapeutics: Biomedical exercises in colloid and surface science. Advances in Colloid and Interface Science. 2003;**106**:147-168

[75] Pantarotto D, Briand J-P, Prato M, Bianco A. Translocation of bioactive peptides across cell membranes by carbon nanotubes. Chemical Communications. 2004;(1):16-17

[76] Kam NWS, Jessop TC, Wender PA, Dai H. Nanotube molecular transporters: Internalization of carbon nanotube-protein conjugates into mammalian cells. Journal of the American Chemical Society. 2004;**126**:6850-6851

[77] Pantarotto D, Singh R, McCarthy D, Erhardt M, Briand JP, Prato M, Kostarelos K, Bianco A. Functionalized carbon nanotubes for plasmid DNA gene delivery. Angewandte Chemie (International Ed. in English). 2004;**43**:5242

[78] Wu W, Wieckowski S, Pastorin G, Benincasa M, Klumpp C, Briand J-P, Gennaro R, Prato M, Bianco A. Targeted delivery of amphotericin B to cells by using functionalized carbon nanotubes. Angewandte Chemie, International Edition. 2005;**44**:6358-6362

[79] Benincasa M, Pacor S, Wu W, Prato M, Bianco A, Gennaro R. Antifungal activity of amphotericin B conjugated to carbon nanotubes. ACS Nano. 2011;**5**:199-208

[80] Gu L, Elkin T, Jiang X, Li H, Lin Y, Qu L, Tzeng T-RJ, Joseph R, Sun Y-P. Single-walled carbon nanotubes displaying multivalent ligands for capturing pathogens. Chemical Communications. 2005;(7):874-876

[81] Brady-Estévez AS, Kang S, Elimelech M. A single-walled-carbon-nanotube filter for removal of viral and bacterial pathogens. Small. 2008;**4**:481-484

[82] Welsher K, Liu Z, Daranciang D, Dai H. Selective probing and imaging of cells with single walled carbonnanotubes as near-infrared fluorescent molecules. Nano Letters. 2008;**8**:586-590

[83] Liu ZA, Li X, Tabakman SM, Jiang K, Fan S, Dai H. Multiplexed multicolor Raman imaging of live cells with isotopically modified single walled carbonnanotubes. Journal of the American Chemical Society. 2008;**130**:13540-13541

[84] Hadjiev VG, Arepalli S, Nikolaev P, Jandl S, Yowell L. Enhanced Raman microprobe imaging of single-wall carbonnanotubes. Nanotechnology. 2004;**15**:562-567

[85] Available from: http://nobelprize.org/nobel_prizes/medicine/laureates/1905/koch.html

[86] Glaziou P, Floyd K, Raviglione M. Global burden and epidemiology of tuberculosis. Clinics in Chest Medicine. 2009;**30**:621-636

[87] Available from: http://www.searo.who.int/tb/documents/annual_tb_repot_2017/en/

[88] Schmit KM, Wansaula Z, Pratt R, Price SF, Langer AJ. Tuberculosis-United States, 2016. MMWR. Morbidity and Mortality Weekly Report. 2017;**66**:289-294

[89] Cegielski JP. Extensively drug-resistant tuberculosis: There must be some kind of way out of here. Clinical Infectious Diseases. 2010;**50**:S195-S200

[90] WHO. Anti-tuberculosis Drug Resistance in the World: The WHO/IUATLD Global Project on Anti-tuberculosis Drug Resistance Surveillance 1994-1997, WHO/TB/97. 229; 1997

[91] Cole ST, Riccardi G. New tuberculosis drugs on the horizon. Current Opinion in Microbiology. 2011;**14**:570

[92] Petrella S, Gelus-Ziental N, Maudry A, Laurans C, Boudjelloul R, Sougakoff W. Crystal structure of the Pyrazinamidase of mycobacterium tuberculosis: Insights into natural and acquired resistance to pyrazinamide. PLoS One. 2011;**6**:e15785

[93] Saikia N, Rajkhowa S, Deka RC. Density functional and molecular docking studies towards investigating the role of single-wall carbon nanotubes as nanocarrier for loading and delivery of pyrazinamide antitubercular drug onto pncA protein. Journal of Computer-Aided Molecular Design. 2013;**27**:257-276

[94] Saikia N, Jha AN, Deka RC. Interaction of pyrazinamide drug functionalized carbon and boron nitride nanotubes with pncA protein: A molecular dynamics and density functional approach. RSC Advances. 2013;**3**:15102-15107

[95] Zanella I, Fagan SB, Mota R, Fazzio A. Ab initio study of pristine and Si-doped capped carbon nanotubes interacting with nimesulide molecules. Chemical Physics Letters. 2007;**439**:348-353

[96] Liu H, Bu Y, Mi Y, Wang Y. Interaction site preference between carbon nanotube and nifedipine: A combined density functional theory and classical molecular dynamics study. Journal of Molecular Structure (THEOCHEM). 2009;**901**:163

[97] Burger KNJ, Staffhorst RWHM, De Vijlder HC, Velinova MJ, Bomans PH, Frederir PM, De Kruijff B. Nanocapsules: Lipid-coated aggregates of cisplatin with high cytotoxicity. Nature Medicine. 2002;**8**:81-84

[98] Hilder TA, Hill JM. Modeling the encapsulation of the anticancer drug cisplatin into carbon nanotubes. Nanotechnology. 2007;**18**:1

[99] Saikia N, Deka RC. Ab initio study on the noncovalent adsorption of Camptothecin anticancer drug onto graphene, defect modified graphene and graphene oxide. Journal of Computer-Aided Molecular Design. 2013;**27**:807-821

[100] Helland A, Wick P, Koehler A, Schmid K, Som C. Reviewing the environmental and human health Knowledge Base of carbon nanotubes. Environmental Health Perspectives. 2007;**115**:1125

Chapter 14

Ultralight Paper-Based Electrodes for Energy Applications

Chuan-Pei Lee

Additional information is available at the end of the chapter

http://dx.doi.org/10.5772/intechopen.74098

Abstract

In this chapter, we briefly introduce our recent work related to the topic of ultralight paper-based electrodes for energy applications. Herein, the ultralight paper-based counter electrodes containing commercial poly(3,4-ethylenedioxythiophene):polystyrenesul fonate (PEDOT:PSS) and homemade graphene dots (GDs) are synthesized for preparing flexible dye-sensitized solar cells (DSSCs). Because the GDs/PEDOT:PSS composite can well fill the porosity of paper substrate, the flexible DSSC with GDs/PEDOT:PSS-coated paper electrode exhibits much higher cell efficiency than that of DSSC using paper electrode with Pt. The features of lightweight, low-cost, space-saving (high flexibility), high machinability (easy-cutting) and environmental friendly would make the GDs/PEDOT:PSS-coated paper electrodes highly potential in portable/wearable electronic applications.

Keywords: conducting polymer, flexible electronic, graphene dot, paper electrode

1. Introduction

Flexible and lightweight electronics have attracted much attention because of their high potential to be integrated into wearable commercial products [1]. Among all the flexible substrates, the printed paper substrate, composed of cellulose fibers, has been considered as the most promising one due to the following features: environmental friendly, low-cost, lightweight and easy for roll-to-roll processing [2]. Currently, the printed paper has been widely used as a substrate for transistors [3], displays [4], sensors [5], memories [6], batteries [7] and supercapacitors [8], etc. Furthermore, Professor Wang's group reported a paper-based triboelectric nanogenerator having origami configurations for harvesting mechanical energy [9].

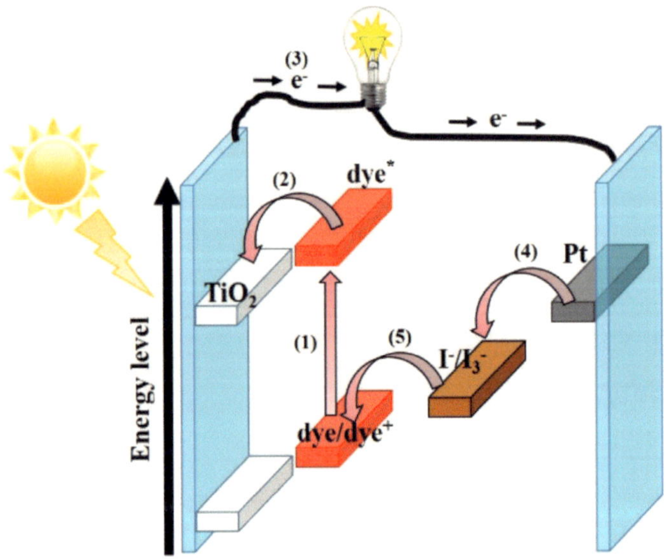

Figure 1. Schematic sketch of a DSSC.

Professor Hu's group developed highly flexible organic light-emitting diodes (OLEDs) on biodegradable and transparent paper substrates [10]. They also demonstrated a range of electronic devices on transparent paper substrate, including organic solar cell [11], touch screen [12] and thin film transistor [13].

For the applications in electrochemical devices, the main challenges of using printed paper as electrode substrates are how to make it to become conductive and electrochemical active via nonsintering processes, as well as to make a conductive layer on it uniformly and continuously because paper-based substrates are nonconductive, porous and rough due to its fibrous nature [14]. Up to now, the printed paper has not been used as the electrode substrates for the third-generation solar cell, namely dye-sensitized solar cell (DSSC). DSSCs have been extensively studied and widely developed in the last two decade since they do not rely on expensive fabrication processes and can be possibly prepared by using roll-to-roll methods [15]. Whereas, flexible DSSCs are handy, convenient for transportation and suitable for setting up at complex environments, which endow them high competitiveness in solar cell markets [16].

As shown in **Figure 1**, a traditional DSSC is consisted of a counter electrode (CE) with Pt catalyst, an I^-/I_3^--based electrolyte and a dye-sensitized TiO_2 photoanode. Briefly, the basic sequence of the working principle of a DSSC is shown below:

Activation

$$TiO_2 - dye \rightarrow TiO_2 - dye^* \tag{1}$$

Electron injection

$$TiO_2 - dye^* \rightarrow TiO_2 - dye^+ + e^-(TiO_2 film) \tag{2}$$

$$e^{-}(TiO_2 film) \rightarrow e^{-}(TCO) \quad (3)$$

Electron reception

$$I_3^{-} + 2\,e^{-}\,(Pt\;CE) \rightarrow 3I^{-} \quad (4)$$

Interception reaction

$$1.5\,I^{-} + TiO_2 - dye^{+} \rightarrow 0.5\,I_3^{-} + TiO_2 - dye \quad (5)$$

Under light illumination, a photo-excited electron is injected from the excited state of the dye (dye*) into the conduction band of the TiO_2. The injected electron percolates through the TiO_2 film by a driving chemical diffusion gradient, and is collected at a conductive glass substrate. After passing through an external circuit, the electron is reintroduced into the DSSC at the Pt CE, where I_3^- is reduced to I^-. Immediately, I^- regenerates the oxidized dye (dye$^+$) to complete the circle of a DSSC and to refresh the DSSC without other chemical side reactions.

However, the most used electrocatalyst on the CE of DSSCs is Pt, which is an expensive noble metal and is rare on earth. To further reduce the cost of fabrication of DSSCs on industrial scale, it is better to develop metal-free electrocatalytic materials for the CEs of DSSCs. Accordingly, conducting polymers (e.g., poly(3,4-ethylene dioxythiophene) (PEDOT) [17], poly(3,3-diethyl-3,4-dihydro-2H-thieno-[3,4-b][1,4]-dioxepine) (PProDOT-Et_2) [18], polypyrrole (PPy) [19], and polyaniline (PANI) [20]) and carbonaceous materials (such as carbon black (CB) [21], graphite [22], carbon nanotube (CNT) [23] and graphene [24]) have become the most promising electrocatalysts for the CE of DSSCs since they are metal element-free, have low material cost and possess good electrocatalytic activity. A water-dispersible conducting polymer, poly(3,4-ethylene dioxythiophene):poly(4-styrene sulfonate) (PEDOT:PSS), has attracted much attention as the catalytic CEs of DSSCs mainly due to its exceptional advantage of aqueous solution processibility [25]. Nevertheless, pristine PEDOT:PSS films are plagued by low conductivity (i.e., <1 S cm^{-1}) and poor electrocatalytic activity for the reduction of I_3^- due to the nonconductive counter anion, PSS^-, disturbing the conduction path of PEDOT inside the film as well as the poor catalytic surface area of its flat film [26]. Inert solvents [27] or carbon materials [28] have been employed to improve conductivity and catalytic surface areas of PEDOT:PSS films. For example, by introducing CB into the PEDOT:PSS-based CEs for their DSSCs, η can achieve to 7.01% [28]. Multiwall CNT-PEDOT:PSS composite CE for DSSCs exhibits η of 6.50% [29]. A catalytic film composited of graphene and PEDOT:PSS for the use of CE in a DSSC had reached 4.50% efficiency [30]; however, a perfect graphene sheet usually possesses limited active sites for electrocatalytic reaction in spite of its extraordinarily high electrical conductivity [31]. Several strategies are employed to increase the electrocatalytic active sites on graphene sheets, such as chemical functionalization [32], heteroatom doping (e.g., nitrogen-doped graphene) [33] and nanosized graphene pieces (e.g., graphene dots (GDs) [34]). Among these graphene nanostructures, GDs have attracted great attention and been widely applied in bioimaging [35], LEDs [36] and photovoltaics [37] due to their unique properties of quantum confinement and edge effects [38, 39]. Moreover, nanometer size and rich oxygen-containing group of GDs facilitate them to be well dispersed in most solvents [40], which benefits for various solution-processable applications. Therefore, the incorporation of GDs in PEDOT:PSS enable us to fabricate an efficient electrocatalytic film by using a simple solution-coating method under low temperature (<100°C).

In this chapter, an all-metal-free CE-containing GDs-PEDOT:PSS and printed paper is developed for flexible DSSCs, and it exhibits higher performance and bending stability than those of a paper electrode with sputtered Pt. The concurrent advantage in low material cost, simple fabrication processes, highly bending durability, lightweight, space-saving, high machinability and environmental friendly makes the GDs-PEDOT:PSS-coated paper electrode playing a crucial role in lightweight electronic devices. Most importantly, this GDs-PEDOT:PSS composite ink can be used in the printable processes for mass production of flexible electrodes.

2. Experimental

2.1. Synthesis of graphene dots

The GDs solution was prepared by using deionized water and glucose as the solvent and source, respectively. First, the as-prepared glucose solution (2.5 mL) was transferred to a glass bottle with 4 mL volume and a tightened cover. The synthesis reaction was carried out in a microwave oven (595 W) for 9 min; the glucose molecules are pyrolyzed and then converted to GDs as shown in **Figure 2**. Subsequently, the reaction bottle was cooled to ambient temperature, and the water-soluble GDs solution was thus prepared. **Figure 3** is the photographs of GDs solution taken under the illumination of visible light and UV light, showing the excitation wavelength-dependent fluorescence property of GDs.

2.2. Preparation of paper-based counter electrodes

The mixing solutions composed of 50 V% PEDOT:PSS aqueous solution and 50 V% binary solution consisting of GDs (X) and ethanol (Y) (X/Y = 3/2) were used for preparing the GDs-PEDOT:PSS composite inks.

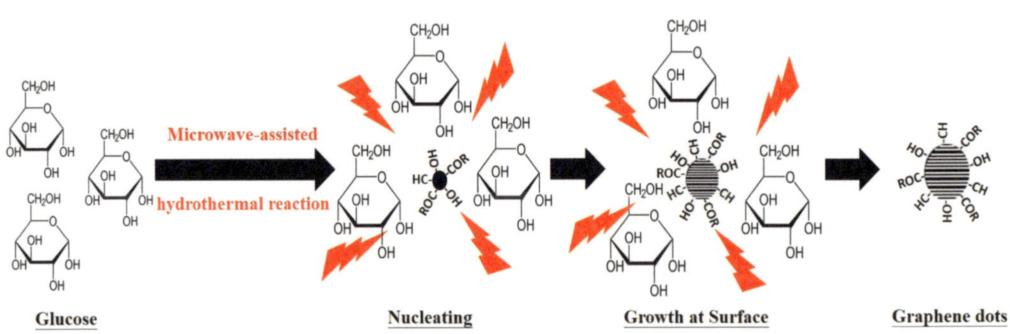

Figure 2. Preparation of graphene dots via the microwave-assisted hydrothermal technique.

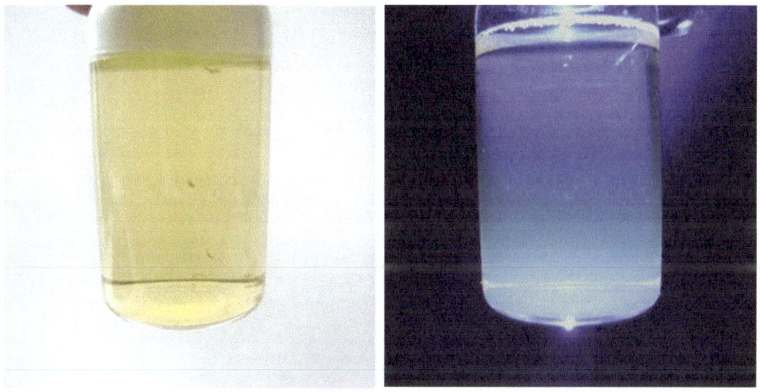

Figure 3. The GDs solutions under ambient light (left) and UV light (right) [41].

For the preparation of paper-based CEs, the commercial printing papers were used as the substrate of counter electrodes. The paper with a fixed coating area was immersed in the GDs-PEDOT:PSS solution (i.e., 30 V% GDs solution content) for 10 min, and it was took out and then dried under 60°C. The thus prepared GDs-PEDOT:PSS/paper electrode (left-hand side of **Figure 4**) was used as the CE for the studies on flexible DSSCs. For comparison purposes, the paper-based CE with sputtered Pt (right-hand side of **Figure 4**) was also prepared as the standard CE.

2.3. Fabrication of flexible dye-sensitized solar cells

For the preparation of the flexible photoanode of DSSCs, the dye-sensitized TiO_2 film was prepared on the conducting plastic substrate (13 Ω $sq.^{-1}$, ITO-PEN) according to the previous reports [42, 43]. First, the binder-free TiO_2 paste was synthesized by mixing 1 g TiO_2 powder (P25) with 6 mL binary solution consisting of tert-butanol and deionized water (volume ratio of 2:1) uniformly. Then, the surface of the ITO-PEN substrate was coated with a TiO_x compact layer by spraying an ethanol solution (10 mL) containing titanium tetraisopropoxide (0.028 g) on it. Using the binder-free TiO_2 paste, a 10-μm-thick film was coated on the treated ITO-PEN substrate through doctor blade technique [44]. Thereafter, an active area of 0.4×0.4 cm^2

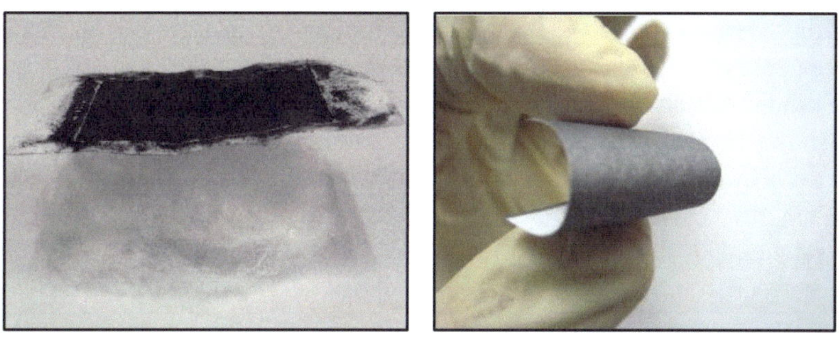

Figure 4. Pictures of CEs with 30 V% GDs-PEDOT:PSS composite (left) and with sputtered Pt (right) on paper substrates.

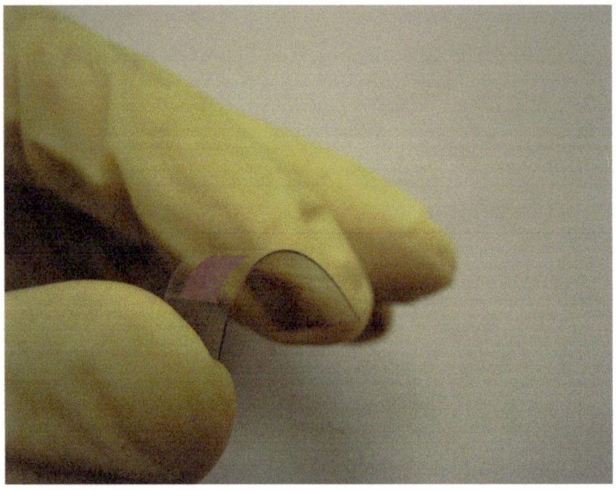

Figure 5. A flexible photoanode with dye-sensitized TiO_2 film on ITO-PEN substrate.

was selected from the TiO_2 films by scrapping. The as-prepared TiO_2/ITO-PEN electrodes were gradually heated to 120°C under ambient conditions, and subsequently annealed at the respective temperatures for 60 min. After annealing and cooling to 80°C, the TiO_2/ITO-PEN electrode was immediately immersed in a 5×10^{-4} M N719 dye solution for 60 min under 55°C. The thus prepared dye-sensitized TiO_2/ITO-PEN photoanode (**Figure 5**) was coupled with paper-based CE (i.e., GDs-PEDOT:PSS or sputtered Pt) as the flexible DSSC.

3. Results and discussion

GDs are edge-bound nanosized graphene pieces and exhibit unique electronic and optical properties due to the quantum confinement and edge effects [38, 39]. **Figure 6(a)** shows the transmission electron microscopy (TEM) image of the monodispersed GDs, which exhibit uniform diameters of ~3.50 nm. As shown in the inset of **Figure 6(a)**, the high-resolution transmission electron microscopy (HRTEM) image indicates high crystallinity of GDs with a lattice spacing of 0.246 nm corresponding to the interplanar separation of graphene (1120). **Figure 6(b)** presents the atomic force microscope (AFM) image of the monodispersed GDs; the inset of **Figure 6(b)** reveals that the average height of GDs is around 2.90 nm. As shown in **Figure 6(c)**, two absorption peaks centered at 228 and 282 nm are observed in the ultraviolet-visible (UV-visible) spectrum of the diluted GD solution, which consists with the result in the previous literature [36]. **Figure 6(d)** shows the photoluminescence (PL) spectrum of the GD solution. The broad emission peaks centered at around 450, 460 and 537 nm are observed when the sample is excited by 300, 400 and 500 nm, respectively, and they show the decrease of PL intensity. The excitation wavelength-dependent intensity and emission wavelength observed here is a common phenomenon for carbonaceous quantum dots [36, 45]. The elemental distribution of the GDs is analyzed by energy-dispersive X-ray spectroscopy (EDS), and the elemental mappings of C and O are shown in **Figure 7**. Obviously, the C content is much higher than O content; the atomic ratio of C/O is 95.32/4.68, which demonstrated C is the dominant element in the GDs.

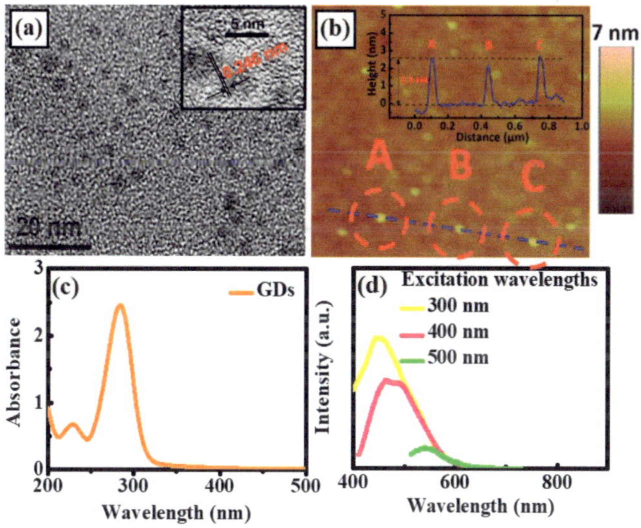

Figure 6. (a) TEM image with the corresponding HRTEM image (inset) of GDs; (b) AFM image of the GDs and its corresponding height profile; (c) absorbance spectrum of the GDs; (d) PL spectra of GDs [41].

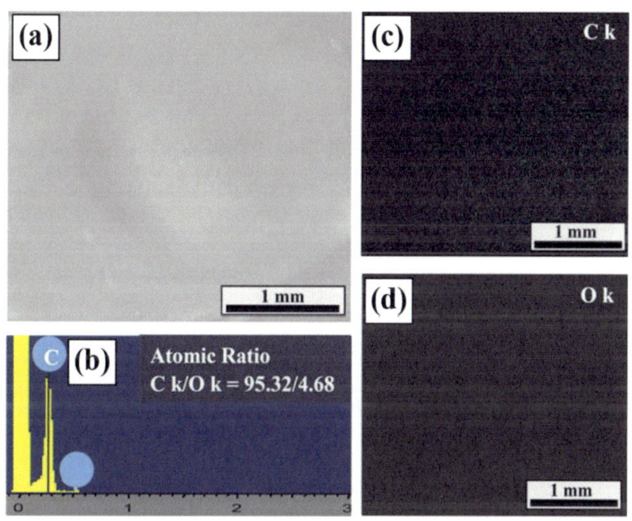

Figure 7. SEM image (a) and EDS spectrum (b) of a GD film. (c) Elemental C mapping of the image shown in (a). The elemental mappings of C (c) and O (d) of the GD film [46].

A mixing solution composed of 50 V% PEDOT:PSS solution, 30 V% GDs solution and 20 V% ethanol were used as the GDs-PEDOT:PSS composite ink. To explore the advantage of the GDs-PEDOT:PSS composite ink for the application in DSSCs, 30 V% GDs-PEDOT:PSS and Pt were separately coated onto the printed papers as the CEs of flexible DSSCs. **Figure 8(a)** and **(b)** shows the top-view SEM images of paper electrodes with sputtered Pt and with 30 V% GDs-PEDOT:PSS, respectively. The SEM images obviously show that the porosity of paper substrate

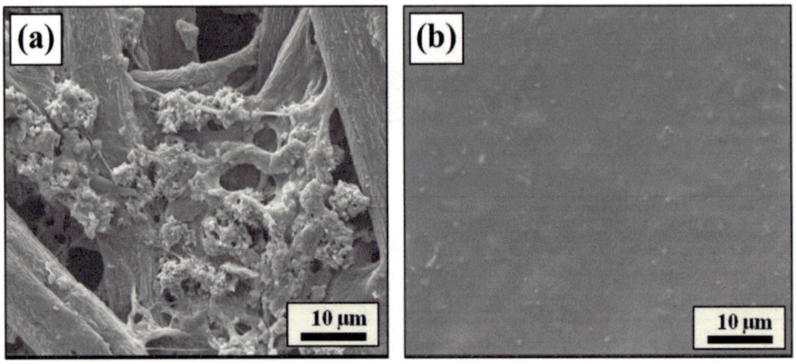

Figure 8. The top-view SEM images of paper substrates with (a) sputtered Pt and with (b) 30 V% GDs-PEDOT:PSS composite [46].

can be perfectly filled by GDs/PEDOT:PSS; however, that cannot be achieved by sputtering Pt, which means that the sputtered Pt layer could not provide a continuous electron transport route in a porous paper substrate. Above two paper, CEs (i.e., 30 V% GDs-PEDOT:PSS and sputtered Pt) are assembled with flexible dye/TiO_2/ITO-PEN photoanodes for studying the pertinent photovoltaic performance.

Figure 9 shows the photovoltaic parameters of the flexible DSSCs using various paper-based CEs with different bending times. As shown in **Figure 9(a)**, the flexible DSSC with GDs/PEDOT:PSS-coated paper CE (η = 4.91%) exhibits three times higher cell efficiency than that of cell using Pt-coated paper CE (η = 1.70%), since the GDs/PEDOT:PSS composite can well fill the porosity of paper substrate. It is worth to mention that the development of an all-metal-free CE is an effective

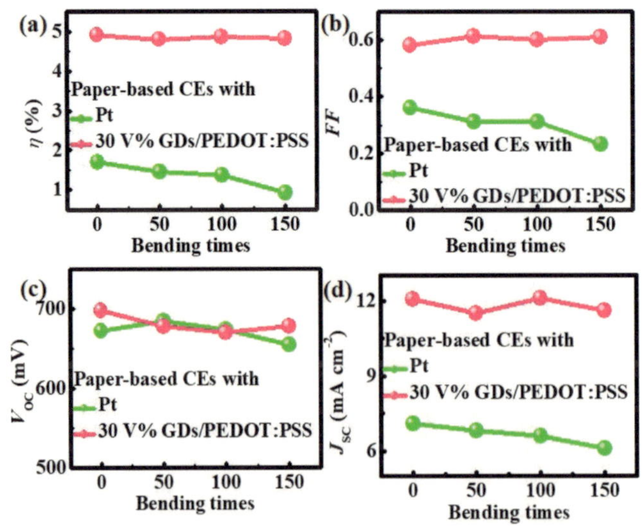

Figure 9. The photovoltaic parameters of the flexible DSSCs using various paper-based CEs with different bending times. (a) Cell efficiency (η); (b) fill factor (FF); (c) open-circuit voltage (V_{OC}); (d) short-circuit current density (J_{SC}) [46].

approach for reducing the cost of the DSSCs. Moreover, the bending test is carried out to study the durability of the paper CEs. Both GDs/PEDOT:PSS-coated paper electrode and Pt-coated paper electrode are bended for several times (0, 50, 100 and 150 times) and then assembled with the flexible photoanodes to measure and record their corresponding photovoltaic performances. **Figure 9(a)** shows the bending time dependence of cell efficiency for the flexible DSSCs with various paper CEs; their corresponding photovoltaic parameters are shown in **Figure 9(b)–(d)**. It obviously shows that the GDs/PEDOT:PSS-coated paper CE shows unfailing performance even though it was bended for 150 times; on the contrary, the Pt-coated paper CE lost its original performance drastically.

4. Conclusion

In summary, a GDs-PEDOT:PSS composite ink was synthesized for preparing the all-metal-free paper-based CEs for flexible DSSCs. The GDs-PEDOT:PSS/paper CE was fabricated through a low-cost and simple coating method (i.e., soak and dry), which could be easily scaled up to mass production. An all-flexible DSSC with GDs/PEDOT:PSS-coated paper CE exhibits much higher cell efficiency (4.91%) than that of cell using paper CE with sputtered Pt (1.70%), since the porosity of paper substrate can be well filled by GDs-PEDOT:PSS, which cannot be achieved by sputtered Pt. After bending for 150 cycles, the performance of GDs/PEDOT:PSS-coated paper CE is still perfectly preserved; on the contrary, the paper CE with sputtered Pt lost its initial performance drastically. In conclusion, this GDs/PEDOT:PSS-coated paper CE is lightweight, low-cost, space-saving (high flexibility), high machinability (easy-cutting) and environmental friendly, which shows the potential for the future applications on portable/wearable electronics.

Acknowledgements

This work was supported by the Ministry of Science and Technology (MOST) of Taiwan.

Conflict of interest

The authors declare no competing financial interests.

Author details

Chuan-Pei Lee

Address all correspondence to: d96524014@ntu.edu.tw

Department of Applied Physics and Chemistry, University of Taipei, Taiwan

References

[1] Bandodkar AJ, You J-M, Kim N-H, Gu Y, Kumar R, Mohan AMV, Kurniawan J, Imani S, Nakagawa T, Parish B, Parthasarathy M, Mercier PP, Xu S, Wang J. Soft, stretchable, high power density electronic skin-based biofuel cells for scavenging energy from human sweat. Energy & Environmental Science. 2017;**10**:1581-1589

[2] Zhu H, Fang Z, Preston C, Li Y, Hu L. Transparent paper: Fabrications, properties, and device applications. Energy & Environmental Science. 2014;**7**:269-287

[3] Yoon J, Lee J, Choi B, Lee D, Kim DH, Kim DM, Moon D-I, Lim M, Kim S, Choi S-J. Flammable carbon nanotube transistors on a nitrocellulose paper substrate for transient electronics. Nano Research. 2017;**10**:87-96

[4] Andersson P, Nilsson D, Svensson PO, Chen M, Malmström A, Remonen T, Kugler T, Berggren M. Active matrix displays based on all-organic electrochemical smart pixels printed on paper. Advanced Materials. 2002;**14**:1460-1464

[5] Khiabani PS, Soeriyadi AH, Reece PJ, Gooding JJ. Paper-based sensor for monitoring sun exposure. ACS Sensors. 2016;**1**:775-780

[6] Lien D-H, Kao Z-K, Huang T-H, Liao Y-C, Lee S-C, He J-H. All-printed paper memory. ACS Nano. 2014;**8**:7613-7619

[7] Aliahmad N, Shrestha S, Varahramyan K, Agarwal M. Poly(vinylidene fluoride-hexafluoropropylene) polymer electrolyte for paper-based and flexible battery applications. AIP Advances. 2016;**6**:065206

[8] Down MP, Foster CW, Ji X, Banks CE. Pencil drawn paper based supercapacitors. RSC Advances. 2016;**6**:81130-81141

[9] Yang P-K, Lin Z-H, Pradel KC, Lin L, Li X, Wen X, He J-H, Wang ZL. Paper-based origami triboelectric nanogenerators and self-powered pressure sensors. ACS Nano. 2015;**9**:901-907

[10] Zhu H, Xiao Z, Liu D, Li Y, Weadock NJ, Fang Z, Huang J, Hu L. Biodegradable transparent substrates for flexible organic-light-emitting diodes. Energy & Environmental Science. 2013;**6**:2105-2111

[11] Fang Z, Zhu H, Yuan Y, Ha D, Zhu S, Preston C, Chen Q, Li Y, Han X, Lee S, Chen G, Li T, Munday J, Huang J, Hu L. Novel nanostructured paper with ultrahigh transparency and ultrahigh haze for solar cells. Nano Letters. 2014;**14**:765-773

[12] Fang Z, Zhu H, Preston C, Han X, Li Y, Lee S, Chai X, Chen G, Hu L. Highly transparent and writable wood all-cellulose hybrid nanostructured paper. Journal of Materials Chemistry C. 2013;**1**:6191-6197

[13] Fang Z, Zhu H, Bao W, Preston C, Liu Z, Dai J, Li Y, Hu L. Highly transparent paper with tunable haze for green electronics. Energy & Environmental Science. 2014;**7**:3313-3319

[14] Leijonmarck S, Cornell A, Lindbergh G, Wagberg L. Single-paper flexible Li-ion battery cells through a paper-making process based on nano-fibrillated cellulose. Journal of Materials Chemistry A. 2013;**1**:4671-4677

[15] Lee C-P, Li C-T, Ho K-C. Use of organic materials in dye-sensitized solar cells. Materials Today. 2017,**20**.267-283

[16] Yugis AR, Mansa RF, Sipaut CS. Review on metallic and plastic flexible dye-sensitized solar cell. IOP Conference Series: Materials Science and Engineering. 2015;**78**:012003

[17] Li Y-C, Jia S-R, Liu Z-Y, Liu X-Q, Wang Y, Cao Y, Hu X-Q, Peng C-L, Li Z. Fabrication of PEDOT films via a facile method and their application in Pt-free dye-sensitized solar cells. Journal of Materials Chemistry A. 2017;**5**:7862-7868

[18] Yeh MH, Lee CP, Lin LY, Nien PC, Chen PY, Vittal R, Ho KC. A composite poly(3,3-diethyl-3,4-dihydro-2H-thieno-[3,4-b][1,4]-dioxepine) and Pt film as a counter electrode catalyst in dye-sensitized solar cells. Electrochimica Acta. 2011;**56**:6157-6164

[19] Sangiorgi N, Sanson A. Influence of electropolymerized polypyrrole optical properties on bifacial dye-sensitized solar cells. Polymer. 2017;**125**:208-216

[20] He Z, Liu J, Khoo SY, Tan TTY. Electropolymerization of uniform polyaniline nanorod arrays on conducting oxides as counter electrodes in dye-sensitized solar cells. ChemSusChem. 2016;**9**:172-176

[21] Wu C-S, Chang T-W, Teng H, Lee Y-L. High performance carbon black counter electrodes for dye-sensitized solar cells. Energy. 2016;**115**:513-518

[22] Li Y-Y, Li C-T, Yeh M-H, Huang K-C, Chen P-W, Vittal R, Ho K-C. Graphite with different structures as catalysts for counter electrodes in dye-sensitized solar cells. Electrochimica Acta. 2015;**179**:211-219

[23] Chang LY, Lee CP, Huang KC, Wang YC, Yeh MH, Lin JJ, Ho KC. Facile fabrication of PtNP/MWCNT nanohybrid films for flexible counter electrode in dye-sensitized solar cells. Journal of Materials Chemistry. 2012;**22**:3185-3191

[24] Casaluci S, Gemmi M, Pellegrini V, Di Carlo A, Bonaccorso F. Graphene-based large area dye-sensitized solar cell modules. Nanoscale. 2016;**8**:5368-5378

[25] Moolsarn K, Tangtrakarn A, Pimsawat A, Duangsa K, Mongkolkachit C, Maiaugree W, Amornkitbamrung V. A dye-sensitized solar cell using a composite of PEDOT:PSS and carbon derived from human hair for a counter electrode. International Journal of Photoenergy. 2017;**2017**:11

[26] Chiang C-H, Wu C-G. High-efficient dye-sensitized solar cell based on highly conducting and thermally stable PEDOT:PSS/glass counter electrode. Organic Electronics. 2013;**14**:1769-1776

[27] Ouyang J, Chu CW, Chen FC, Xu Q, Yang Y. High-conductivity poly(3,4-ethylenediox ythiophene):poly(styrene sulfonate) film and its application in polymer optoelectronic devices. Advanced Functional Materials. 2005;**15**:203-208

[28] Yue G, Wu J, Xiao Y, Lin J, Huang M. Low cost poly(3,4-ethylenedioxythiophene):poly styrenesulfonate/carbon black counter electrode for dye-sensitized solar cells. Electrochimica Acta. 2012;**67**:113-118

[29] Fan B, Mei X, Sun K, Ouyang J. Conducting polymer/carbon nanotube composite as counter electrode of dye-sensitized solar cells. Applied Physics Letters. 2008;**93**:143103

[30] Hong W, Xu Y, Lu G, Li C, Shi G. Transparent graphene/PEDOT–PSS composite films as counter electrodes of dye-sensitized solar cells. Electrochemistry Communications. 2008;**10**:1555-1558

[31] Xue Y, Liu J, Chen H, Wang R, Li D, Qu J, Dai L. Nitrogen-doped graphene foams as metal-free counter electrodes in high-performance dye-sensitized solar cells. Angewandte Chemie International Edition. 2012;**51**:12124-12127

[32] Wang Z, Chen Y, Li P, He J, Zhang W, Guo Z, Li Y, Dong M. Synthesis of silicon-doped reduced graphene oxide and its applications in dye-sensitive solar cells and supercapacitors. RSC Advances. 2016;**6**:15080-15086

[33] Gao Z, Wang L, Chang J, Liu X, Wu D, Xu F, Guo Y, Jiang K. Nitrogen doped porous graphene as counter electrode for efficient dye-sensitized solar cell. Electrochimica Acta. 2016;**188**:441-449

[34] Chang Q, Ma Z, Wang J, Li P, Yan Y, Shi W, Chen Q, Huang Y, Huang L. Hybrid graphene quantum dots@graphene foam nanosheets for dye-sensitized solar cell electrodes. Energy Technology. 2016;**4**:256-262

[35] Sun H, Wu L, Gao N, Ren J, Qu X. Improvement of photoluminescence of graphene quantum dots with a biocompatible photochemical reduction pathway and its bioimaging application. ACS Applied Materials & Interfaces. 2013;**5**:1174-1179

[36] Tang L, Ji R, Cao X, Lin J, Jiang H, Li X, Teng KS, Luk CM, Zeng S, Hao J, Lau SP. Deep ultraviolet photoluminescence of water-soluble self-passivated graphene quantum dots. ACS Nano. 2012;**6**:5102-5110

[37] Kim JK, Park MJ, Kim SJ, Wang DH, Cho SP, Bae S, Park JH, Hong BH. Balancing light absorptivity and carrier conductivity of graphene quantum dots for high-efficiency bulk heterojunction solar cells. ACS Nano. 2013;**7**:7207-7212

[38] Gupta V, Chaudhary N, Srivastava R, Sharma GD, Bhardwaj R, Chand S. Luminescent graphene quantum dots for organic photovoltaic devices. Journal of the American Chemical Society. 2011;**133**:9960-9963

[39] Ponomarenko LA, Schedin F, Katsnelson MI, Yang R, Hill EW, Novoselov KS, Geim AK. Chaotic Dirac billiard in graphene quantum dots. Science. 2008;**320**:356-358

[40] Li M, Ni W, Kan B, Wan X, Zhang L, Zhang Q, Long G, Zuo Y, Chen Y. Graphene quantum dots as the hole transport layer material for high-performance organic solar cells. Physical Chemistry Chemical Physics. 2013;**15**:18973-18978

[41] Lee C-P, Lin C-A, Wei T-C, Tsai M-L, Meng Y, Li C-T, Ho K-C, Wu C-I, Lau S-P, He J-H. Economical low-light photovoltaics by using the Pt-free dye-sensitized solar cell with graphene dot/PEDOT:PSS counter electrodes. Nano Energy. 2015;**18**:109-117

[42] Lin L-Y, Lee C-P, Tsai K-W, Yeh M-H, Chen C-Y, Vittal R, Wu C-G, Ho K-C. Low-temperature flexible Ti/TiO_2 photoanode for dye-sensitized solar cells with binder-free TiO_2 paste. Progress in Photovoltaics: Research and Applications. 2012;**20**:181-190

[43] Lee CP, Lin LY, Tsai KW, Vittal R, Ho KC. Enhanced performance of dye-sensitized solar cell with thermally-treated TiN in its TiO_2 film prepared at low temperature. Journal of Power Sources. 2011;**196**:1632-1638

[44] Ahmadi S, Asim N, Alghoul MA, Hammadi FY, Saeedfar K, Ludin NA, Zaidi SH, Sopian K. The role of physical techniques on the preparation of photoanodes for dye-sensitized solar cells. International Journal of Photoenergy. 2014;**2014**:19

[45] Zhu S, Zhang J, Qiao C, Tang S, Li Y, Yuan W, Li B, Tian L, Liu F, Hu R, Gao H, Wei H, Zhang H, Sun H, Yang B. Strongly green-photoluminescent graphene quantum dots for bioimaging applications. Chemical Communications. 2011;**47**:6858-6860

[46] Lee C-P, Lai K-Y, Lin C-A, Li C-T, Ho K-C, Wu C-I, Lau S-P, He J-H. A paper-based electrode using a graphene dot/PEDOT:PSS composite for flexible solar cells. Nano Energy. 2017;**36**:260-267

Chapter 15

Rationally Fabricated Nanomaterials for Desalination and Water Purification

Rajendra S. Dongre

Additional information is available at the end of the chapter

http://dx.doi.org/10.5772/intechopen.74738

Abstract

Rationally designed nanomaterials from synthetic/biopolymers like chitosan, zeolites, graphene, nanometal/oxides, zerovalent metal/magnetic iron, OMS and nanocarbon/carbon nanotube (CNT) utilized in desalination/purification are thoroughly discussed. Conventional desalination membrane/materials own inherent limitations; nevertheless, designed nanocomposite/hybrid/films address the new challenges/constraints and consequently aid the remediation of environmental/water pollution, thus denoting prospective nanotechnology/science. The morphology and chemical functionality of certain natural/synthetic polymers are altered/controlled rationally yielding advanced membranes/materials, for example, aquaporin, nanochannels, graphene and smart self-assemble block copolymer blends to cater futuristic desalination needs besides superseded conventional membrane limitations too. In a nut shell, advance nanotechnology via electrospinning, track-etching, phase inversion and interfacial polymerization yields structured composites/matrixes that conquer traditional barriers of conventional desalination and supplies treated/purified water. This review confers synthetic strategy and utility of nanomaterials that are procured via ordered/rational designing/self-assembly to be used in separation techniques including RO/FO antifouling membrane, superwet surface, oil-water/emulsion separation and multifunctional desalination nanodevices.

Keywords: nanomaterial, designing, chitosan, OMS, desalination, water purification, electrospun

1. Introduction

The ever growing population and economic expansion put potential crisis in supply, and the availability of fresh water as UN for the decade 2030 forecasts 40% high global water uptake hints intensifies water consumption [1]. Amid estimated 780 million that do not have access to safe water, UN reported that global populations will face water scarcity by 2050. Thus, obligatory purifications of existing polluted/unsafe water were achieved via desalination or any other techniques. Desalination via membrane technique was developed since 1960 which removes soluble salts and micro-pollutants that are not of concern in conventional treatments, nevertheless being expensive over conventional water purifications. It also has drawbacks like high costs/energy inputs, and greenhouse gas pollution puts constraints on the environment to discover rationally developed desalination materials. Conventional polyamide-based membranes own inherent drawbacks viz. limited permeability, less selectivity and low chemical stability, and affecting separation performance [2]. Nanotechnology aids in the design of smart/advanced materials/membranes via fabrication of one-dimensional (1D), two-dimensional (2D), and three-dimensional (3D) nanomaterials owing to improved efficiency for revolutionary desalination/water purification. Rationally designed nanomembranes trounce such demerits, for example, graphene membranes owing to exclusive features like tunable properties, extraordinary robustness, and advantageous leverage with a superior separation efficiency on par with CNT and biomimetic aquaporin membranes [3–5]. Fabricated 2-D membranes own fundamentally different separation capability due to peculiar features like punching nanometer/ultrathin pores, precisely controlled/manipulated shape/size super-strong impermeable monolayers, and facile industrial scale-up since they use cheap/biopolymeric feedstock/raw materials [6–9]. Certain nanomaterials such as 2-D graphenes, zeolites and molybdenum disulfides assembled via layer-stacking and own unique configurations with controlled interlayer spacing have fascinated research in making high-performance desalination membranes with advantages peculiarity like tangible manipulation for permeability and contaminant's selectivity. Further, 2-D materials innovatively converted into 3D nanomembrane to impart enhanced selectivity and minimized fouling, which find potential applicability in solar desalination devices [10]. Versatile pertinence of nanotechnology is exploited in fabrication of many superior organic and/or inorganic based nano-materials owing myriad functionality by virtue of inherent firm controlled shape and pore distribution with facile alterations in porosity, surface area: volume ratio, and optimized dimensions/shapes as designed prior to their synthesis. Desalination materials/membranes have been designed and developed via assorted methodologies reported in the study [6–12]. On the basis of structure, properties, and performance relationships, these smart nanomaterials are fabricated for water treatments [1–13], and so this review signifies/focuses on recent advances in synthetic strategy in its designing/fabrication besides purification applications.

1.1. Assorted techniques for desalination

1.1.1. Adsorptive separation and pervaporation

Desalination extracts/removes salts and mineral components from saline water [1–14]. It is an earliest form of treatment still popular throughout for the conversion of seawater (salinity due to dissolved salts at 10,000 ppm) to drinking water for addressing water scarcity. Desalinated water is better than river/groundwater, as it has less salt and lime-scale contents. This desalination technology has demanded the usage of assorted smart/fabricated materials involving through advancement in nanotechnology. Thermal-based desalination is achieved via adsorption onto fabricated porous silica surfaces owing to double-bond surface force affinity for pollutants, resulting in high permeability as demonstrated for seawater at a substantially lower cost than conventional membrane technologies [10–14]. Pervaporation is performed via a designed membrane owing to preferential higher affinity with one component with a faster diffusion rate responsible for the separation of oil-water/mixtures. Pervaporation technique is advantageous in desalination due to 100% salt rejections and low-energy consumption attenuated by combined membrane permeation and evaporation. Several sophisticated advanced membranes were synthesized, for example, polyvinyl alcohol-based membrane owing to peculiar features viz. excellent film formation, hydrophilicity, and hydroxyl-induced swelling accountable for the effective desalination/separation of pseudo-liquid blend/mixtures [13, 14]. Nanosilica in maleic acid-polyvinyl alcohol yielding composite membranes shows enhanced water flux/diffusion coefficient and 99.9% salt rejections as desired for an efficient desalination [10]. Polyethylene terephthalate grafted in styrene obtains a membrane for pervaporation with an improved toluene selectivity [15]. PEBAX matrix of 100-μm dimension obtained for working at 50°C feed water temperature showed enhanced diffusion and viscosity reduction ruled by vacuum and thickness, which vitally determines pervaporation performance [10]. Commercial coalescing water filtration and adsorptive difficult emulsion separations for oily contaminants were achieved via fabricated nanofibers found to get partially deactivated besides membrane fouling, thus increasing their treatment cost [1–15]. Coalescing filtrations break stable/difficult oil-water and surfactant emulsions [16–18] that are viable on a coalescing medium/material used, as large droplets downstream settlement require less residence time. In this context, electrospun nanofibrous-based membranes own a larger surface area, vitally enhancing its coalescing filtration performance [1].

1.2. Fabrication methods of nanomaterials/membranes

Assorted nanoporous membranes/materials yield via diversified synthetic techniques [19] including phase inversion, interfacial polymerization, track-etching, and electrospinning as described in the below section:

1.2.1. Phase inversion

Chemical stratification is performed to remove the solvent from liquid-polymer solution and converts homogeneous solution into porous solid membrane/films in a controlled fashion [1, 19]. This phase inversion is vastly reliant on both the types of solvent and polymers, besides accomplished via captivate precipitation and/or heat/vapor/evaporation-induced phase separations. Immersion precipitation and thermo-induced phase separation are applied for nanofiltration (NF), ultra-filtration (UF), and reverse osmosis (RO) membrane fabrications [10]. Membrane morphology and porosity are found to be controlled by the nature of solvent/oil-water mixtures. Phase inversion technique is used to yield superoleophobic poly(acrylic acid)-graft PVDF membrane that induces efficient desalination and major emulsion/oil-water separations.

1.2.2. Interfacial polymerization

A step-growth polycondensation occurs in immiscible solvents (aqueous solution, impregnates one monomer and organic solution containing a second monomer), for example, diamine and di-acid chloride solution yield polyamide membrane [1, 10]. This technique is used to fabricate ultra-thin films (10 nm to μm) to be used for RO and NF membranes [1–10]. A membrane skeletal morphology or a layer-by-layer build-up of a composite can be controlled by many factors like the type/concentration of monomer and solvents, and reaction time, besides posttreatment conditions [10, 19]. The study reported interfacial polymerization of diamine and acyl chloride onto cellulose nanocrystal layer yielding a triple-layered composite/membrane for nanofiltration/desalination [1]. Interfacial polymerized MCM-41-silica and graphene oxide onto polyamide surface yield ultra-films [10] that showed huge water flux and salt rejections in desalination.

1.2.3. Track-etching

Track-etching involves energetic heavy ion irradiation onto a substrate resulting in a linear-damaged track across irradiated polymer surfaces that yield nanoporous membrane [19]. This technique precisely augmented a pore-size distribution of nm to μm dimension and a pore density of $1–10^{10}$ cm^{-2} [10], for example, nanoporous silicon nitride membranes using porous nanosilicon templates [1, 10].

1.2.4. Electrospinning

Electrospinning technique uses high-voltage treatment to polymers that is outsized to overcome the surface tension of solution droplets, and the resulted charged liquid jet is then converted into ultrafine/nanofibrous at collection drums. The morphology and skeletal parameters like porosity, shape/size distribution, and ratios of the corresponding nanomaterials/membranes are controlled by adjusting the electric voltage and treatment conditions viz. polymeric solution's viscosity and flow of solution [19]. Illustrious nanofibrous PVDF and 2-D nanosheets (Ti3C2Tx MXene) are prepared for sieving cations and dyes from contaminated water [1, 10].

1.3. Myriad nanomaterials/membranes for desalination

Assorted nanoporous material-based membranes are categorized as inorganic, organic, and inorganic-organic hybrid as per the material compositions. Various inorganic membranes include Al_2O_3-, TiO_2-, ZrO_2-, SiO_2-, TiO_2-SiO_2-, TiO_2-ZrO_2-, and Al_2O_3-SiC-based ceramics, and 2-D matter like graphenes and carbon nanotubes. Organic membranes are obtained from polymers like polyvinyl alcohol, polyimide, polypropylene, polyethersulfone, cellulose acetate, cellulose nitrates, polysulfone, polyvinylidene fluoride, polyacrylonitrile, polytetrafluoroethylene, and biopolymers like chitin, chitosan, and so on. Chitosan-blended dendrimers showed highly efficient anionic dyes, heavy metals, and organic contaminants removal from water [1]. Hybrid membranes yield, using inorganic metal/metal oxide, carbon materials into a polymeric matrix [1–19].

1.3.1. Electrospun membranes

Electrospinning technique is found to control many parameters like porous tortuosity, pores size and/or shape deviations (straight or cylindrical) which are crucially reduced in nano-fibers synergistic amalgamation. Polystyrene-based nanofibrous smooth surface of 452 nm owns a peculiarity viz. uniform porosity with low tortuosity, no bead formation, and least surface roughness, establishing futuristic coalescing filtration media for oily emulsion separations.

Morphological/topographical features of such electrospun membranes can be altered by numerous methodologies namely molecular bonding, in situ polymerization, and dopants encroachment technology. Akin to surface modifications achieved via nanoparticle coating, chemical/heat treatments, grafting, and interfacial polymerization were found to be efficient in coalescing filtration across commercial treatments. During the last decade, electrospun technique utilizes myriad polymeric feedstocks for devising nanofibrous membranes for pressure/thermal-driven microfiltration (MF), UF, NF, forward osmosis (FO) and coalescing filtrations besides adsorptive desalination [19].

Electrospun membranes own 3D interconnected skeletons as critical for improved desalination media which is advantageous over traditional membranes in regard to the efficiency, price, and energy. Microporous polymeric progressive desalination membranes are fabricated via assorted techniques like film lithography, stretching, phase inversion, electrospinning owing to peculiarity viz. huge interconnective 3D porosity, adjustable pore-size distribution, high water flux, high surface area with molecular orientations and facile fiber-axis directional macroscales. These films/composites fascinated prominent field like water treatment/purification/separation, pressure-driven distillation, oil-water/marine oil spill cleanups, and RO/NF pretreatment feeds. Functionalized polymeric membranes can act as methanol fuel cell, separators for rechargeable lithium-ion batteries, pressure-retarded/driven/osmosis used for bacteria/fungus culture media/suspended particles micro-filtration, dye solutions, and ultra-filtration [1–19]. Tailored polyvinylidenefluoride-polyacrylonitrile nanofiber membranes are characterized for chemical adsorption, liquid filtration, and extraction of harmful chemicals from contaminated water. Nanochitosan owns fast adsorption kinetics, high arsenic sorption capacity, and facile arsenic and other such anionic removal [20–23]. Electrospun polystyrene-polydopamine/PDA fibrous cross-linked with β-cyclodextrin coating was found to overcome existing limitations for the removal of anionic pollutants from water compared to non β-cyclodextrin/mere PDA fiber [21, 22]. Biohybrid membranes from polyvinyl alcohol and hydrocolloidal natural gum yield via electrospinning remove assorted nanometal like Ag, Au, Pt, Fe_3O_4, and CuO from water [1, 21]. PVA/GK, amidoxime webs yield via methane plasma treatments and two-nozzle electrospinning, respectively, exhibited altered porosity and hydrophobicity besides elevated sorption capacity for the selective adsorption of uranium and vanadium, besides offering huge utility in tissue engineering and drug-delivery systems [1, 10, 19]. Self-assembly/phase separation techniques for nanomaterial synthesis control 3-D porosity without fiber orientation, while templating controls for fiber orientation over dimension arrangement using sacrificing agents [20]. Electrospinning fabricates tunable morphology and diameter in 1D to 3D nanofiber pores with a facile modification achieved via chemical grafting of rough surface [1, 19]. Chemical compositions, mechanical features, patterns, and membrane pore areas are dominant factors in desalinations that are attenuated by electrospinning, for example, polymethyl methacrylate fiber axis with 0.97-μm diameter found to affect cross-link adherence or extended orientation that ultimately regulates water flux. Thus, nanofiber augmented/controlled via electrospinning to manage fiber alignment yields aligned membranes for desalination utilities as shown in **Figure 1**. Still, many challenges require balancing the degree of orientation, nanofiber thickness, and productivity to be achieved by effective strategic controls like pore area, gap width, and orientation. Electrospinning coats nanofiber onto polymers/ceramic, imparting a high-specific surface and tunable porosity needs in separation [1, 19].

1.3.2. Advanced nanomaterials

Novel nanocomposites, films, hybrids, matrixes, and membranes were fabricated via constitutional morphological alterations to owe substantial water flux/permeability and salt rejection sought in desalination [1, 19]. Lucratively, R & D has designed nanoporous materials that accede to high water flux to keep salt/other contaminates away, thus paving the path for futuristic overwhelmed desalination. Single-layer nanoporous sulfur-coated molybdenum (MoS_2) sheets exhibited 70% higher water flux than graphene RO membranes due to unusual features like a fish-bone/hourglass architecture owing to a nozzle subnanoporosity withstanding necessary water pressure/volumes and robustness as energetic and economic than other counterparts [24]. Certain single-layer nanosheet (pore area of 20–60 Å^2), MFI-zeolite, polymeric high-flux RO, and graphene membranes owing to nanofiltration are significant due to vital modulated parameters like porosity, velocity distributions, permeation, and water density imparting 90% ion rejection with huge water transportation than conventional membranes [24]. Modern nanotechnology aids in designing opportunistic energy-efficient membranes/sheets/mats for efficient desalinations, for example, few A^0 to several nanoporous dimensionally drilled resultant molecular sieves/membranes that *control mass transportations* [1, 19]. *Boron nitride-doped carbon nanotubes and* single-atom-thick *graphene* are augmented for assorted hydrated ions/salt rejections performance than conventional zeolitic membranes in desalination [1]. Hydroxyl functionality gets altered via

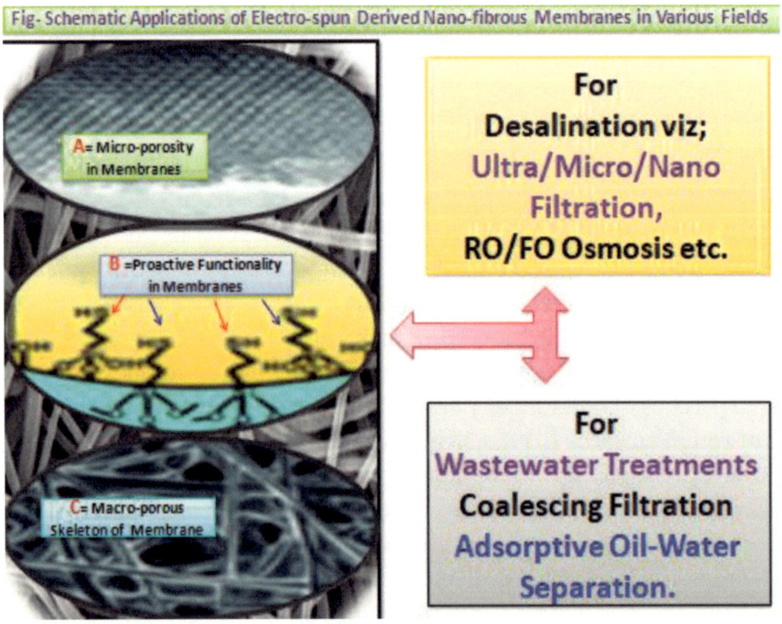

Figure 1. Schematic application of electrospun nanofibrous membranes in various fields.

chemical treatment, for example, graphene nanopore provides hydrophilic sites with enhanced water flux achieved via precise nanopore edges with tunable properties and hydrophilicity in certain advanced desalination [19, 24]. Chemical vapor deposition is combined with focused electron beam to get sculptured single-layer assorted metal dichalcogenide ($MoSe_2$, $MoTe_2$, WS_2, WSe_2)-based membranes that carry effective ion separations as a function of pore size, chemistry, geometry, and hydrostatic pressure [1]. Mechanistic separations via reverse osmosis and capacitive deionization utilized many nanomaterials including zeolites, carbon nanotubes, and graphene to develop highly efficient and capable futuristic desalinations [1, 19, 24]. Advanced designing of nanomaterials paved new avenues and ability to manipulate nanomaterials like carbon nanotubes, nanowires, graphene, quantum dots, super-lattices, and nanoshells [1, 24]. Nanotechnology enables unique nanofabrication that controls macroscopic nontortuous ~1-nm pores specifically designed in MCM, carbon nanotubes, and graphene to offer well-designed size-selective, filtration membranes superior to conventional polymeric membranes (rigidity owing to size-selective nondesign porosity) for desalination [1, 19, 24].

1.3.3. Zeolites

Aluminosilicates are commonly termed as zeolites that possess 3–8-nm pore dimensions found to control morphological features that are well exploited for adsorption/ion exchangers in water treatments. Molecular dynamics stimulates a tight pore distribution as vital for absolute salt rejections and high water permeability, for example, ZK-4 molecular sieve-based RO membrane (4.4-nm) solvates salt and allows the passage of water to flow [1, 25, 26]. Mordenite-coated α-alumina zeolite (MFI, 5.6-nm) RO membranes exhibited hydrophobicity and lowers salt/ionic transport due to fabricated interstitial-defective surfaces so as to control major ion transportations. Zeolites/molecular sieves can be directly coated onto ceramics and incorporated via laser-induced fragmentation to yield RO/FO membranes, for example, Linde-zeolite-A (4.4-nm) interfacial-embedded composites [25, 26] as shown in **Figure 2**. Further rising of zeolite weight % was found to enhance water permeability and salt rejections in resultant membranes compared to experimental thin-film membranes (without zeolite). Zeolite-coated membrane permeates high salt/water throughout since specific transport limits intrinsic pores, and it is difficult to establish its exact role in water transport and salt rejection [10, 25, 26].

1.3.4. Nanocarbons

Assorted nanocarbon materials get vast popularity due to their specific morphology, physicochemical properties, and varied significant utilities. Thus, carbon-based materials are developed as nanoparticles, nanoonions, peapods, nanofibers, nanorings, nanowires, nanotubes, and fullerenes, owing to extensive analytical explicabilities. Intrinsic surface defects of nanostructured carbons were found to affect their stability, and mechanical and physicochemical properties, which further aids in rationally designed requisite materials like zero-dimensional fullerenes and diamond clusters, 1-D nanotubes, 2-D graphenes, 3D nanodiamond (<1 μm), and ultra-hard fullerite [1, 19, 24]. Nanocarbon materials have special advantages viz. facile functionality alterations, high carrier capacity, hydrophilic/hydrophobic incorporations besides high chemical stability [1, 15, 24].

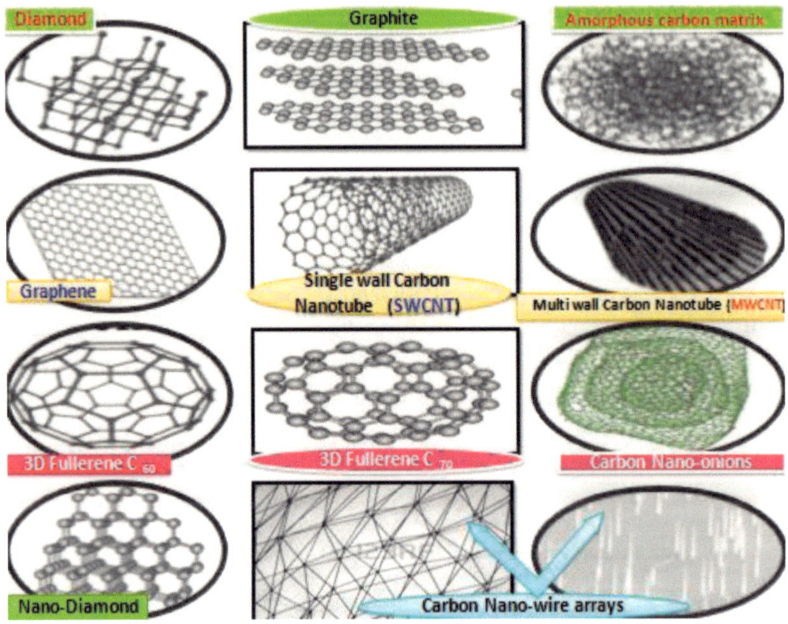

Figure 2. Some nanocarbon materials used in desalination/water purifications.

Allotropic carbon nanotube (CNT) contains rolled-up/cylindrical graphite layers arising as fascinated structural materials due to excellent thermal-electrical conductance, strength, and adsorption properties [1, 12]. CNTs are characterized as single-walled and multi-walled based on the built-up route. A high specific area and elevated available adsorption sites of CNT aid its adaptable surface chemistry. Hydrophobic CNT surfaces are benignly stabilized to avoid persistent contaminant adsorption/aggregation and for preconcentration/detections [15, 19] besides metals/ions too adsorbed onto CNTs through electrostatic attraction and chemical bonding. CNT-based desalinations are facile for point-of-use water purification, for example, plasma-modified ultra-long CNT-based membranes own specific enhanced salt, organic, and metal adsorptions compared to conventional carbon [19]. Futuristic techniques can equip with CNT so as to achieve superior desalination, disinfection, and filtration. USA developed CNT-boron sponges to separate oil-water/emulsion besides oil spills removal aiding oil remediation [1]. Less CNT amount is available in mitigations of degradable antibiotics/pharmaceutical contaminants in point-of-use purification systems [19]. Facile, cheap, and compatibility features of CNT, nanometal, and zeolite nanomaterial-based pellet/bead are utilized for arsenic removal from water [1, 22]. Electrochemically active CNT-based membrane can remove salt, proteins, virus, dyes, and phenol from water [1]. Advanced nanotechnology controls carbon nanotube diameter [19] that aids fabrication of CNT-coated high flux RO membrane owing to labile hydrophobicity and surface roughness [13–17]. Rapid mass transport occurs in CNT-coated polystyrene/silicon nitride membranes that are designed via carboxylation offering high salt ions rejection due to salt anionic and carboxyl repulsions. Some CNT composites owe a low surface area of 210 m^2/g than the activated carbon of 1500 m^2/g; instead, the adsorptive capacity is more due to ordered nanoporosity, which enhances the adsorption of pollutants

[1]. Graphene is utilized potentially in the fabrication of RO/FO membrane fabrications [1] due to high breaking strength and impermeability aiding to create tunable pore size and ultrathin high flux membranes akin to molecular sieves. Molecular dynamic and impregnation techniques simulate the designing of pores (3–25 nm) in ordered carbon, subterminal pores in CNT that allows facile ionic transportation within electrolyte interfaces [20, 21]. The porosity of nanocarbon, CNT, and graphene can be tailored to direct specific adsorption/ultrafast ion transports vital in separation/desalination [1, 19].

1.3.5. *Nanometal/metal oxides*

Nanometals/metal oxides owe peculiar features like high specific surface area, short intraparticle diffusions, and facile compressibility, which are devoid of activated carbon; hence they are preferred for use in the remediation of metals, arsenic, radionuclides, and organics [1, 19]. Nanometal/oxide gets easily compressed into porous pellets, bead, and powder, for example, Solmete-X, Inc., USA, commercialized nanoselective resin *Arsen-Xnp* from nanoiron oxide-coated organic polymer for the removal of arsenate from water. Nanozerovalent iron adsorbs chlorinated polycyclic aromatic hydrocarbons and perchlorate from water due to its high specific surface area than granular iron [30]. Magnetic nano-Fe_3O_4 separates arsenic from contaminated water [1] besides being injected directly into systems as it gets easily removed by increased osmotic pressure in FO osmosis [1, 19]. Nanosilver-coated TiO_2 has high stability and low toxicity availability in disinfectants in water treatments [19]. Assorted properties, applications, and approach of nanomaterials in treatments [19] are summarized in **Table 1**.

1.3.6. *Rationally designed composites/hybrid/biomimetics*

Rational designing is done to get ultra-thin, dense active membranes/composites which own less salt rejections and ambient water flux utilized in much FO/RO desalination [1]. Nanomaterial's intrusion in contemporary polymeric skeleton offers extra physical/permeable barriers so as to attain two bulk phases as utilized in promising oil-water separations and desalination [1, 19]. Nanomaterial matrix/blend, for example, nano-SiO_2, CNT, aquaporin-rapped polyamide, and polysulfone-zeolite's superior sieving fabricated myriad smart membranes [25] (**Figure 1**). Functionalized multi-wall carbon nanotube (MWCNT) polyamide-based RO membrane reported 200% water flux with at par salt rejection, amplified water flux [1–17], high hydrophilicity, and thermal stability [27]. Mesosilica MCM-41 incorporated polyamide-based RO membranes augmented water permeability/flux [16]. Biomimic amphiphilic tri-block vesicles enclosed aquaporin-Z and interfacial polymerize RO/FO mem brane impart complete salt rejection and 800 times water flux for seawater desalinations [15, 17]. Conventionally intense polarized membranes own depleted water flux and operation efficiency as serious issues in RO/FO separations [1]. Smart nanosupported bottleneck RO/FO membranes are proficient due to nanoporosity, low tortuosity, high mechanical strength [17], and huge water flux than conventional under similar conditions. Zeolite/nanosilica-polysulfone hydrogel (10 nm) was found to alleviate conventional membrane clogged owing [1, 15] to water flux enhancement and anti-fouling mitigation as achieved due to anti-adhesive surface hydrophilicity [15]. Grafted polyethylene glycol-based RO/FO membrane has great

SN	Nanoporous materials	Characteristics		Applications
		Useful properties	Adverse properties	
1	Nanofiltration (RO/FO)	Reliable, automated, charge-based repulsion, high selectivity, low pressure, costly	High energy, costly, membrane jam, concentrated polarization, few nanoscale pore dimensions	Water hardness, color, odor reduction and heavy metal removal, sea water desalination, wastewater
2	Nanocomposite membranes	Large hydrophilicity, water permeability/flux, thermal/mechanical robust, fouling resistant	Leakage of nanoparticle, bulk nanomaterials needed for oxidation, and composite dependency	Reverse osmosis and removal of micropollutants, bionanocomposite membrane utility
3	Self-assembled membranes	Homogeneous nanoporosity, tunable designed/tailoring	Applied on laboratory/small scales	Ultra-filtrations, process scale-up.
4	Nanocarbon membranes	High porous, high permeable, bacteriosidal, superhydrophobic, tailored electrospun and sustain high salinity, hydrophobic. Survive filtration under high pressure/vacuum	Pore blockages/chocking, leakages of nanofibers	Ultra-filtrations, filters, cartridges, nanofiber composite membranes in water treatments and high-performing direct contact membrane distillations
5	Aquaporin membranes	Assort molecular transportation, highly selective, tailored to dense polyfilm, mechanically stable, regenerative/self-healed	Low pressure desalination, mechanical weakness	Low pressure desalination, biomimetic membrane for RO and FO filtrations, and surface imprint and embed membrane filtrations
6	CNT: carbon nanotubes	Assessable adsorption sites, vast reusability, high cost, health risk	Production cost is very high, own health risks	Degradation of antibiotics, organic, pharmaceutics and own high specific salt adsorption
7	Dendrimers/dendrons (arborol cascade species)	Water-soluble bifunctional (inner hydrophobics and outer hydrophilic absorptions), encapsulate molecules, reusable, bioactive mimics, handy toxicity, nontoxic	Dendrimer production steps are complex and multistage processes	Organic and heavy metal removals, biodegradable, biocompatible, for example, chitosan/dendrite-based
8	Zeolites/aluminosilicate	Highly microporous molecular sieves, lodge variety of cations, selective molecule sorts at size exclusions, control ion release, regular molecular porosity	Reduction in active surface through immobilization	Water disinfections, ion-exchange beds in water purification/softening, solar thermal collector, adsorptive refrigeration
9	Nanozerovalent iron (1–100 nm)	Permeable reactive barrier to filter out contaminants, high specific surface area, and nZVI high mobility/reactivity	Stabilization is needed, that is, surface modifications	As permeable reactive barriers, sediment cap, groundwater remediation like PCB and PAH degradations

http://dx.doi.org/10.5772/intechopen.74738

SN	Nanoporous materials	Characteristics		Applications
		Useful properties	Adverse properties	
10	Nano-TiO_2	High reactivity, stable, durable, toxic, needs UV activation, controllable nanosizes controlling by process conditions like calcinations	Removes suspended fine particles, needs ultraviolet activation	Water disinfections, antifouling processes
11	Metals magnetic nanoparticles/alloys	Nanoparticle manipulations via magnetism/magnetically tunable colloidality and shows superparamagnetism, highly recyclable, easy magnetic separation, own very large surface to volume ratio and biocompatible	Stabilization is needed to enhance its potential	Environmental remediation, cation sensors, nanobeads adsorbents in water treatments, groundwater remediations
12	Nanometals/ nanometal oxides	Short intraparticle diffusivity, tunable pore size/surface chemistry, compressible, abrasion resistive, magnetic, reusable	Needs variant nanodopant to enhance capturing potential by interconnections	Heavy metallic anion removal, filters, slurry reactors, palates, powders

Table 1. Characteristics of the utility of nanomembrane/materials in water treatment processes.

surface tensions, which are utilized for more wetability for hydrophobic foulants [1]. Certain smart nanomembranes, for example, graphene oxide-coupled polyamide, nanosilver-coated polysulfobetaine and polypeptide-grafted single-walled carbon nanotube owing peculiar features viz. bendable size/porosity, high defect density, irreversible bacterial cell adhesion, and antibacterial activity exploited in antimicrobial RO/FO and desalination [1–20].

Thus, nanotechnology improved flux efficiency via ordered pore/size/shape variations and physical barrier/selective charge-base repulsion as requisite for specific/establish emulsion separation/desalination [1–20]. Argonide corporation-USA developed nanoceram interlinked nanofibrous of 2–100 nm diameter and surface area of 300–600 m^2/g to work as an electropositive filter cartridge [27]. Cellulose polymers tailored on nonwoven glass sheets are utilized for ultra-filtration of dirt, bacteria, viruses, and proteins [18]. Certain nanomaterials like fluorocarbon-coated tetramethyl orthosilicate and polyurethane/polylactic acid/polyethylene oxide-coated bio-films' superior features have exploited in desalination due to its mechanical strength, hydrophobicity, biocompatibility, and nontoxicity [1–19]. Biomimetic membranes akin to a*quaporin* are embedded in polymeric matrix/nanofilters [20] to withstand 10 bar high pressures and huge water flux of >100 L/(hm^2) involved in brackish water desalination. Nanotechnology aids in the fabrication of specific aquaporin-based membranes competitive with conventional membranes withstanding under all critical conditions like operating pressures of reverse osmosis, high temperature, acidic/alkaline range, and fouling-based corrosion [1]. Nano-Al_2O_3/TiO_2 zeolite, CNT, photo-catalytic, and nanobimetallic incorporation improve hydrophilicity, raise water permeability, enhance foul resistance, and elevate mechanical and thermal stability which impart high water permeation in RO/FO membrane desalination [1, 20]. Carbon intrude nanopolymeric matrices-based semi-permeable membranes raised the hydrophilicity and increased water permeability/salt permeation as utilized for reverse osmosis [1]. Trimesoyl chloride-metaphenylenediamine interfacial-coated polyethersulfone yields nano-NaX zeolite with 40–150 nm dimensions catering to effective RO/FO membrane [25, 26]. Nano-H_2O Company-USA has commercialized *Quantum-Flux/WO-2006/098872-A3* matrix owing to more permeable efficiency with low fouling and no clogging for membrane-based reverse osmosis [27]. Such coated matrix membranes maintained a surface profile that carries immobilization of potential harmful nanoparticles, for example, P25-Evonik robust membrane [1, 27]. Smart material-based membranes overcome inherent conventional material limitations and address global challenges of water scarcity besides combating environmental pollution [1–20]. Molecular designing of CNT and aquaporin membranes highlighted surface modification and interfacial interactions with enhanced fouling resistivity as shown in **Figure 3**.

1.3.7. Multifunctional nanodevices

Advancement in nanoscience aids in the design of proactive/flexible synchronous and synergistic functionality in desired water treatments as achieved via nanodevice concepts like Fenton nanofiltration and self-floated solar device using well-ordered CNT, nanogold-derived hydrophobic membranes [1, 17]. Polypyrrole-coated stainless steel-based hydrophobic membranes enhance water evaporation than natural solar heating/point-of-use seawater desalination [44]. Molecular simulation theoretically designed smart nanomaterials/devices/systems for best ground-breaking solutions to the existing desalination/purification problems [1, 19].

http://dx.doi.org/10.5772/intechopen.74738

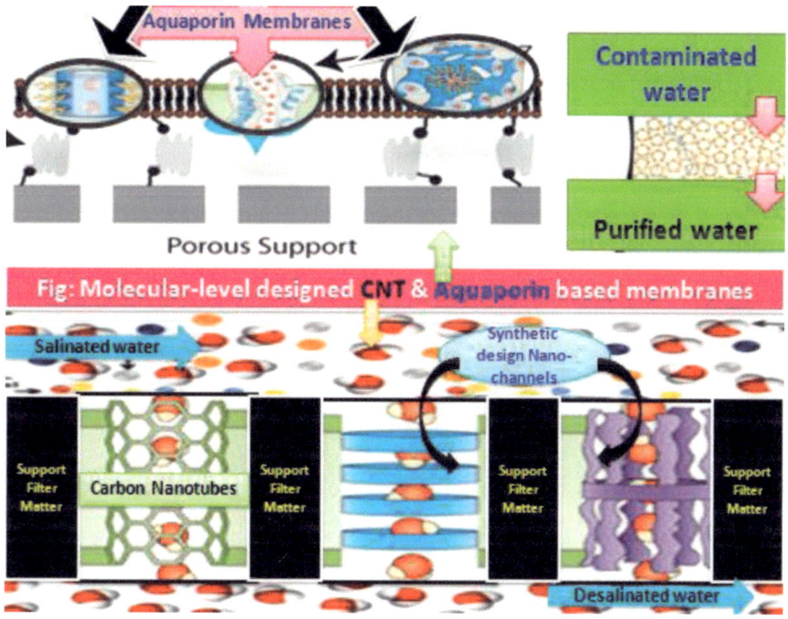

Figure 3. Molecular-level designed CNT and aquaporin-based membranes.

1.3.8. Ordered mesosilica

Ordered mesosilica (OMS) are designed via eco-friendly synthetic paths to offer controlled pore-size network as explored in industrial applications [28]. Kuroda-Mobil Oil Company in 1990 had prepared ordered mesosilica with huge surface areas of >1000 m^2/g and a controlled porosity of 2–20 nm via amphiphilic co-polymeric pore/structure directing agent/precursor [1, 28]. Eco-sustainable mesosilica has achieved via greener syntheses from raw feedstock, and thus sol-gel is preferred [28]. Research on various applications of OMS (last decade) is given in **Figure 4**.

OMS has an ordered chemical/structural/textural distinctiveness which permits to exhibit pollutant adsorption selectivity in water purifications/desalinations [29]. Conventional petroleum-derived surfactants are replaced by renewable amphiphilic polysaccharides that impart an efficient porosity and a high pore volume in resultant OMS [28, 29]. Hydro-soluble polyionic micelles afford wide-ordered porosity needed for oil-water/emulsion separations as targeted OMS applicability differs by means of porosity and template removal/thermal calcinations/chemical extractions [1, 29]. Calcination temperature vitally shrinks shape/size, porosity, and morphology and affects micro-skeleton of OMS as mentioned in **Table 2**.

OMS manufacturing is expensive due to costly precursor usage like silicon-alkoxide/TEOS, so, it is developed in a more eco-synthetic way for large scale [28, 29]. Environmental impacts are lessened if biomass/recycled waste silica sources are used viz. fly/rice husk's ashes extracted silica. Several templating surfactants are recognized for lyotropic liquid crystalline phase formations. However, classical surfactants are substituted by new hydro-soluble/dissociable and recoverable porogens. The ecodesigned industrial ordered meso-silica synthesis

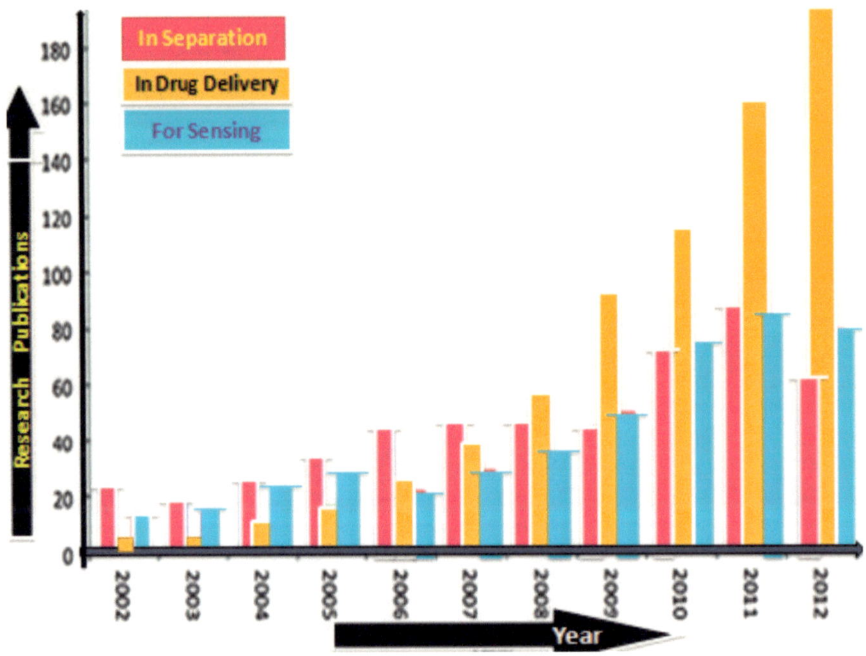

Figure 4. Research publications on various field applications of mesoporous silica (last decade).

Silica solids	Particle shape	Size dimensions	Pore-size ordering
MTAB-MSN	Spherical	50–100 nm	Low-stretched mesoporous channels
CTAB-MSN	Spherical	70–100 nm	Well-ordered mesoporous channels
STAB-MSN	Rod-like	100–500 nm length 50–75 nm width	Well-ordered mesoporous channels
DDAB-MSN	Irregular	80–100 nm	Ink-bottle porosity
TDTHP-NS	Spherical	100–500 nm	Microporosity

Table 2. Shape, size, pore-order morphology, and microskeletons in mesosilica.

which uses assisted self-assembly path for removal of template own profound impacts on surface, chemical and textural properties of OMS and thus affects their synthetic applications [29]. Calcination and solvent extraction eliminate organic templates during synthesis, and numerous template removals are studied to lessen process time and solvent usage. Assorted varied surfactants vitally fix sustainable synthetic paths as shown with waste generation (E-factor = kg waste/kg product) in **Table 3**.

These developed sustainable OMS syntheses rely on the choice of silica precursor, environmental impact, byproduct, and cost [29]. Green chemistry synthesis aspects escort silica from natural mineral deposits, biomass, or industrial wastes, and constitute huge resources that are commercially compared in **Table 4**.

Nomenclature	Surfactant used	Structure	E-factor [30]
MCM-41/48: mobile composition of matter	Alkyltrimethyl ammonium salt $C_nH_{2n+1}N^+(CH_3)_3X^-$ (with *n* = 12, 14, 16 or 18 and X = Cl or Br)	*P6mm*, hexagonal for MCM-41 and *Ia3d*, cubic MCM-48	41.9/21–2.0 and 45.3/1 = 45.3
FSM-16[a] Folder sheet mesoporous	Alkyltrimethyl ammonium $C_nH_{2n+1}N^+(CH_3)_3X^-$ (with *n* = 12, 14, 16 or 18 and X = X or Br)	*P6mm*, hexagonal	31.9/10
HMS Hexagonal mesoporous silica	Uncharged amine surfactant $C_nH_{2n+1}NH_2$	Wormhole framework structure	29.9/12
SBA-15 and 16 Santa Barbara amorphous	P123 and F127, respectively	*P6mm*, hexagonal and *Im3m*, cubic, respectively,	16/2.3–7.0
KIT-6[b] Korea Advanced Institute of Science and Technology	P123	*Ia3d*, cubic	25/4–6.2
FDU-1[c] FuDan University	Polyethylene oxide-polybutylene oxide-polyethylene oxide tri-block copolymer B50-6600 ($EO_{39}BO_{47}EO_{39}$, Dow)	*Fm3m*, cubic	–
COK-12[d] Centrum voor Oppervlaktechemie & Katalyse	P123	*P6m*, hexagonal	–

[a]FSM-16 yield from layered silicate kanemite

[b]KIT-6 by tri-block copolymer ($EO_{20}PO_{70}EO_{20}$)-butanol.

[c]FDU-1 in NaCl salt.

[d]COK-12 in citrate/citric acid surfactant, E-factors are found to be lower than nanomaterials and bulk chemicals.

Table 3. Ordered mesoporous materials, used surfactant, and their crystallographic structure.

Origin	Type	Advantages	Drawbacks
Natural	Natural clays, diatomite/kieselgur, siliceous sedimentary rock and minerals, besides zeolite	Copious economical amicable, mesoporous materials without the need of organic surfactants rather than metal cation impurities, residual natural lignins are helpful	Polycondensation is uncontrolled under neutral/acidic conditions, strong acids and high temperatures needed purification
Synthetic	Silicon alkoxides Si(OR)4, for example, TEOS, TMOS, besides soluble silicates, for example, sodium silicate, colloidal and fumed silica	Silicon alkoxides, uniform oligomers impart highly organized mesostructure at any pH, facile and cheap protocol, water as solvent	High energy input and expensive procedures, toxic precursors, needs catalysts and organic solvents only, byproduct alcohol during hydrolysis. Silicate oligomers with varied degrees of polymerization, polycondensation is uncontrolled under neutral/acidic conditions

Origin	Type	Advantages	Drawbacks
Recycling wastes	Industrial ashes, for example, coal, rice husk, and packaging resin waste, glasswares	Plentiful cheap and nontoxic, acidity conferred by residual metal ions within matrix besides waste disposal solutions	Has lower surface area and pore volume than other precursors, porous silica material regenerations as hard templates used for nanocasting

Table 4. Comparison of different silica sources for the synthesis of silica-based mesoporous material.

Eco-designed OMS synthesis focused on assisted self-assembly involving assorted [28, 29] treatments via precipitation, liquid phase reactions, solid recovery, washing, and drying as shown in **Figure 5**.

1.4. Major smart materials/membranes for desalination and oil-water/emulsion separations

Nanomaterial-aided modus operandi have simulated membranes especially designed onto advanced nanostructures like CNT, graphene, graphene oxide and reduced graphene oxide for desalination [1, 19]. Graphene/graphene oxide/rGO showed unique performance and promising direction in futuristic water treatment/desalination [1–20]. Such stimulated membrane penetrates water via nanometer pores in single-layer graphene and offers salt nanofiltration with enhanced permeability than conventional RO [17, 19]. Graphene oxide membrane exhibits

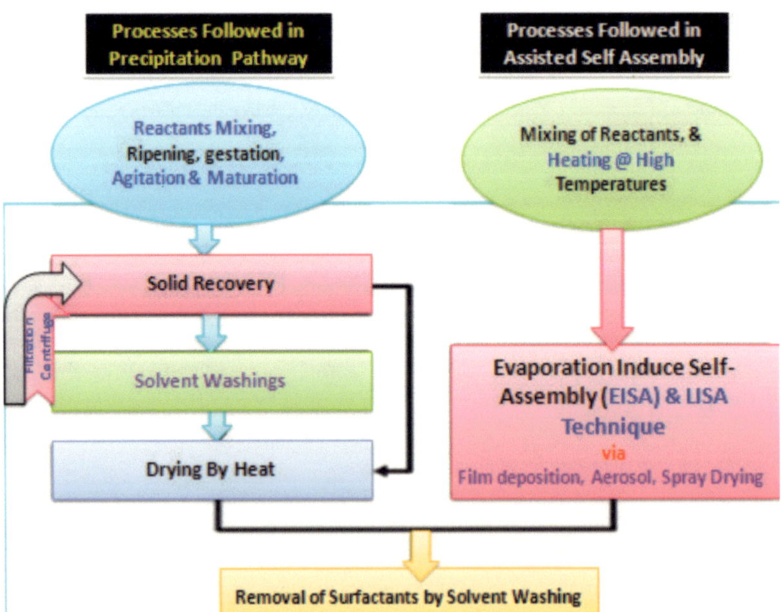

Figure 5. Steps involved in precipitation and assisted self-assembly routes for OMS synthesis.

anomalous liquid fast permeation into capillary high pressure created inside inter-layers and used for seawater desalinations as it contains varied ions including Na^+, K^+, Mg^{2+}, and Cl^-. Such ultrafast transportation via precise casing imparts single/double-layered graphene as enthusiastic representative in desalination [18] which has 100% LiCl, NaCl, and KCl salt rejections than traditional membranes [17 19]. Akin, two-dimensional inorganic matrixes like MXenes possess atomic-layered skeletons with controllable compositions as potential nanofiltration membranes. Nanofiltration assembly/devices were developed with specified and smartly designed nano-TiO_2, CNT, and noble-metals and Fenton agents [1, 19].

Oil-water separations are required if fuel spillage especially gasoline/diesel/petrol during transportation failures releases oil in aqueous systems to pose environmental hazardous, and thus emulsion separations are highly sought [1, 19]. Interface science and bionic knowledge have developed some 2-D membranes via surface micronanohierarchical structure grafting so as to impart unique/super-wetting characters viz. superhydrophobicity, superoleophobicity, and superamphiphilicity to be utilized for oil-water separation [1, 15]. Bio-inspired spatial hierarchical polytetrafluoroethylene-coated stainless steel mesh has self-cleaned superhydrophobic surfaces that impart a water contact angle of >150° and a diesel contact angle of ~0° and responsibly perform excellent oil-water separations [15]. Such hydrogel-coated mesh is superior due to extraordinary water passage selectivity over oil, thus preventing oil-material contacts and avoiding membrane clogging caused by viscous oils besides allowing gravity-driven separation. Surface chemistry designed membranes like polyvinylpyridine-polydimethylsiloxane-polyurethane sponges, grafted polyacrylic acid, and sodium silicate-TiO_2 stainless steel mesh have irreversible encapsulation of low-surface-energy species, flexible wet-ability, and toggled superoleophilicity as benefited for emulsion/oil-water separation [15].

1.5. Futuristic nanomaterials for desalination

Advanced nanotechnologically designed/engineered nanoadsorbents, nanometals, nanomembranes, and photo-catalysts have vulnerable flexibility and adjustability with water treatment systems encompassing assorted micro-pollutants [1]. The compatible existing water treatment processes can be integrated simply in conventional modules. Nanomaterials are advantageous due to their ability to be integrated to various multifunctional membranes that enable both particle retention and contaminant mitigations compared to conventional materials used in water technologies [20]. Auxiliary usages of nanomaterials impart higher process efficiency and higher sorption rates. However, in order to minimize the health risk of nanomaterial's usage in water treatments, several regulatory norms need to be prepared for being adaptable to mass/large-scale utility. Still, nanostructured materials have offered potential innovations in serious contaminants degradation, decentralized water treatments, and point-of-use devices [1–20]. Nanotechnology assists in wastewater treatments for the mitigation of pathogens, organic and inorganic, heavy metals, and other toxic contaminants using nano-ZnO RO/FO nanofilms and polyrhodanine-encapsulated magnetic nanoparticle that are removed by contaminants up to ppb level [19]. There emerge innovative technologies in nanoscience; yet, many challenges posed by water purification need to be resolved by future scientists.

1.6. Conclusions

Rationally designed smart nanomaterials provide myriad scientific and technology growth to desalination/water purification. Still, biomaterials must be explored in this perspective that owes high permeation and low salt rejections in futuristic desalination. Thus, advanced nanotechnology aided commercially viable products/solutions that enhance/replace existing desalination/purification. Certain functionalized nanoporous biopolymeric membranes were found to cater to inherent challenges. Bio-polymer cross-linking fixes usual instability and imposes functionally cost-effective nanoporous biomaterials to be used in desalinations. Overall, rational fabrication highlighted "*design-for-purpose*" unlike *trial-and-error approach* starts with scientific perceptions by knowing inherent barriers, and thus the conceptual design of nanomaterial is proposed, as fed back to desalination problems for thorough usage in water treatments. Nanomaterials' performance must unambiguously be defined in water purifications and need redesigning under failure conditions with "thinking-outside-the-box" prospective as confronted by desalination. Till now, nanomaterials' rational designing offered more unprecedented opportunities for solving desalination challenges in a sustainable manner. A prospective-ordered designing should owe few aspects namely molecular dynamics/simulation tools to extend problem definition and theoretically needs more multi-functional/all-in-one nanomaterial for effective desalination/water purifications. Therefore, progressive and versatile nanomaterials/devices, which can work under ambient conditions with paramount desalinations/water purification performance, are expected in the near future. Such design-sophisticated material surfaces reversibly counter stimuli via innovative and promising exterior/interior changes and eminent environmental adaptability which displayed myriad functions viz. interfacial pollutant adsorption, omniphobic slippery coatings, responsive particle-stabilized emulsions, and self-healed surface membranes. This chapter described the strategic and valuable rational-designing idea for assorted biomimetic membranes/materials owing to environmental utilities.

Acknowledgements

The author is thankful to the Head, Department of Chemistry, R.T.M. Nagpur University, Nagpur, for laboratory facilities and to the Vice Chancellor, Nagpur University, Nagpur, for the sanction of a research project under University Research Project Scheme, No. Dev./RTMNURP/AH/1672 (9), dated 24 September 2016.

Author details

Rajendra S. Dongre

Address all correspondence to: rsdongre@hotmail.com

Department of Chemistry, R.T.M. Nagpur University, Nagpur, Maharashtra, India

References

[1] Li R, Zhang L, Wang P. Rational design of nanomaterials for water treatment. Nanoscale. 2015;**7**:17167-17194. DOI: 10.1039/C5NR04870B

[2] Eijkel JCT, van den Berg A. Nanofluidics: What is it and what can we expect from it? Microfluidics and Nanofluidics. 2005;**1**:249-267

[3] Humplik T, Lee J, O'Hern SC, Fellman BA, Laoui T, Wang EN. Nano-structured materials for water desalination. Nanotechnology. 2011;**22**(29):1-19

[4] Han JY, Fu JP, Schoch RB. Molecular sieving using nanofilters: Past, present and future. Lab on a Chip. 2008;**8**:23-33. DOI: 10.1039/b714128a

[5] Sint K, Wang B, Kral P. Selective ion passage through functionalized graphene nanopores. Journal of the American Chemical Society. 2008;**130**:16448-16449. DOI: 10.1021/ja804409f

[6] Murad SM, Lin JC. Using thin zeolite membranes and external electric fields to separate supercritical aqueous electrolyte solution. Industrial and Engineering Chemistry Research. 2002;**41**:1076-1083

[7] Jeong B et al. Interfacial polymerization of thin film nano-composites: A new concept for reverse osmosis membranes. Journal of Membrane Science. 2007;**294**:1-7

[8] Wang HZ, Huang ZP, Carnahan D. Reversible transformation of hydrophobicity & hydrophilicity of aligned carbon nanotube arrays & buckypapers by dry processes. Carbon. 2010;**48**(3):868-875

[9] Ji L, Zhu D. Mechanisms for strong adsorption of tetracycline to carbon nanotubes: Comparative study using activated carbon & graphite as adsorbents. Environmental Science & Technology. 2009;**43**:2322-2327

[10] Qu X. Applications of nanotechnology in water treatment. Water Research. 2013;**47**:3931-3946

[11] Li LX et al. Ordered mesoporous carbons synthesized by a modified sol-gel process for electrosorptive removal of sodium chloride. Carbon. 2009;**47**:775-781

[12] Striemer CC et al. Charge- and size-based separation of macromolecules using ultrathin silicon membranes. Nature. 2007;**445**:749-753

[13] Hajeh M, Laurent S, Dastafkan K. Nanoadsorbents: Classification, preparation, and applications (with emphasis on aqueous media). Chemical Reviews. 2013;**113**:S7728-S7768

[14] Li J-J, Zhou Y-N, Luo Z-H. Smart fiber membrane for pH-induced oil/water separation. ACS Applied Materials & Interfaces. 2015;**7**:19643-19650

[15] Liu H, Gao S-W, Cai J-S, He C-L, Mao J-J, Al-Deyab SS, Lai Y-K. Recent progress in fabrication & applications of superhydrophobic coating on cellulose-based substrates. Materials. 2016;**9**(3):124

[16] Lee A, Elam JW, Darling SB. Membrane materials for water purification: Design, development, and application. Environmental Science: Water Research & Technology. 2016; **2**:17-42

[17] Lalia BS, Kochkodan V, Hashaikeh R, Hilal N. A review on membrane fabrication: Structure, properties and performance relationship. Desalination. 2013;**326**:77-95

[18] Dervin S, Dionysiou DD, Pillai SC. 2D nanostructures for water purification: Graphene and beyond. Nanoscale. 2016;**8**:15115-15131

[19] Wang Z, Ciacchi LC, Wei G. Recent advances in nanoporous membranes for water purification: Review. Nanomaterials. 2018;**8**(2):65

[20] Davis ME. Ordered porous materials emerging application. Nature. 2002;**417**:813-821

[21] Diallo MS, Goddard WA. Dendrimer enhanced ultrafiltration. Recovery of Cu(II) from solutions PAMAM dendrimer with ethylenediamine core-terminal NH_2. Environmental Science & Technology. 2005;**39**:S1366-S1377

[22] Aredes S, Klein B, Pawlik M. The removal of arsenic from water using natural iron oxide minerals. Journal of Cleaner Production. 2012;**29-30**:208-213

[23] Torrey JD et al. Oxidation behavior of zero-valent iron nanoparticles in mixed matrix water purification membranes. Environmental Science: Water Research & Technology. 2015;**1**(2):146-152. DOI: 10.1039/c4ew00068d

[24] Lin SC, Buehler MJ. Mechanics and molecular filtration performance of graphyne nanoweb membranes for selective water purification. Nanoscale. 2013;**5**:11801

[25] Humplik T, Wang EN. Framework water capacity & infiltration pressure of MFI Zeolites. Microporous & Mesoporous Materials. 2014;**190**:84-91

[26] Gehrke I, Geiser A, Somborn-Schulz A. Nano-techn innovations for water treatment. Nanotechnology, Science Applications. 2015;**8**:1-17. DOI: 10.2147/NSA.S43773

[27] Porada S et al. Direct prediction of the desalination performance of porous carbon electrodes for capacitive deionization. Energy & Environmental Science. 2013;**6**:3700

[28] Corine G'r, Reboul J, Bonneb M, Lebeau B'n'd. Ecodesign of ordered mesoporous silica materials. Chemical Society Reviews. 2013;**42**:4217

[29] Fernandes FM, Coradin T, Aimé C. Self-assembly in biosilicification and biotemplated silica materials: Review. Nanomaterials. 2014;**4**:792-812

[30] Sheldon RA. E factors, green chemistry & catalysis: Odyssey. Chemical Communications. 2008;(9):3352-3365. DOI: 10.1039/B803584A

Chapter 16

Noble Metal-Based Nanocomposites for Fuel Cells

Hongpan Rong, Shuping Zhang,
Sajid Muhammad and Jiatao Zhang

Additional information is available at the end of the chapter

http://dx.doi.org/10.5772/intechopen.71949

Abstract

Noble metal-based nanocomposites are attractive for a rich variety of electrocatalytic applications as they can exhibit not only a combination of the properties associated with each component but also synergy due to a strong coupling between different constituents. Using noble metal as the base component, a plenty of methods have been recently demonstrated for the synthesis of noble metal-based nanocomposites with novel structures (e.g., alloys, core-shell, skin and 1D/2D structures). In this chapter, an account of recent advances of synthetic approaches to noble metal-based nanocomposites with controlled structures, compositions and sizes are reviewed. The relationship between structures and electrochemical properties of these nanocomposites in fuel cell field is discussed. The potential future directions of research in the field are also addressed.

Keywords: noble metal, nanocomposites, electrochemical, fuel cell

1. Introduction

With the global rapid increase of energy demand and the depletion of fossil fuels, research on environment-friendly energy sources has attracted considerable attention in recent years. The real commercialization of fuel cells is a promising solution for the global problems of energy supply and clean environment. A fuel cell can convert chemical energy into electric energy by an electrochemical reaction of hydrogen-containing fuel with oxidant. Based on the electrolyte type, fuel cells are classified to be several kinds: proton exchange membrane fuel cells (PEMFCs), phosphoric acid fuel cells (PAFCs), solid acid fuel cells (SAFCs), alkaline fuel cells (AFCs) and high-temperature fuel cells. Among all kinds of fuel cells, only hydrogen PEMFC has been used in commercial vehicles (the Toyota Mirai) from 2014, due to its short start-up time, high-energy density and low working temperature. Thus, we will focus on PEMFCs in this chapter. Besides hydrogen, other fuels which are suitable for PEMFCs include alcohols (methanol, ethanol, glycol etc.) and formic acid. Compared with hydrogen, they have lower energy density [1]. But they are cheap, plentiful, easily stored and transported. Therefore, direct methanol fuel cell (DMFC), direct ethanol fuel cell (DEFC), and direct formic acid fuel cell (DFAFC) are also promising in commercial applications.

At the anode of a fuel cell, hydrogen-containing fuel is oxidized to produce electrons that are transferred to the cathode through an external circuit. At the cathode, oxygen is reduced to be water. Thus, catalysts are essential at both electrodes to promote the fuel oxidation reaction and oxygen reduction reaction (ORR). Most fuel oxidation and oxygen reduction reactions catalyzed by noble metal-based nanocomposites are performed in acidic solutions. The reason contains two aspects: (1) for DMFCs, DEFCs and DFAFCs, carbonates can form in alkaline electrolyte [1]; (2) hydrogen oxidation reaction on platinum (Pt) in acid is 2 orders of magnitude quicker than in alkaline electrolytes [2]. When the fuel is hydrogen, the hydrogen oxidation reaction (HOR) rate on Pt is extremely fast, and 0.05 mg cm^{-2} Pt loading at the anode is enough. However, to achieve a desirable catalytic performance, the slow reaction rate of ORR at the cathode even on the best Pt-based catalyst needs much more Pt loading amount (~0.4 mg cm^{-2}) [3]. According to the data reported by Vesborg in 2012, Pt is one of the most expensive metals [4]. The large-scale application of fuel cell technology in transportation field or portable power generation is limited by the high cost of electrode catalysts. Numerous researches have been undertaken to improve the intrinsic activity of electrocatalysts to reduce the noble metal loading of the electrodes without compromising fuel cell performance.

The catalytic activity, selectivity and stability of noble metal-based nanocomposites are closely related to the factors including size, shape/morphology and composition [5–8]. In the past decade, various experimental methods have been developed to synthesize size-dependent, high catalytic performance noble metal-based nanocomposites with diverse morphologies, such as polyhedron, concave, wire, plate, belt/ribbon, dendrite/branch and cage/frame structures [9–21]. Smaller size means higher surface-area-to-volume ratio, higher atomic utilization efficiency and more catalytic active sites. During electrochemical process, small-size noble metal-based nanocomposites provide high electrocatalytic activity, but the challenge is the aggregation that could happen under electrocatalytic conditions, resulting in the poor performance of long-term stability test (e.g., commercial Pt/C). Alloying one or two kinds of transition metals with noble metal has become a valid strategy to develop excellent electrocatalysts [22–25]. The addition of the second/third metal not only changes the surface active sites by ensemble effect, but also alters the binding strength of reactants, intermediates, and products by electronic/strain effect [26–28]. Different noble metals polyhedrons/concave expose atoms with different coordination numbers, thus have different surface energy and exhibit different catalytic performances [29–31]. 1D (wire) and 2D (plate, belt/ribbon) noble metal-based nanostructures have high surface area (high atom utilization efficiency), high conductivity and large interfacial area contacting with the support in electrochemical reactions [12, 15, 32]. Dendrite and branch structures not only exhibit large surface area and high active sites, but also significantly relieve the aggregation happened in electrocatalytic stability test [9, 33, 34]. Highly open noble metal nanocages and nanoframes exhibit enhanced electrocatalytic properties due to their three-dimensional accessible surface atoms [20, 35, 36]. A nanosegregated noble metal skin on these open structures can further improve the catalytic performance [18, 37]. Thus, numerous shape-controlled synthetic strategies are established in colloidal reaction system, such as facet-selective capping (CO [38, 39], halide anions [40], amines [9], formaldehyde [30, 41], etc.), seed growth [42, 43] and oxidative etching [44].

In this chapter, we focus on the key factors behind the novel synthesis strategies and provide the discussions along the topic of composition. The very recent achievements of the controllable synthesis of noble metal-based nanocomposites are given and discussed on the structure-property relations in electrochemical reactions.

2. Pt-based nanocomposites

At present, owing to the outstanding catalytic activity and superior resistant characteristics to corrosion, Pt-based nanoparticles (NPs) supported on porous carbon are still the most efficient catalysts in PEMFCs [15, 16, 18, 27]. In the past several years, to develop advanced Pt-based nanocatalysts for electrochemical reaction, researchers have proposed several synthesis strategies, which mainly include (1) alloying, (2) core-shell, (3) Pt-skin, (4) 1D or 2D structures and (5) porous, caged, hollow and frame structures. The integration of two or more strategies into one kind of Pt-based nanocomposites is beneficial for the construction of super-electrocatalysts with excellent catalytic activity and stability.

2.1. Pt-based alloys and intermetallics

In the past decade, Pt-based nanostructures have been widely used to prepare excellent catalysts for diverse chemical reactions [45–47]. Compared with pure Pt nanocatalysts, Pt-based alloy and/or intermetallics exhibit higher activity and stability [48]. As since the surface electronic structure and *d*-band center position of Pt were changed by incorporating a second and/or third metal into Pt lattice. Owing to higher oxidation potentials of transition metals than Pt, they are always removed from the surface of Pt-based alloy by electrochemical leaching. A pure Pt shell with a thickness of more than several atomic layers is produced after the dealloying process [49]. Although the dealloyed Pt-based nanocatalysts show a higher electrochemical activity compared with Pt/C [50], they perform a poor long-term durability since the re-deposition of leached transition metal onto the anode [51].

For bimetallic and multi-metallic Pt-based nanocomposites, the exposed crystal facet has a close relationship with their electrocatalytic performance. Tian et al. and Stamenkovic et al. have demonstrated that the ORR activities on Pt decrease in the order of high-index facet (hkl) > (111) ≫ (100) in $HClO_4$ solution [52]. Well-defined shapes of Pt-based nanocrystals (NCs) are, thus, of great interest for boosting electrochemical performance. It is believed that the surface energy of a fcc Pt NCs exposed with different facets increases in an order of (111) < (100) < (110) < (hkl) [53]. Accordingly, high-energy crystal facets and surface defects vanish quickly during the NC growth process due to the adatom incorporation on them. Therefore, it is a challenge to prepare NCs with high-index facets and high density of surface defects. Compared with alloy, intermetallic compounds show outstanding structure stability owing to the ordered atom arrangements. Rong et al. developed a kinetically controlled method to tune the surface defect of Pt-based intermetallics [24]. Based on their proposed growth mechanism, large electronegativity difference, etching and diffusing processes are necessary for the synthesis of defect-rich cubic intermetallic Pt_3Sn NCs. In this protocol, N,N-dimethylformamide (DMF) is solvent and poly(vinylpyrrolidone) (PVP) is surfactant. As shown in **Figure 1**, rich surface defects endow the defect-rich cubic Pt_3Sn NCs with excellent catalytic activity for formic acid oxidation reaction, excellent stability caused by the structure stability of intermetallic compounds.

2.2. Ultrathin Pt-based nanocomposites

2.2.1. Ultrathin Pt-based nanowires (NWs)

Ultrathin Pt-based NWs with a diameter of few atomic layers present high ratio of surface atoms to bulk atoms, which could increase Pt utilization efficiency greatly. During the synthesis process of bimetallic and multi-metallic Pt-based NWs, there are two obstacles to overcome: (1) controlling the reduction and nucleation process of different kinds of metal precursors with different reduction potentials; (2) confining the growth along a certain direction and inhibiting the other two directions. While diverse efforts to develop such materials have been made, some achievements of the fabrication of ultrathin Pt-based NWs with few nanometers have been reported. Oleylamine (OAm) was widely used as soft template in the preparation of ultrathin Pt-based NWs. In 2007, Sun's group synthesized ultrathin Pt-Fe NWs with a diameter of 2–3 nm in a mixed solution of 1-octadecene (ODE) and OAm [54]. The length of Pt-Fe NWs ranged from 20 to 200 nm with the tuning volume ratio of OAm/ODE. They claimed that OAm self-organizes into reverse-micelle-like structure in ODE solution and higher ratio of OAm/ODE resulted in the formation of longer NWs.

In the same OAm/ODE system, Li et al. prepared Pt/NiO core-shell NW first, subsequently reduced it into PtNi alloy NWs by thermal annealing and finally converted it into jagged Pt NWs through electrochemical dealloying (**Figure 2**). The jagged Pt NWs had shorter Pt—Pt bond length than regular Pt NWs and/or bulk Pt. Short Pt—Pt bond length resulted in a compressive strain on the surface, which produced the world record ORR properties. The specific and mass activity for ORR at 0.9 V vs. reversible hydrogen electrode (RHE) are 11.5 mA cm^{-2} and 13.9 A/mg_{Pt}, respectively [16].

As shown in **Figure 2**, Huang's group used OAm as the solvent, reductant and surfactant, cetyltrimethylammonium chloride (CTAC) as structure-directing and glucose as reducing reagent to synthesize hierarchical platinum-cobalt (Co) NWs [14]. These PtCo NWs had high-index, Pt-rich facets and ordered intermetallic structure. The novel structure enabled the PtCo NWs excellent performance toward alcohol oxidation reactions and ORR (39.6/33.7 times of the specific/mass activities than commercial Pt/C for ORR). They used CTAC, $Mo(CO)_6$ and Ni $(acac)_2$ (acac = acetylacetonate) as structure-directing reagent to prepare subnanometer Pt NWs with a diameter of 0.8 nm [17]. They claimed that the carbonyl decomposed from $Mo(CO)_6$, Ni^{2+} and proper amount of CTAC were important for the formation of 1D structure. With addition of a stronger reductant (glucose), they fabricated PtNi, PtCo and PtNiCo NWs with similar diameter. These subnanometer Pt alloy NWs showed enhanced ORR activity and stability (mass and specific activities of 4.20 A mg_{Pt} and 5.11 mA cm^{-2} at 0.9 V vs. RHE). They

http://dx.doi.org/10.5772/intechopen.71949

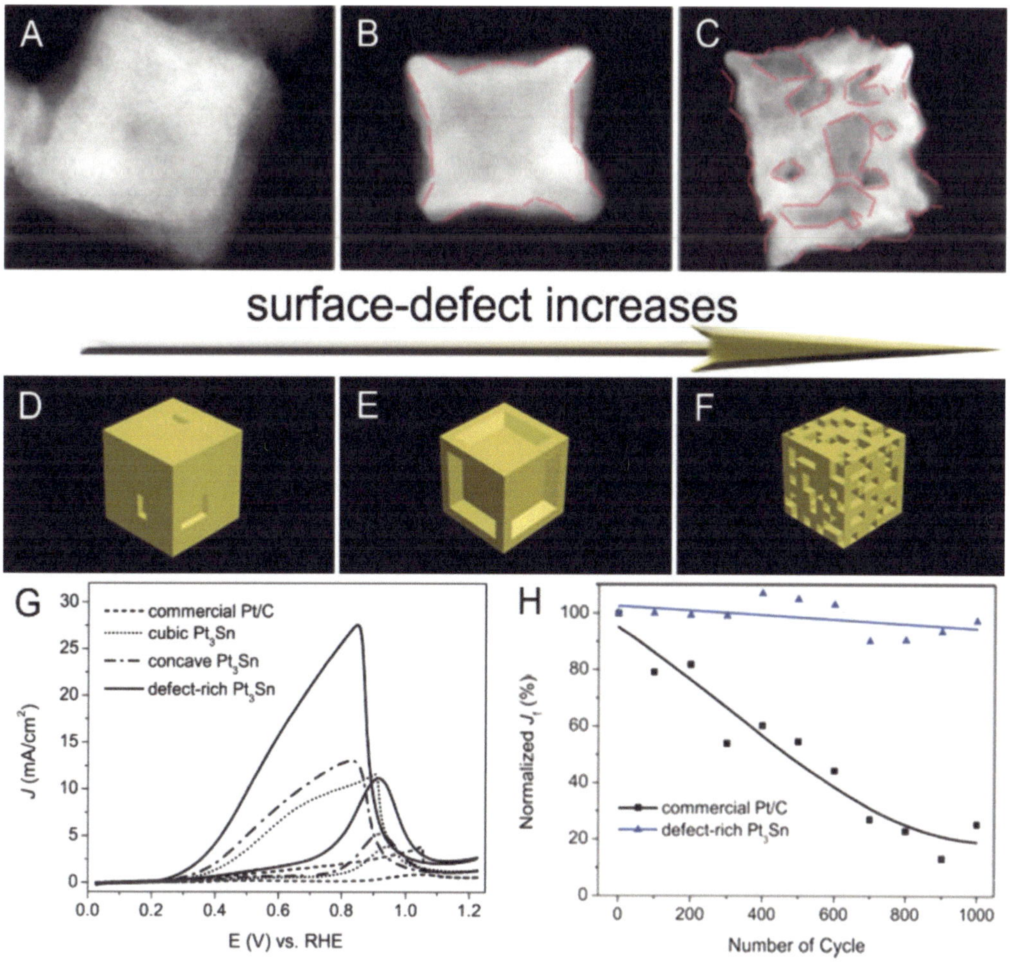

Figure 1. (A–C) HAADF-STEM images and (D–F) corresponding structure model of (A and D) cubic, (B and E) concave cubic, and (C and F) defect-rich cubic intermetallic Pt_3Sn NCs. (G) Cyclic voltammograms of formic acid oxidation in 0.1 M $HClO_4$ + 1 M HCOOH (scan rate: 50 mV/s). (H) Loss of peak current density in forward scans as a function of cycling numbers (0.65–1.23 V vs. RHE, scan rate: 50 mV/s). Reprinted with permission from Ref. [24]. Copyright 2016 Wiley-VCH Verlag GmbH & Co. KGaA, Weinheim.

used this effective reaction system to synthesize a series of Pt-based NWs for electrochemical reactions: (1) PtNiPd core-shell NWs showed superior glycoloxidation reaction (EGOR), glycerol oxidation reaction (GOR) and ORR performances [55], (2) porous Pt_3Ni NWs with extraordinary catalytic performance toward methanol oxidation reaction (MOR) and ORR [56], (3) hierarchical PtPb NWs exhibited higher activity and stability for MOR and ethanol oxidation reaction (EOR) than commercial Pt/C [57], (4) screw thread-like PtCu NWs showed excellent properties for MOR and EOR [58].

Recently, Li's group realized the reduction of Mo in this reaction system by a hydrogen assisted solution route (HASR) [32]. Ultrathin Pt-Mo-Ni NWs with a diameter of ~2.5 nm were

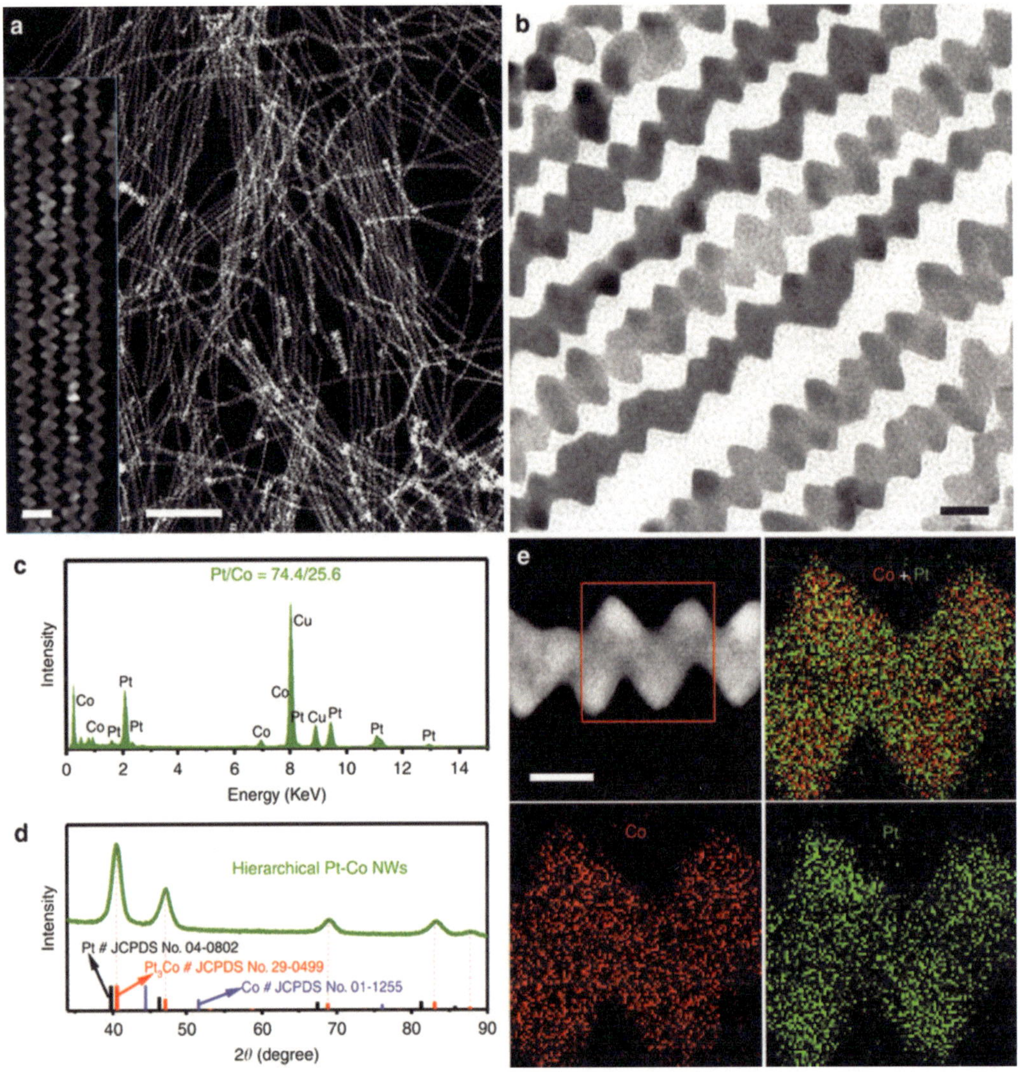

Figure 2. Representative (a) STEM image, (b) TEM image, (c) TEM-EDS, (d) PXRD pattern and (e) STEM-ADF image and EDS elemental mappings of the hierarchical Pt_3Co NWs. Inset in (a) is an enlarged STEM image. The composition is Pt/Co = 74.8/25.2, as revealed by ICP-AES. The scale bars in (a), inset of (a), (b) and (e) are 200, 20, 10 and 10 nm, respectively. Reprinted with permission from Ref. [14]. Copyright 2016 Nature Publishing Group.

successful synthesized in OAm. Hydrogen served not only as reductant, but also as structure-directing species. A series of Pt-Mo-M (M = Fe, Co, Mn, Ru) NWs could be synthesized by this HASR method. The Pt-Mo-Ni NWs exhibited higher maximum power density than Pt/C in DEFC experiments.

In the above mentioned cases, soft template method is very effective in OAm system to construct ultrathin Pt-based NWs. The largest extent exposure of Pt atoms results in the extraordinary electrochemical activity. The presence of the second and third transition metal can not only relieve the undesirable overbinding of oxygen atoms on catalyst, but also stabilize the undercoordinated active sites, further improving the electrochemical activity and stability. While the synthesis of ultrathin Pt-based NWs has many achievements, more innovative works are still desired to enrich this field.

2.2.2. *Core-shell structures with ultrathin Pt shell*

Since catalytic reactions take place on the surface of nanocatalysts, the alteration of surface composition and electronic structure directly causes different catalytic properties [59]. The physical and chemical properties of core-shell nanostructures can be altered by the design of core, shell and interface. The construction of Pt shell on the surface of nanocatalysts is another possible solution to prepare enhanced Pt-based catalysts with minimum usage of Pt. The catalytic properties of Pt based core-shell NCs have close relationship with the Pt shell thickness. The charge transfer between core and shell components changes the band structure of shell, thus influencing the catalytic performance [60]. The electronic and geometric properties of the top surface layer are affected by the core, and the influence decreases with the increased thickness of shell. For core-shell structures with thin shell, the different lattice constant between core and shell components results in the lattice strain and brings geometric effect [61]. Strong electrochemical activity enhancements have been observed for monolayer or a few atomic layer Pt shells. Thanks to the development of instrumental techniques, researchers can study ultrathin Pt shells closely and clearly.

So far, diverse synthetic strategies have been developed to prepare diverse inorganic core-shell structures: semiconductor@semiconductor [62], metal@metal [63, 64], semiconductor@metal [65], metal@semiconductor [66–68], and even multishell structures [69].

Several synthetic strategies have been developed to fabricate core-shell structures, such as seed-mediated growth, dealloying, galvanic replacement and one-pot synthesis. The most effective and widely used synthetic methodology is seed-mediated growth. Adzic and co-workers have fabricated various core-shell electrocatalysts with monolayer and ultrathin Pt shell via Cu under potential deposition (UPD) and pulse electrodeposition (PED). The core materials contain intermetallics (PtPb, PdPb, PdFe) [70], core-shell (Pd@PdAu) [71], alloy (AuNi) [72] and transition metal nitride (TiCuN) [73]. As shown in **Figure 3**, in an UPD process, they prepared core first, then dropped onto a flat glassy electrode to form a thin film. The electrodes were deposited Cu monolayer in a H_2SO_4 + $CuSO_4$ solution, then rinsed in a K_2PtCl_4 + H_2SO_4 solution to replace Cu with Pt. The Cu deposition and replacement operations were conducted under an Ar atmosphere [70]. Due to the limited electrode area and critical synthesis conditions, UPD and PED are not suitable for the large-scale preparation of core-shell nanocatalysts with ultrathin Pt shell. Zhang's group reported the synthesis of Ag@Pt and Au@Ag@Pt particles with a controllable thin Pt shell by interface-mediated galvanic replacement. These sub-ten-nanometer core-shell NCs showed better ORR performance than Ag [74].

Utilizing the different reduction potentials of different metal precursors, bimetallic core-shell structure could be prepared by one-pot synthetic process [75, 76]. However, it is still a challenge to control this process since alloys or heterostructures with uncontrolled morphologies are usually produced. Bu et al. synthesized PtPb@Pt core-shell nanoplate in OAm/ODE mixture and this

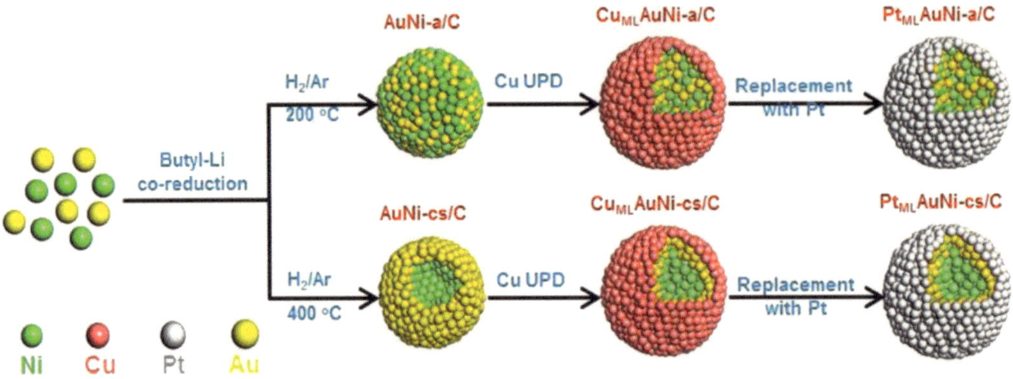

Figure 3. Schematic illustration of the stepwise synthesis of monolayer Pt core-shell electrocatalysts. Reprinted with permission from Ref. [72]. Copyright 2016 American Chemistry Society.

nanoplate exhibited about eight-fold increase in mass activity for MOR and EOR and enhanced ORR performance compared with Pt catalyst. In their synthesis process, L-ascorbic acid was used as the reducing reagent. The Pt shell thickness was 0.8–1.2 nm (equal to four to six atomic layers). According to the well-established *d*-band model, tensile lattice strain upshifts the *d*-band center and thus results in a stronger bonding with adsorbates and a lower reaction rate. In the case of PtPb@Pt core-shell nanoplate, although Pb atom in the core is larger than Pt, the tensile strain did not decrease the electrochemical properties. They claimed that the surface strain effect on binding behavior was facet dependent, and the tensile strain on the exposed Pt(110) helped to decrease the Pt—O bond strength.

Compared to monolayer Pt shell, a few atomic layer Pt shell can protect the transition metal atoms from leaching under electrochemical operation and have better stability. However, the stability of core-shell structured nanocatalysts with ultrathin Pt shell is still far away from satisfactory.

3. Other noble metal-based nanocomposites

3.1. Pd-based nanocomposites

Palladium (Pd)-based nanocomposites have exhibited extraordinary performance in diverse applications such as hydrogenation reactions [5, 77], photothermal therapy [78], cross-coupling reactions [79] and electrochemical reactions [80–82]. Pd and Pt are in the same group and Pd-Pt nanostructures have showed excellent electrocatalytic properties [37, 83]. Considering Pt-based nanocomposites that have been discussed above, Pd-Pt nanostructures are not reviewed in this subsection.

The electrochemical reactions on Pd NCs are structure sensitive. For the low index planes of Pd, Kondo et al. reported that reduction current density of ORR at 0.9 V (RHE) increased in the order of (110) < (111) < (100) [84]. Shao et al. also demonstrated that the specific activity of Pd nanocubes exposed with {100} facets was about 10 times higher than that of Pd octahedral enclosed with {111} facets toward ORR [85]. Compared with hydrophilic solvents such as water, ethylene glycol and DMF, systematic shape control of Pd NCs in hydrophobic solvent remains a challenging task. As shown in **Figure 4**, Niu et al. synthesized icosahedral, decahedral, octahedral, tetrahedral and triangular plates Pd NCs in the mixture of OAm, formaldehyde and toluene [30]. In this protocol, $Pd(acac)_2$ was used as the metal precursor and the morphology of Pd was determined by the quantity of OAm. Different intermediates were formed with different amount of OAm under room temperature stirring and reduction rate decreased with the increase amount of OAm. Seeds were formed with the addition of formaldehyde at room temperature, which also acted as selective surfactant in this reaction

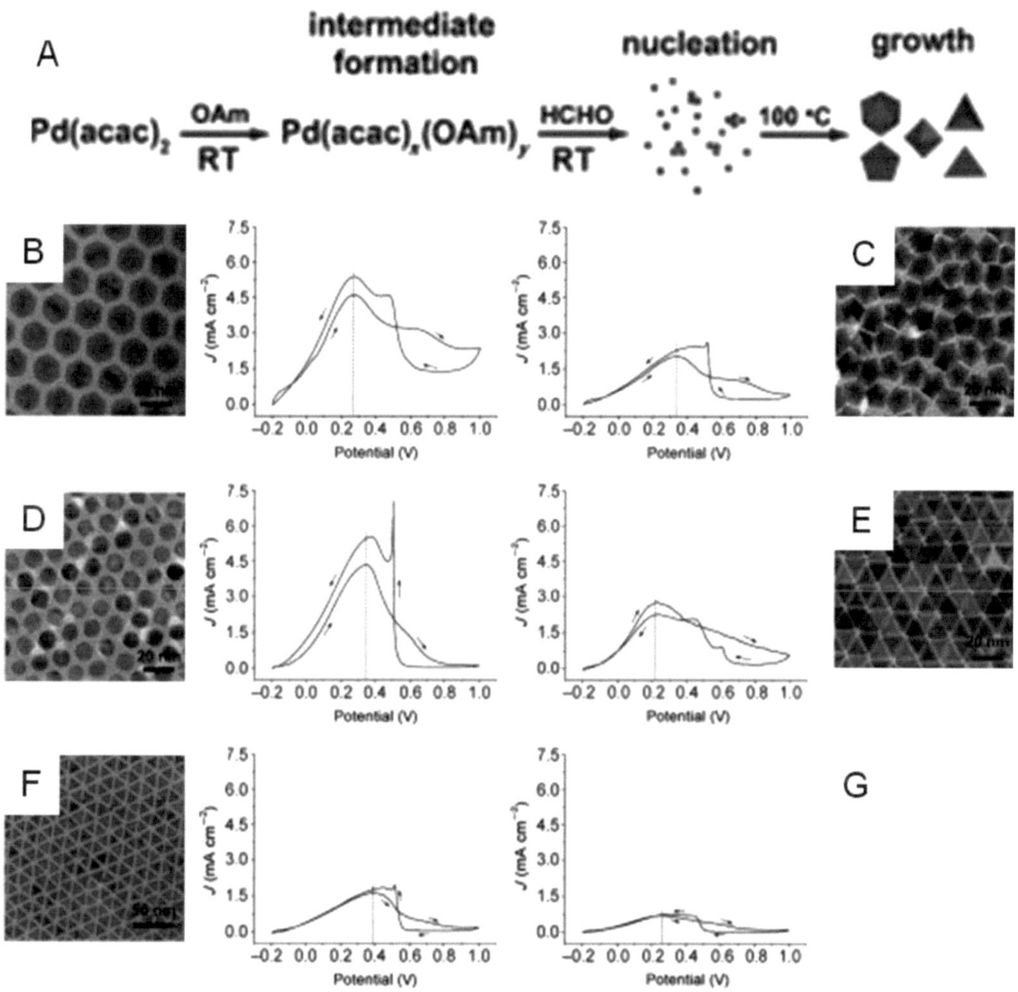

Figure 4. (A) Proposed mechanism for formation of Pd NCs. (B–F) TEM images and (B–G) cyclic voltammograms of Pd catalysts with different shapes in 0.1 M $HClO_4$ and 2 M HCOOH solution at a scan rate of 50 mV s^{-1}. (B) Icosahedral Pd, (C) decahedral Pd, (D) octahedral Pd, (E) tetrahedral Pd, (F) triangular platelike Pd, and (G) commercial Pd. Adapted with permission from Ref. [30]. Copyright 2011 Wiley-VCH Verlag GmbH & Co. KGaA, Weinheim.

system. Growth process was observed at elevated temperature and the final morphology was decided by the co-action of surface energy, elastic strain energy and twin boundary energy. The maximum current densities of the prepared Pd NCs were in the order of icosahedral ≈octahedral > tetrahedral > decahedral > triangular plates toward formic acid oxidation reaction.

Among bimetallic Pd-base nanocatalysts for electrochemical reactions, PdCu NCs have received a great deal of attention due to the low cost and good performance. Yin et al. prepared PdCu NPs by emulsion-assisted ternary ethylene glycol system, which showed superior methanol oxidation activity than Pd in alkaline solution [86]. Mao et al. also synthesized PdCu NCs with tunable compositions in octadecylamine. $Pd_{0.5}Cu_{0.5}$ exhibited the highest mass activity for ethanol electro-oxidation in alkaline solution [8]. Gao et al. prepared PdCu nanocubes in OAm and the PdCu nanocubes showed 2 times mass activity than commercial Pt/C catalysts [87].

Although numerous Pd-based nanocomposites have been proven to show excellent electrocatalytic activities, but the stability is still a major issue for Pd-based electrocatalysts even in alkaline solution. Better performance of Pd-based nanocomposites is essential for the substitute of Pt in the application of fuel cell.

3.2. Other noble metal-based nanocomposites

Generally speaking, besides Pt and Pd, noble metals also contain gold (Au), silver (Ag), rhodium (Rh), ruthenium (Ru), osmium (Os) and iridium (Ir). They are always used to modify the electronic structure of Pd and Pt in the fuel cell application and these nanocomposites will not be discussed below [88–92].

Since Haruta et al. reported the oxidation of CO by supported Au at low temperature, Au-based nanocomposites have received a great deal of attention in catalytic field [93]. Surfactants being on the surface may decrease the electrochemical activities. Yin et al. used reduced grapheme oxide (rGO) as reductant and support to confine the growth of Au so they prepared surfactant-free Au/rGO hybrids for ORR in alkaline solution [94]. In this method, utilizing the electrostatic interaction between negative-charged rGO and positive Au(III) ions also the coordination effects between heteroatoms at the defects of rGO and Au(III) ions, Au clusters with a diameter of 1.8 nm were reduced and anchored by rGO. Although the catalytic activity was not as good as Pt/C, Au/rGO exhibited relatively high onset potential, methanol tolerance, and extraordinary stability (**Figure 5**).

By introducing other metal components, the cost of Au can be reduced, the modification of crystallography and electronic structure may increase the catalytic performance. With abundant tip, edge and surface atoms, 2D Au-based nanocomposites have attracted increasing attention. Xu et al. prepared AuCu bimetallic NCs with tunable composition via a simple wet-chemical method [95]. The 2D AuCu triangular porous nanoprisms exhibited high electrocatalytic activity. The mass activities of Au_1Cu_1 for ethylene glycol and glycerol electrooxidation reactions were 3.0- and 3.9-fold improvements over those of pure Au, respectively.

Owing to the poor stability in acidic solution, ORR catalyzed by Ag was always demonstrated in alkaline solution [96, 97]. Tammeveski et al. prepared the Ag-based nanocomposites by sputter deposition of Ag on the surface of multi-walled carbon nanotube [96]. They analyzed

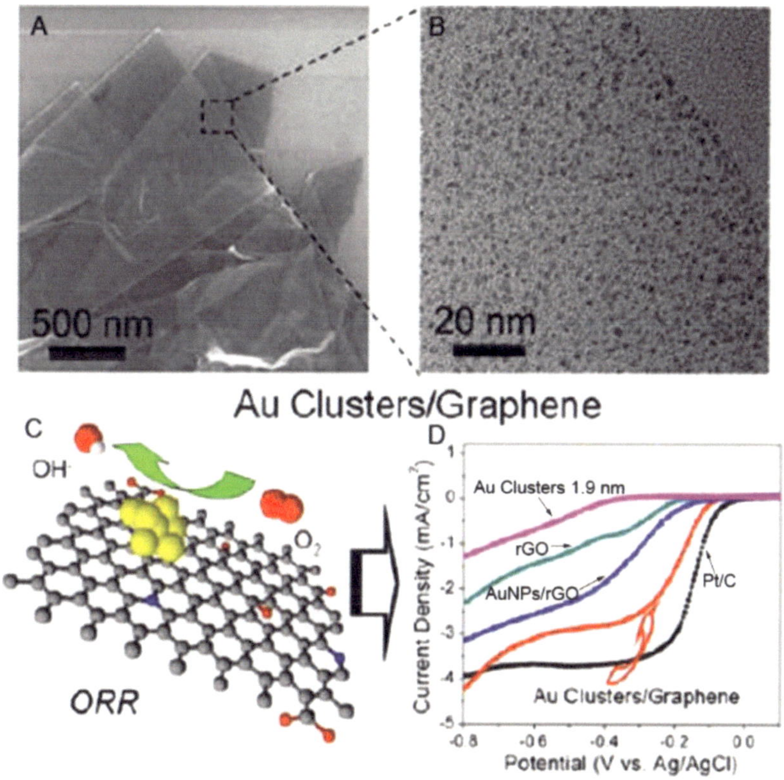

Figure 5. (A) SEM, (B) TEM images of as-prepared Au/rGO hybrids. (C) Scheme of the ORR happened on Au/rGO hybrids. (D) RDE curves of commercial Pt/C, Au/rGO hybrids, Au NP/rGO hybrids, rGO sheets, Au clusters in O_2-saturated 0.1 M KOH at a scanning rate of 50 mV/s at 1600 rpm. Reprinted with permission from Ref. [94] Copyright 2012 American Chemistry Society.

rotating disk electrode (RDE) data by Koutecky-Levich equation and demonstrated that O_2 was reduced by a four-electron pathway. Gupta et al. studied the influence of anisotropic Ag on ORR activity in alkaline media [97]. They prepared Ag NPs in ethylene glycol with different ratio of anisotropic NPs by adding different amount of $Cu(CH_3COO)_2 \cdot H_2O$, which acted as etchant. They reported that sample containing higher ratio of anisotropic Ag NPs showed better ORR performance.

Zheng et al. studied the size effect of Ir/C toward hydrogen oxidation/evolution reaction (HOR/HER) in alkaline solution [98]. They prepared Ir/C with different sizes by heat treating commercial Ir/C at different temperature. Higher temperature resulted in larger size. The particle sizes of Ir/C changed from 3 to 12 nm. Different from the law of Pt NCs with the same size range, larger Ir/C had higher HOR/HER activities. Qin et al. utilized CeO_2 and C to disperse Ir NPs; the mass and specific activities of Ir/CeO_2-C toward HOR in alkaline conditions are 2.8- and 1.8-fold higher than Ir/C. They claimed that the addition of CeO_2 decreased the agglomeration extent of Ir NPs, and the particle size changed from 1.1 to 3.4 nm after 2000 potential cycles [99].

Lin et al. reported a salt-templated method to prepare Rh (1.6 nm)/ultrathin carbon nanosheets, which displayed a comparable ORR activity and better durability to commercial Pt/C under basic conditions [100]. The salt template was green and easy to remove. In this approach, Na_2SO_4 particles acted as template and poured into prefabricated rhodium oleate. During the subsequent heat-treatment under N_2, Rh NPs were reduced and anchored on the ultrathin carbon nanosheets transformed from oleate. After water rinsing, Na_2SO_4 particles were removed and Rh/ultrathin carbon nanosheets were obtained.

Owing to the strong Ru—O bond strength, metallic Ru show poor ORR activity. The addition of Se can transfer charge from Ru to Se, which reduces the oxygen binding energy on Ru and increases the ORR activity [3]. Cao et al. prepared Se-modified Ru NPs by heating the Ru black and Se in xylene under refluxing condition with Ar bubbling [101]. The ORR activity of Se-modified Ru NPs in H_2SO_4 was higher than that of clean and Se-modified Ru disk. Zaikovskii et al. fabricated Se-modified Ru/C by reacting Ru/C NPs with SeO_2 followed by annealing [102]. They prepared RuSe/C with different Se coverage by tuning the ratio of Se/Ru. Higher amount of Se caused lower extent of Ru oxidation, thus resulting in higher ORR activity. However, Ru is the active center for ORR and blocked shell of RuSe decreased ORR activity. Neergat et al. synthesized RuSe NPs by reducing $RuCl_3{\cdot}xH_2O$ and SeO_2 by $NaBH_4$ in water at 80°C [103]. They investigated the ORR performance of several nanocatalysts in the presence of methanol, and the ORR activities were in the order of Ir < Pt < Rh < IrSe < RhSe < RuSe.

4. Conclusions and outlooks

This chapter has demonstrated the significant developments that have been accomplished in the noble metal-based nanocomposites for fuel cell. For the past decade, great progress has been made in developing Pt- and Pd-based nanocomposites for electrochemical reactions in fuel cell. Surface electronic structure of Pt and Pd can be modified by incorporating a second and/or third metal, which significantly influences the electrochemical performance. Constructing Pt ultrathin structure is an effective strategy to improve the utilization of Pt and the specific activity of Pt-based nanocatalysts. Significant progresses have been made on Ir-, Rh-, and Ru-based nanocatalysts. However, their activities are still not comparable to those of Pt- and Pd-based catalysts, especially those of Pt-based nanocomposites. For the practical application, fuel cell is still a long way to go.

At present, most Pt-based catalysts are characterized by RDE technique, which is a fast screening technique. They need to be evaluated in a fuel cell environment before large-scale application. As a result, simple preparation method is very important for scale production of catalysts. Wet-chemical method is suitable for scale preparation. Researchers need to optimize the fabrication technology of the state-of-the-art noble metal-based electrocatalysts, and find a simple, green and low-cost preparation method. Developing in situ characterization technique for the electrochemical process will provide a basis for mechanism exploration. According to the excellent catalytic performance of binary noble metal structures, general preparation strategies and catalytic property studies for tri- and multi-metallic nanocomposites are urgent.

Author details

Hongpan Rong, Shuping Zhang, Sajid Muhammad and Jiatao Zhang*

*Address all correspondence to: zhangjt@bit.edu.cn

Beijing Key Laboratory of Construction-Tailorable Advanced Functional Materials and Green Applications, School of Materials, Beijing Institute of Technology, Beijing, PR China

References

[1] Kakati N, Maiti J, Lee SH. Anode catalysts for direct methanol fuel cells in acidic media: Do we have any alternative for Pt or Pt-Ru? Chemical Reviews. 2014;**114**:12397-12429. DOI: 10.1021/cr400389f

[2] Elbert K, Hu J, Ma Z. Elucidating hydrogen oxidation/evolution kinetics in base and acid by enhanced activities at the optimized Pt shell thickness on the Ru core. ACS Catalysis. 2015;**5**:6764-6772. DOI: 10.1021/acscatal.5b01670

[3] Shao M, Chang Q, Dodelet J-P. Recent advances in electrocatalysts for oxygen reduction reaction. Chemical Reviews. 2016;**116**:3594-3657. DOI: 10.1021/acs.chemrev.5b00462

[4] Vesborg PCK, Jaramillo TF. Addressing the terawatt challenge: Scalability in the supply of chemical elements for renewable energy. RSC Advances. 2012;**2**:7933-7947. DOI: 10.1039/c2ra20839c

[5] Rong H, Cai S, Niu Z. Composition-dependent catalytic activity of bimetallic nanocrystals: AgPd-catalyzed hydrodechlorination of 4-chlorophenol. ACS Catalysis. 2013;**3**:1560-1563. DOI: 10.1021/cs400282a

[6] Rong H, Niu Z, Zhao Y. Structure evolution and associated catalytic properties of PtSn bimetallic nanoparticles. Chemistry—A European Journal. 2015;**21**:12034-12041. DOI: 10.1002/chem.201501442

[7] Cai S, Rong H, Yu X. Room temperature activation of oxygen by monodispersed metal nanoparticles: Oxidative dehydrogenative coupling of anilines for azobenzene syntheses. ACS Catalysis. 2013;**3**:478-486. DOI: 10.1021/cs300707y

[8] Mao J, Liu Y, Chen Z. Bimetallic Pd-cu nanocrystals and their tunable catalytic properties. Chemical Communications. 2014;**50**:4588-4591. DOI: 10.1039/c4cc01051e

[9] Niu Z, Wang D, Yu R. Highly branched Pt-Ni nanocrystals enclosed by stepped surface for methanol oxidation. Chemical Science. 2012;**3**:1925-1929. DOI: 10.1039/c2sc00004k

[10] Ge J, He D, Chen W. Atomically dispersed Ru on ultrathin Pd nanoribbons. Journal of the American Chemical Society. 2016;**138**:13850-13853. DOI: 10.1021/jacs.6b09246

[11] Wu Y, Cai S, Wang D. Syntheses of water-soluble octahedral, truncated octahedral, and cubic Pt-Ni nanocrystals and their structure-activity study in model hydrogenation reactions. Journal of the American Chemical Society. 2012;**134**:8975-8981. DOI: 10.1021/ja302606d

[12] Bu L, Ding J, Guo S. A general method for multimetallic platinum alloy nanowires as highly active and stable oxygen reduction catalysts. Advanced Materials. 2015;**27**:7204. DOI: 10.1002/adma.201502725

[13] Liu H, An W, Li Y. In situ probing of the active site geometry of ultrathin nanowires for the oxygen reduction reaction. Journal of the American Chemical Society. 2015;**137**: 12597-12609. DOI: 10.1021/jacs.5b07093

[14] Bu L, Guo S, Zhang X. Surface engineering of hierarchical platinum-cobalt nanowires for efficient electrocatalysis. Nature Communications. 2016;**7**. DOI: 10.1038/ncomms11850

[15] Bu L, Zhang N, Guo S. Biaxially strained PtPb/Pt core/shell nanoplate boosts oxygen reduction catalysis. Science. 2016;**354**:1410-1414. DOI: 10.1126/science.aah6133

[16] Li M, Zhao Z, Cheng T. Ultrafine jagged platinum nanowires enable ultrahigh mass activity for the oxygen reduction reaction. Science. 2016;**354**:1414-1419. DOI: 10.1126/science.aaf9050

[17] Jiang K, Zhao D, Guo S. Efficient oxygen reduction catalysis by subnanometer Pt alloy nanowires. Science Advances. 2017;**3**. DOI: 10.1126/sciadv.1601705

[18] Chen C, Kang Y, Huo Z. Highly crystalline multimetallic nanoframes with three-dimensional electrocatalytic surfaces. Science. 2014;**343**:1339-1343. DOI: 10.1126/science.1249061

[19] He DS, He D, Wang J. Ultrathin icosahedral Pt-enriched nanocage with excellent oxygen reduction reaction activity. Journal of the American Chemical Society. 2016;**138**:1494-1497. DOI: 10.1021/jacs.5b12530

[20] Ding J, Zhu X, Bu L. Highly open rhombic dodecahedral PtCu nanoframes. Chemical Communications. 2015;**51**:9722-9725. DOI: 10.1039/c5cc03190g

[21] Wu Y, Wang D, Zhou G. Sophisticated construction of Au islands on Pt-Ni: An ideal trimetallic nanoframe catalyst. Journal of the American Chemical Society. 2014;**136**: 11594-11597. DOI: 10.1021/ja5058532

[22] Wang D, Zhao P, Li Y. General preparation for Pt-based alloy nanoporous nanoparticles as potential nanocatalysts. Scientific Reports. 2011;**1**. DOI: 10.1038/srep00037

[23] Mao J, Chen Y, Pei J. Pt-M (M = Cu, Fe, Zn, etc.) bimetallic nanomaterials with abundant surface defects and robust catalytic properties. Chemical Communications. 2016;**52**:5985-5988. DOI: 10.1039/c6cc02264b

[24] Rong H, Mao J, Xin P. Kinetically controlling surface structure to construct defect-rich intermetallic nanocrystals: effective and stable catalysts. Advanced Materials. 2016;**28**:2540. DOI: 10.1002/adma.201504831

[25] Yu X, Li L, Su Y. Platinum-copper nanoframes: One-pot synthesis and enhanced electrocatalytic activity. Chemistry—A European Journal. 2016;**22**:4960-4965. DOI: 10.1002/chem.201600079

[26] Liu H-l, Nosheen F, Wang X. Noble metal alloy complex nanostructures: Controllable synthesis and their electrochemical property. Chemical Society Reviews. 2015;**44**:3056-3078. DOI: 10.1039/c4cs00478g

[27] Huang X, Zhao Z, Cao L. High-performance transition metal-doped Pt3Ni octahedra for oxygen reduction reaction. Science. 2015;**348**:1230-1234. DOI: 10.1126/science.aaa8765

[28] Bu L, Shao Q, Bin E. PtPb/PtNi intermetallic core/atomic layer shell octahedra for efficient oxygen reduction electrocatalysis. Journal of the American Chemical Society. 2017;**139**:9576-9582. DOI: 10.1021/jacs.7b03510

[29] Liu X, Wang W, Li H. One-pot protocol for bimetallic Pt/Cu hexapod concave nanocrystals with enhanced electrocatalytic activity. Scientific Reports. 2013;**3**. DOI: 10.1038/srep01404

[30] Niu Z, Peng Q, Gong M. Oleylamine-mediated shape evolution of palladium nanocrystals. Angewandte Chemie-International Edition. 2011;**50**:6315-6319. DOI: 10.1002/anie.201100512

[31] Wu B, Zheng N. Surface and interface control of noble metal nanocrystals for catalytic and electrocatalytic applications. Nano Today. 2013;**8**:168-197. DOI: 10.1016/j.nantod.2013.02.006

[32] Mao J, Chen W, He D. Design of ultrathin Pt-Mo-Ni nanowire catalysts for ethanol electrooxidation. Science Advances. 2017;**3**. DOI: 10.1126/sciadv.1603068

[33] Wang W, Wang D, Liu X. Pt-Ni nanodendrites with high hydrogenation activity. Chemical Communications. 2013;**49**:2903-2905. DOI: 10.1039/c3cc40503f

[34] Nosheen F, Zhang Z, Xiang G. Three-dimensional hierarchical Pt-Cu superstructures. Nano Research. 2015;**8**:832-838. DOI: 10.1007/s12274-014-0565-1

[35] Ding J, Bu L, Guo S. Morphology and phase controlled construction of Pt-Ni nanostructures for efficient electrocatalysis. Nano Letters. 2016;**16**:2762-2767. DOI: 10.1021/acs.nanolett.6b00471

[36] Shan A, Cheng M, Fan H. NiPt hollow nanocatalyst: Green synthesis, size control and electrocatalysis. Progress in Natural Science-Materials International. 2014;**24**:175-178. DOI: 10.1016/j.pnsc.2014.03.013

[37] Zhang Y, Bu L, Jiang K. Concave Pd-Pt core-shell nanocrystals with ultrathin Pt shell feature and enhanced catalytic performance. Small. 2016;**12**:706-712. DOI: 10.1002/smll.201502755

[38] Huang X, Tang S, Mu X. Freestanding palladium nanosheets with plasmonic and catalytic properties. Nature Nanotechnology. 2011;**6**:28-32. DOI: 10.1038/nnano.2010.235

[39] Dai Y, Mu X, Tan Y. Carbon monoxide-assisted synthesis of single-crystalline Pd tetrapod Nanocrystals through hydride formation. Journal of the American Chemical Society. 2012;**134**:7073-7080. DOI: 10.1021/ja3006429

[40] Xiong Y, Cai H, Wiley BJ. Synthesis and mechanistic study of palladium nanobars and nanorods. Journal of the American Chemical Society. 2007;**129**:3665-3675. DOI: 10.1021/ja0688023

[41] Duan H, Yan N, Yu R. Ultrathin rhodium nanosheets. Nature Communications. 2014;**5**. DOI: 10.1038/ncomms4093

[42] Xia X, Zeng J, McDearmon B. Silver nanocrystals with concave surfaces and their optical and surface-enhanced Raman scattering properties. Angewandte Chemie International Edition. 2011;**50**:12542-12546. DOI: 10.1002/anie.201105200

[43] Mahmoud A, Tabor CE, El-Sayed MA. A new catalytically active colloidal platinum nanocatalyst: The multiarmed nanostar single crystal. Journal of the American Chemical Society. 2008;**130**:4590-4591. DOI: 10.1021/ja710646t

[44] Wu Y, Wang D, Niu Z. A strategy for designing a concave Pt-Ni alloy through controllable chemical etching. Angewandte Chemie-International Edition. 2012;**51**:12524-12528. DOI: 10.1002/anie.201207491

[45] Bai Q, Li D, He L. Solvent-free selective hydrogenation of o-chloronitrobenzene to o-chloroaniline over alumina supported Pt nanoparticles. Progress in Natural Science: Materials International. 2015;**25**:179-184. DOI: https://doi.org/10.1016/j.pnsc.2015.05.005

[46] Lin L, Zhou W, Gao R. Low-temperature hydrogen production from water and methanol using Pt/alpha-MoC catalysts. Nature. 2017;**544**:80. DOI: 10.1038/nature21672

[47] Li N, Tang S, Meng X. Preparation of Pt-GO composites with high-number-density Pt nanoparticles dispersed uniformly on GO nanosheets. Progress in Natural Science-Materials International. 2016;**26**:139-144. DOI: 10.1016/j.pnsc.2016.03.011

[48] Wu J, Yang H. Platinum-based oxygen reduction electrocatalysts. Accounts of Chemical Research. 2013;**46**:1848-1857. DOI: 10.1021/ar300359w

[49] Gan L, Heggen M, Rudi S. Core–shell compositional fine structures of dealloyed PtxNi1−x nanoparticles and their impact on oxygen reduction catalysis. Nano Letters. 2012;**12**:5423-5430. DOI: 10.1021/nl302995z

[50] Koh S, Strasser P. Electrocatalysis on bimetallic surfaces: Modifying catalytic reactivity for oxygen reduction by voltammetric surface dealloying. Journal of the American Chemical Society. 2007;**129**:12624-12625. DOI: 10.1021/ja0742784

[51] Neyerlin KC, Srivastava R, Yu C. Electrochemical activity and stability of dealloyed Pt–Cu and Pt–Cu–Co electrocatalysts for the oxygen reduction reaction (ORR). Journal of Power Sources. 2009;**186**:261-267. DOI: https://doi.org/10.1016/j.jpowsour.2008.10.062

[52] Tian N, Zhou Z-Y, Sun S-G. Synthesis of tetrahexahedral platinum nanocrystals with high-index facets and high electro-oxidation activity. Science. 2007;**316**:732-735. DOI: 10.1126/science.1140484

[53] Vitos L, Ruban AV, Skriver HL. The surface energy of metals. Surface Science. 1998;**411**:186-202. DOI: https://doi.org/10.1016/S0039-6028(98)00363-X

[54] Wang C, Hou Y, Kim J. A general strategy for synthesizing FePt nanowires and nanorods. Angewandte Chemie-International Edition. 2007;**46**:6333-6335. DOI: 10.1002/anie.200702001

[55] Zhang N, Feng Y, Zhu X. Superior bifunctional liquid fuel oxidation and oxygen reduction electrocatalysis enabled by PtNiPd core shell nanowires. Advanced Materials. 2017;**29**. DOI: 10.1002/adma.201603774

[56] Jiang K, Shao Q, Zhao D. Phase and composition tuning of 1D platinum-nickel nanostructures for highly efficient electrocatalysis. Advanced Functional Materials. 2017;**27**. DOI: 10.1002/adfm.201700830

[57] Zhang N, Guo S, Zhu X. Hierarchical Pt/PtxPb core/shell nanowires as efficient catalysts for electrooxidation of liquid fuels. Chemistry of Materials. 2016;**28**:4447-4452. DOI: 10.1021/acs.chemmater.6b01642

[58] Zhang N, Bu L, Guo S. Screw thread-like platinum-copper nanowires bounded with high index facets for efficient electrocatalysis. Nano Letters. 2016;**16**:5037-5043. DOI: 10.1021/acs.nanolett.6b01825

[59] Yang L, Vukmirovic MB, Su D. Tuning the catalytic activity of Ru@Pt core–shell nanoparticles for the oxygen reduction reaction by varying the shell thickness. The Journal of Physical Chemistry C. 2013;**117**:1748-1753. DOI: 10.1021/jp309990e

[60] Oezaslan M, Hasche F, Strasser P. Pt-based core-shell catalyst architectures for oxygen fuel cell electrodes. Journal of Physical Chemistry Letters. 2013;**4**:3273-3291. DOI: 10.1021/jz4014135

[61] Brodsky CN, Young AP, Ng KC. Electrochemically induced surface metal migration in well-defined core–shell nanoparticles and its general influence on electrocatalytic reactions. ACS Nano. 2014;**8**:9368-9378. DOI: 10.1021/nn503379w

[62] Ji M, Li X, Wang H. Versatile synthesis of yolk/shell hybrid nanocrystals via ion-exchange reactions for novel metal/semiconductor and semiconductor/semiconductor conformations. Nano Research. 2017;**10**:2977-2987. DOI: 10.1007/s12274-017-1508-4

[63] Wu Y, Wang D, Li Y. Understanding of the major reactions in solution synthesis of functional nanomaterials. Science China-Materials. 2016;**59**:938-996. DOI: 10.1007/s40843-016-5112-0

[64] Chen Y, Yu Z, Chen Z. Controlled one-pot synthesis of RuCu nanocages and Cu@Ru nanocrystals for the regioselective hydrogenation of quinoline. Nano Research. 2016;**9**:2632-2640. DOI: 10.1007/s12274-016-1150-6

[65] Zhang J, Tang Y, Weng L. Versatile strategy for precisely tailored Core@Shell nanostructures with single shell layer accuracy: The case of metallic shell. Nano Letters. 2009;**9**:4061-4065. DOI: 10.1021/nl902263h

[66] Ji M, Xu M, Zhang W. Structurally well-defined au@Cu2-xS core-shell nanocrystals for improved cancer treatment based on enhanced photothermal efficiency. Advanced Materials. 2016;**28**:3094-3101. DOI: 10.1002/adma.201503201

[67] Huang L, Zheng J, Huang L. Controlled synthesis and flexible self-assembly of monodisperse Au@semiconductor core/shell hetero-nanocrystals into diverse superstructures. Chemistry of Materials. 2017;**29**:2355-2363. DOI: 10.1021/acs.chemmater.7b00046

[68] Zhang J, Tang Y, Lee K. Nonepitaxial growth of hybrid core-shell nanostructures with large lattice mismatches. Science. 2010;**327**:1634-1638. DOI: 10.1126/science.1184769

[69] Cai S, Hu H, Li H. Design of multi-shell Fe_2O_3@MnOx@CNTs for the selective catalytic reduction of NO with NH_3: Improvement of catalytic activity and SO_2 tolerance. Nanoscale. 2016;**8**:3588-3598. DOI: 10.1039/c5nr08701e

[70] Ghosh T, Vukmirovic MB, DiSalvo FJ. Intermetallics as novel supports for Pt monolayer O-2 reduction electrocatalysts: Potential for significantly improving properties. Journal of the American Chemical Society. 2010;**132**:906. DOI: 10.1021/ja905850c

[71] Xing Y, Cai Y, Vukmirovic MB. Enhancing oxygen reduction reaction activity via Pd—Au alloy sublayer mediation of Pt monolayer electrocatalysts. The Journal of Physical Chemistry Letters. 2010;**1**:3238-3242. DOI: 10.1021/jz101297r

[72] Chen G, Kuttiyiel KA, Su D. Oxygen reduction kinetics on Pt monolayer shell highly affected by the structure of bimetallic AuNi cores. Chemistry of Materials. 2016;**28**:5274-5281. DOI: 10.1021/acs.chemmater.6b00500

[73] Tian X, Tang H, Luo J. High-performance core–shell catalyst with nitride nanoparticles as a core: Well-defined titanium copper nitride coated with an atomic Pt layer for the oxygen reduction reaction. ACS Catalysis. 2017;**7**:3810-3817. DOI: 10.1021/acscatal.7b00366

[74] Deogratias N, Ji M, Zhang Y. Core@shell sub-ten-nanometer noble metal nanoparticles with a controllable thin Pt shell and their catalytic activity towards oxygen reduction. Nano Research. 2015;**8**:271-280. DOI: 10.1007/s12274-014-0664-z

[75] Lee YW, Kim M, Kim ZH. One-step synthesis of Au@Pd core-shell nanooctahedron. Journal of the American Chemical Society. 2009;**131**:17036, 10.1021/ja905603p

[76] Yan J-M, Zhang X-B, Akita T. One-step seeding growth of magnetically recyclable Au@Co core-shell nanoparticles: Highly efficient catalyst for hydrolytic dehydrogenation of ammonia borane. Journal of the American Chemical Society. 2010;**132**:5326. DOI: 10.1021/ja910513h

[77] Feng Q, Zhao S, Wang Y. Isolated single-atom Pd sites in intermetallic nanostructures: High catalytic selectivity for semihydrogenation of alkynes. Journal of the American Chemical Society. 2017;**139**:7294-7301. DOI: 10.1021/jacs.7b01471

[78] Chen M, Tang S, Guo Z. Core-shell Pd@Au nanoplates as theranostic agents for in-vivo photoacoustic imaging, CT imaging, and photothermal therapy. Advanced Materials. 2014;**26**:8210-8216. DOI: 10.1002/adma.201404013

[79] Niu Z, Peng Q, Zhuang Z. Evidence of an oxidative-addition-promoted Pd-leaching mechanism in the Suzuki reaction by using a Pd-nanostructure design. Chemistry—A European Journal. 2012;**18**:9813-9817. DOI: 10.1002/chem.201201224

[80] Yin Z, Lin L, Ma D. Construction of Pd-based nanocatalysts for fuel cells: Opportunities and challenges. Catalysis Science & Technology. 2014;**4**:4116-4128. DOI: 10.1039/c4cy00760c

[81] Koenigsmann C, Sutter E, Adzic RR. Size- and composition-dependent enhancement of electrocatalytic oxygen reduction performance in ultrathin palladium gold ($Pd_{1-x}Au_x$) nanowires. The Journal of Physical Chemistry C. 2012;**116**:15297-15306. DOI: 10.1021/jp306034d

[82] Shao M. Palladium-based electrocatalysts for hydrogen oxidation and oxygen reduction reactions. Journal of Power Sources. 2011;**196**:2433-2444. DOI: 10.1016/j.jpowsour.2010.10.093

[83] Zhang H, Jin M, Xia Y. Enhancing the catalytic and electrocatalytic properties of Pt-based catalysts by forming bimetallic nanocrystals with Pd. Chemical Society Reviews. 2012;**41**:8035-8049. DOI: 10.1039/c2cs35173k

[84] Kondo S, Nakamura M, Maki N. Active sites for the oxygen reduction reaction on the low and high index planes of palladium. The Journal of Physical Chemistry C. 2009;**113**:12625-12628. DOI: 10.1021/jp904278b

[85] Shao M, Yu T, Odell JH. Structural dependence of oxygen reduction reaction on palladium nanocrystals. Chemical Communications. 2011;**47**:6566-6568. DOI: 10.1039/C1CC11004G

[86] Yin Z, Zhou W, Gao Y. Supported Pd–Cu bimetallic nanoparticles that have high activity for the electrochemical oxidation of methanol. Chemistry–A European Journal. 2012;**18**:4887-4893. DOI: 10.1002/chem.201103674

[87] Gao Q, Ju Y-M, An D. Shape-controlled synthesis of monodisperse PdCu nanocubes and their electrocatalytic properties. ChemSusChem. 2013;**6**:1878-1882. DOI: 10.1002/cssc.201300404

[88] Wang Y, Xiong Z, Xia Y. Branched PdAu nanowires with superior electrocatalytic formic acid oxidation activities. RSC Advances. 2017;**7**:40462-40469. DOI: 10.1039/c7ra02115a

[89] Huang JJ, Yang H, Huang QH. Methanol oxidation on carbon-supported Pt-Os bimetallic nanoparticle electrocatalysts. Journal of the Electrochemical Society. 2004;**151**:A1810-A1815. DOI: 10.1149/1.1785932

[90] Saleem F, Ni B, Yong Y. Ultra-small tetrametallic Pt-Pd-Rh-Ag nanoframes with tunable behavior for direct formic acid/methanol oxidation. Small. 2016;**12**:5261-5268. DOI: 10.1002/smll.201601299

[91] Zhang T, Li S-C, Zhu W. Shape-tunable Pt-Ir alloy nanocatalysts with high performance in oxygen electrode reactions. Nanoscale. 2017;**9**:1154-1165. DOI: 10.1039/c6nr08359e

[92] Wu Y-N, Liao S-J, Guo H-F. Ultralow platinum-loading PtPdRu@PtRuIr/C catalyst with excellent CO tolerance and high performance for the methanol oxidation reaction. Rare Metals. 2014;**33**:337-342. DOI: 10.1007/s12598-013-0063-8

[93] Masatake H, Tetsuhiko K, Hiroshi S. Novel gold catalysts for the oxidation of carbon monoxide at a temperature far below 0°C. Chemistry Letters. 1987;**16**:405-408. DOI: 10.1246/cl.1987.405

[94] Yin H, Tang H, Wang D. Facile synthesis of surfactant-free Au cluster/graphene hybrids for high-performance oxygen reduction reaction. ACS Nano. 2012;**6**:8288-8297. DOI: 10.1021/nn302984x

[95] Xu H, Yan B, Wang J. Self-supported porous 2D AuCu triangular nanoprisms as model electrocatalysts for ethylene glycol and glycerol oxidation. Journal of Materials Chemistry A. 2017;**5**:15932-15939. DOI: 10.1039/c7ta04598k

[96] Tammeveski L, Erikson H, Sarapuu A. Electrocatalytic oxygen reduction on silver nanoparticle/multi-walled carbon nanotube modified glassy carbon electrodes in alkaline solution. Electrochemistry Communications. 2012;**20**:15-18. DOI: 10.1016/j.elecom.2012.04.003

[97] Gupta RK, Verma AD, Sinha I. Oxygen reduction reaction on anisotropic silver nanoparticles in alkaline media. Chemical Physics Letters. 2017;**680**:6-9. DOI: 10.1016/j.cplett.2017.05.031

[98] Zheng J, Zhuang Z, Xu B. Correlating hydrogen oxidation/evolution reaction activity with the minority weak hydrogen-binding sites on Ir/C catalysts. ACS Catalysis. 2015;**5**:4449-4455. DOI: 10.1021/acscatal.5b00247

[99] Qin B, Yu H, Chi J. A novel Ir/CeO2-C nanoparticle electrocatalyst for the hydrogen oxidation reaction of alkaline anion exchange membrane fuel cells. RSC Advances. 2017;**7**:31574-31581. DOI: 10.1039/c7ra03675b

[100] Lin C, Wu G, Li H. Rh nanoparticles supported on ultrathin carbon nanosheets for high-performance oxygen reduction reaction and catalytic hydrogenation. Nanoscale. 2017;**9**:1834-1839. DOI: 10.1039/c6nr09739a

[101] Cao DX, Wieckowski A, Inukai J. Oxygen reduction reaction on ruthenium and rhodium nanoparticles modified with selenium and sulfur. Journal of the Electrochemical Society. 2006;**153**:A869-A874. DOI: 10.1149/1.2180709

[102] Zaikovskii VI, Nagabhushana KS, Kriventsov VV. Synthesis and structural characterization of Se-modified carbon-supported ru nanoparticles for the oxygen reduction reaction. Journal of Physical Chemistry B. 2006;**110**:6881-6890. DOI: 10.1021/jp056715b

[103] Neergat M, Gunasekar V, Singh RK. Oxygen reduction reaction and peroxide generation on Ir, Rh, and their selenides—A comparison with Pt and RuSe. Journal of the Electrochemical Society. 2011;**158**:B1060-B1066. DOI: 10.1149/1.3604744

Chapter 17

Capillary Electrophoresis in Nanotechnologies versus Nanotechnologies in Capillary Electrophoresis

Vojtech Adam and Marketa Vaculovicova

Additional information is available at the end of the chapter

http://dx.doi.org/10.5772/intechopen.72015

Abstract

Nanomaterials are attracting an interest of many researches. All this attention is due the unique physical and chemical properties of nanomaterials differing significantly from the bulk materials mainly due to their size in range of nanometers. Capillary electrophoresis (CE) is a powerful, well-established analytical technique that provides numerous valuable benefits over other separation methods including high-performance liquid chromatography. The connection between CE and nanotechnology can be approached by two strategies: (i) CE analysis of nanomaterials and (ii) nanomaterials for CE improvement. The first perspective focuses on uses of CE as a method for characterization employed during nanomaterial production and modification as well as for monitoring their properties and interactions with other molecules. The second viewpoint deals with applications of nanomaterials for improving CE performance, mainly by enhancing efficiency of separation using nanomaterials as a stationary or pseudo-stationary phase and by enhancing detection sensitivity and/or selectivity in both optical and electrochemical detection. Moreover, applications of nanomaterials for sample preparation before CE analysis will be mentioned. This chapter aims at highlighting the symbiosis of CE and nanotechnology as a combination of modern, progressive field with well-known and reliable analytical method.

Keywords: capillary electrophoresis, nanomaterials, separation efficiency, sensitivity

1. Introduction

The very first nano-scientist was a Roman potter, who made the Lycurgus Cup as the oldest known application of nanomaterials (fifth-fourth century B.C.). This cup was made from so-called "gold-ruby glass" containing tiny gold droplets (5–60 nm in size). Therefore, the glass appeared green in daylight (reflected light), but red when light was transmitted from the inside of the vessel [1]. In spite of the fact that we do not know the name of the potter, which is based on nano-research without the knowledge of nanoparticles, Richard Feynman opened "nano-window of twenty-first century" with his lecture "There's plenty of room at the bottom". This lecture came to be looked upon as the starting point of nanoscience as we are already living in [2]. In 1974, Taniguchi used the term "Nanotechnology" for the first time. This term was defined as the technology, where dimensions, within the range of 0.1–100 nm, play a key role. At the nanolevel, gravity is less an issue while the strength of materials is a bigger one and also quantum size effect is a key aspect. Due to the unique size-dependent spectroscopic, electronic, and thermal features and also chemical properties, and ability to be functionalized, arising from the small sizes and large surface-to-volume ratio, nanomaterials found their applications not only in electronics, physics, and engineering but also in natural sciences. Although nanomaterials are greatly affecting numerous scientific fields, it can be perceived differently. In chemistry, the range of sizes has been associated with colloid solutions, micelles, polymeric molecules, and also large molecules, or aggregates of number of molecules. Recently, structures such as carbon nanotubes, silicon nanorods, and semiconductor quantum dots have been emerged as particularly interesting classes of nanomaterials. In physics and electrical engineering, nanoscience is most often associated with quantum behavior, and the behavior of electrons and photons in nanoscale structures. Biochemistry and biology is interested in nanostructures such as cells components. The most widely investigated biological structures including DNA, viruses, and subcellular organelles can be considered as nanostructures [3].

1.1. From nano-pottery to modern analytical tools

It is obvious that "nano" influences the whole scientific world including instrumental analytical chemistry. Due to above-mentioned unique properties, not only new approaches and assays are being developed, but also standard techniques have been upgraded and capillary electrophoresis (CE) belongs to the group of these highly affected methods. In 1981, and since then, this powerful analytical technique progressed significantly not only in instrumentation, but also in method development, data acquisition, and processing. The group of applications has also widened markedly. The applications of CE are covering huge number of analytes from inorganic ions [4–8] and organic molecules [9–11] to biomolecules such as proteins [12–14] and DNA [15–17]. The golden era of CE was in 1990s, during the Human Genome Project [18]. The sequencing of the whole human genome was successfully finished in 2006 identifying all 20,000–25,000 genes (approximately) in human DNA and determining the sequences of 3 billion base pairs that make up human DNA.

Next great boom of CE begun due to the micro-total analysis system (μTAS) concept [19]. Due to the relatively simple instrumentation and ease of miniaturization of CE, the fast growth of attention in microfluidics and particularly in chip-based CE [20–23] was observed. Even though CE provides rapid results with high efficiency and resolution, and sample consumption is low; advantages of high number of theoretical plates can be diminished by relatively low sensitivity of commonly used photometric detection systems [24]. Therefore, new approaches improving these weak sites are investigated and the use of nanomaterials is widely tested.

http://dx.doi.org/10.5772/intechopen.72015

2. Capillary electrophoresis

Capillary electrophoresis is an extremely powerful microcolumn separation technique, separating molecules based on their mobilities in the electric field. Its main advantages include high separation efficiency, short time of analysis, and low consumption of chemicals.

Classical CE separation takes place in a fused silica capillary with internal diameter of 20–100 μm, where the voltage of up to ±30 kV is applied. The scheme of the setup is shown in **Figure 1**.

Diversity of detection modalities applicable in CE is very wide. Optical detection methods including photometric and fluorimetric detection are probably the most common ones currently used in CE; however, electrochemical detection including amperometric and contactless conductometric detection have unparalleled advantages. Especially for analysis of small inorganic ions, which do not absorb light, electrochemical detection is an alternative to the indirect optical approach. Also, popularity of mass spectrometric (MS) detection coupled to CE increased rapidly in past few decades.

From the very beginning, electromigration methods benefited from the use of certain sieving media, such as paper or gel. Moreover, since the separations have been transferred to capillary and/or chip, the addition of some kind of "stationary phase", sieving media or pseudo-stationary phase increased the number of applications. This step led to rise of several electrophoretic methods such as capillary electrochromatography or capillary micellar electrokinetic chromatography.

Low sensitivity, which is probably the main weakness of CE in connection with the most wide spread detection method—spectrophotometric detection—is caused by the short optical path-length (given by the capillary diameter) and the low sample volume that is injected.

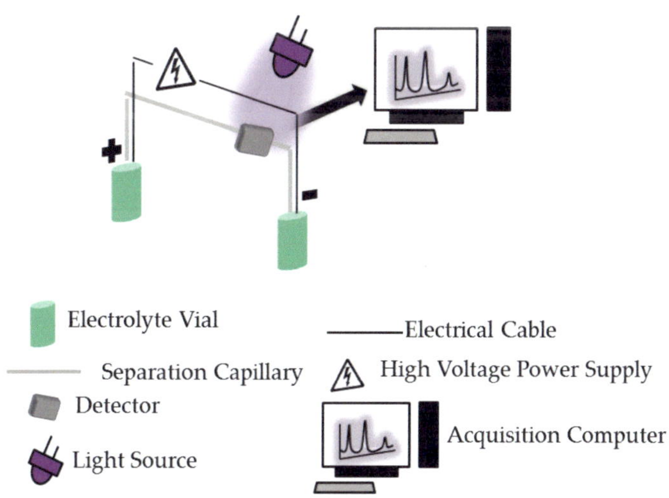

Figure 1. Scheme of CE setup.

Therefore, a preconcentration step is usually used for analysis of relatively diluted sample solutions. The injected sample solution should have low salt concentration to enable the sample stacking process. Otherwise the electro-osmotic flow can be altered, and the unfavorable detector background signal may be observed. Two main methods are used for sample pretreatment and preconcentration: (i) an electrophoretic method and (ii) a chromatographic technique. Electrophoretic techniques are based on differences in the electrophoretic mobilities of the analytes. Four main types of these techniques have been presented: transient isotachophoresis [25, 26], stacking [27–29], sweeping [30–32], and dynamic pH junction [33, 34]. On the other hand, chromatographic techniques rely on analyte sorption on a surface of the stationary phase. Chromatographic techniques allow for loading of large volumes followed by preconcentration on a sorbent and subsequent elution into a small volume of solvent, resulting in lower limits of detection.

For such applications, nanomaterials are excellent candidates because, compared to bulk materials, nanomaterials offer significantly higher surface-to-volume ratios, and therefore provide higher sorption capacity and thus better extraction efficiencies. As an example, may serve the comparison of surface area of carbon microparticles with 60 μm in diameter of (0.01 mm^2) and the surface area of carbon nanoparticles with 60 nm in diameter (11.3 mm^2). Similar to the increase in the surface area, the reactivity is also increased by approximately three orders of magnitude. Not only the surface area, but also the chemical affinity may be beneficial. For example, gold nanoparticles provide excellent extraction power due to their high affinity for thiol-containing compounds.

Similarly, magnetic separation is a method using magnetism for the efficient separation mediated by paramagnetic and superparamagnetic particles. This technique takes advantage of the option of surface modification of magnetic nanoparticles to enable so-called immunoextraction. Such particles may be modified by either antibodies for specific capture of the target analyte or by oligonucleotide fragments having sequences complementary to the desired nucleic acid. Magnetic particles can be immobilized using an external magnetic field and interfering compounds are removed from the solution [35].

3. Capillary electrophoresis for analysis of nanomaterials

The field of nanomaterials (e.g. metal or polymeric nanoparticles and carbon nanomaterials) is one of the most attractive and quickly emerging, while these materials have often valuable properties for various applications. The synthesis, however, is problematic especially from batch-to-batch repeatability point of view and, sometimes, techniques enabling characterization of nanomaterial properties and composition are absent. Even within a single batch, the polydispersity of the particles and the variability of their properties may present insurmountable problem for reliable application [36].

The conventional methods evaluating the size distribution are transmission electron microscopy and/or size exclusion chromatography. However, these methods have disadvantages including high instrumental costs, time consuming and laborious sample preparation, and high requirements on an operator, because of interpretation of the results. Therefore conventional or microfluidic CE is a good alternative for characterization of colloids and nanomaterials. Review articles focusing on electrophoretic separation of nanoparticles has been published in 2004 by Rodriguez and Armstrong [37], later by Surugau and Urban [38], Pyell [39], Lopez-Lorente et al. [40], and in 2017 by Aleksenko et al. [41]. Microfluidic format used in nanoparticle separation was reviewed by Salafi et al. [42]. More focused review article about CE analysis of poly(amidoamine) dendrimeric structures was prepared by Shi et al. [43]. Paper summarizing the application of separation techniques (including CE) of gold nanoparticles [44, 45] and QDs [46, 47] analysis are also accessible.

One of the main advantages of nanomaterials is that their surface can be easily functionalized and modified with various molecules potentially applicable for interactions with other molecules. Therefore, CE can (i) monitor the interaction between nanomaterial and analyte, (ii) monitor the interaction between two analytes facilitated by the nanomaterial, and (iii) monitor the interaction between two analytes expressed as a change in the signal of the nanomaterial.

4. Nanomaterials enhancing performance of capillary electrophoresis

4.1. Enhancement of separation

Nowadays, an increasing number of researchers perform CE separations in short capillaries (units of centimeters) instead of microfluidic chips [48, 49]. In such capillaries, fast and efficient separations are carried out without complicated chip preparation requiring expensive facilities (e.g. clean rooms and lithography). Compared with microchip-based high-speed CE systems, short capillary-based high-speed CE systems take advantage of simple structure, easy fabrication, and low costs.

The disadvantage, however, is in lowered resolution connected with short separation length. This obstacle can be solved either by injection of extremely low sample volumes (picoliters) or by additional selectivity given by stationary of pseudo-stationary phases of various natures (e.g. micelles, nanoparticles, nanostructures, etc.), which significantly eliminate the adsorption of highly abundant proteins on the capillary wall [50–53]. Nanomaterials have been proven to be effective (pseudo)stationary phase due to their beneficial properties, such as large surface–to-volume ratio and easy modification. The most commonly used nanomaterials are carbon nanotubes [54]. However, other structures including nanoparticles [55, 56], nanofibres [57], and/or nanorods [58, 59] have been utilized for these purposes. On the other hand, surface-binding method uses interaction between analyte with the surface of fixed nanostructures such as monoliths, nanopillars, immobilized nanoparticles, and/or other nanomaterials. All of these types have already been employed in coupling with in either capillary-based CE or microfluidic CE. Immobilized nanomaterials, either deposited on capillary wall as a thin layer coating or packed within the capillary, are commonly utilized as stationary phases for capillary electrochromatography. The (pseudo)stationary phases enable a broad range of functionalities offering a variety of interactions [60] (**Figure 2**).

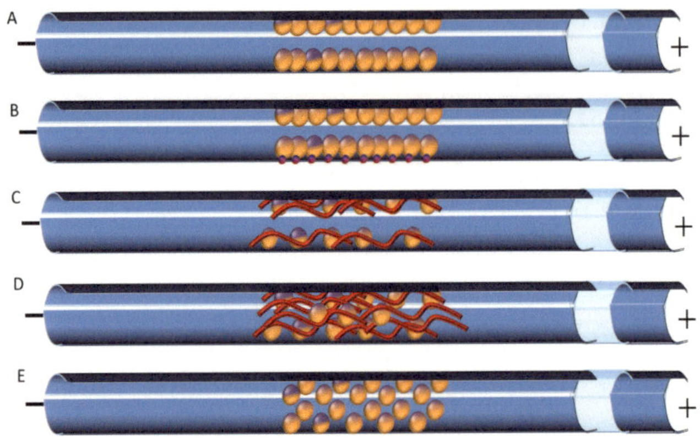

Figure 2. Approaches for separation performance enhancement by nanomaterials. (A) Physically adsorbed opentubular phase, (B) covalently bound opentubular phase, (C) co-polymerized opentubular pahse, (D) full-filling sieving matrix, (E) mobile pseudo-stationary phase.

The efficiency of separation of two compounds is defined as resolution (R_S). It can be affected by alternating the electrophoretic mobility of the analytes and their electrophoretic mobilities. R_S can be calculated according to the equation (1):

$$R_S = \sqrt{N}\frac{\mu_{elA} - \mu_{elB}}{4(\mu_{av} + \mu_{EOF})} \tag{1}$$

where N is the number of theoretical plates, μ_{elA} and μ_{elB} are electrophoretic mobilities of the analytes, μ_{av} is the mean electrophoretic mobility of analyte A and B, and μ_{EOF} is mobility of electro-osmotic flow.

4.2. Enhancement of detection

Laser-induced fluorescence detection is (and most likely will be also in the future) the most sensitive detection technique among the optical detection modes following chemical separation. It has exceptionally low limits of detection (10^{-13} M) [61] and good detection selectivity in cases of sample analysis with rather complex matrices. Simultaneously, this selectivity could be perceived as a limitation because the majority of analytes lacks the desired fluorescent properties, and therefore, derivatization by some fluorescent label is needed.

Optical detection in association with nanomaterials is mainly connected with quantum dots due to their application as a fluorescent labels in laser-induced fluorescence detection [62–65]. An indirect laser-induced fluorescence mode of detection by means of CdTe quantum dots has been demonstrated and therefore determination of small organic acids in food with detection limits in the range of tenths of mg/L was enabled [64]. Moreover, determination of pesticide and antibiotic residues in vegetables [66, 67] and in foods [68] has been described. Chen and Fung presented laser-induced fluorescence detection using immobilized QDs to determine organophosphate pesticides (mevinphos, phosalone, methidathion, and diazinon) in vegetable samples [69]. Detection limits of the method were in the range of tens of μg/kg.

Besides photoluminescence detection, chemiluminescence (CL) and electrochemiluminescence detection are also benefiting from properties of nanomaterials [70]. CL detection is based on measurement of electromagnetic radiation released after excitation of the electron by chemical reaction. The main advantages of CL are in the absence of undesired background signals, improved sensitivity, and wide linear dynamic range. Moreover, no excitation sources and/or optical filters are essential; therefore the instrumentation is simple, robust, and relatively low cost. In such instruments, metal-based nanoparticles can be used as catalysts, reductants, fluorophores, or acceptors of energy. Metal nanoparticles, such as gold, silver, platinum, semiconductors, and magnetic types, provide beneficial properties for CL detection [71].

Electrochemical detection in CE can be carried out in three modes: potentiometric, amperometric, and conductometric. Potentiometric and conductometric detectors provide good sensitivity and on contrary, amperometric detection is selective and can be tuned to the analyte of interest. One of the main differences of this approach compared to the optical detection modes is that the electrochemical detection is mostly performed by off-column, end-capillary, and therefore, in destructive arrangement.

The use of nanomaterials for electrochemical detection covers a remarkably broad field. Due to their electrochemical properties, nanomaterials have been applied for electrochemical analysis of many analytes, comprising of nucleic acids [72–74], proteins [75, 76], secondary metabolites [77, 78], and/or metals [79]. The key roles delivered by nanoparticles include biomolecule immobilization, catalysis of electrochemical reactions, enhancement of electron transfer between electrode surfaces and proteins, biomolecule labeling, and even use as a reactant [80]. In addition to the relatively low financial demands of electrochemical detection in comparison to optical instrumentation, advantages such as the possibility of miniaturization and in-field applications are vital. Number of reviews covering this topic has been published. For example, Pumera and Escarpa [81] summarized the different approaches for constructing nanomaterial-based detectors for conventional CE and microchip electrophoresis and mostly focused on three main types of nanomaterials, that is, carbon nanotubes, nanoparticles, and nanorods, in various designs. The work by Garcia-Carmona focusses on highlighting the electrochemical detection enhancement in CE, chip electrophoresis, and paper-based microfluidic devices [82].

In our opinion, it is highly unlikely that nanomaterials will wholly substitute such well-established approaches as organic dyes for fluorescent labeling. However, nanomaterials offer new options for a broad range of applications. The electrochemical detection particularly benefits from use of nanomaterials that enable increasingly sensitive detection.

5. Conclusion

There is no doubt that nanomaterials are extremely valuable tool for analytical applications enhancing highly the efficiency of extraction techniques, increasing significantly the resolution of separations, and improving greatly the capabilities of detection systems. There are a lot of key features of instruments used for clinical purposes including being easy to use and robust. In spite of the great advantages of capillary electrophoresis, robustness and repeatability of measurements belong to its weaknesses, which represent an obstacle for using of capillary electrophoretic instrumentation in clinical practice with one exception represented by DNA sequencer. Utilization of nanomaterials in capillary electrophoresis is opening new perspective in the field of clinical usage because these advanced materials can lower detection limits on one side and enhance the separation effectiveness on the other side. Nevertheless, this is at the beginning and waiting for exploration.

From the CE point of view, NMs such as liposomes and dendrimes have abilities to improve the separation part of the CE analysis and QDs, on the other hand, they can significantly improve the detection part. However, there are several members of the nanomaterial family, which can improve both of those—carbon nanotubes and metal nanoparticles. At the same time, CE is an effective technique for NMs characterization, evaluation, and/or observation (**Figure 3**).

The symbiosis of CE and nanomaterials is beneficial not only for analytical chemists and material scientists, but also for biochemists and molecular biologists, because it leads to the development of new, more effective, and more sensitive methods.

The combination with the simplicity of miniaturization is opening the opportunities for portable and point-of-care applicable instrumentation suitable for personalized diagnostics. Moreover the separation power of electrophoretic analysis, even increased by nanomaterial-based stationary and pseudo-stationary phased in combination with advances in microfluidics, promises effective analyses of complex biological samples.

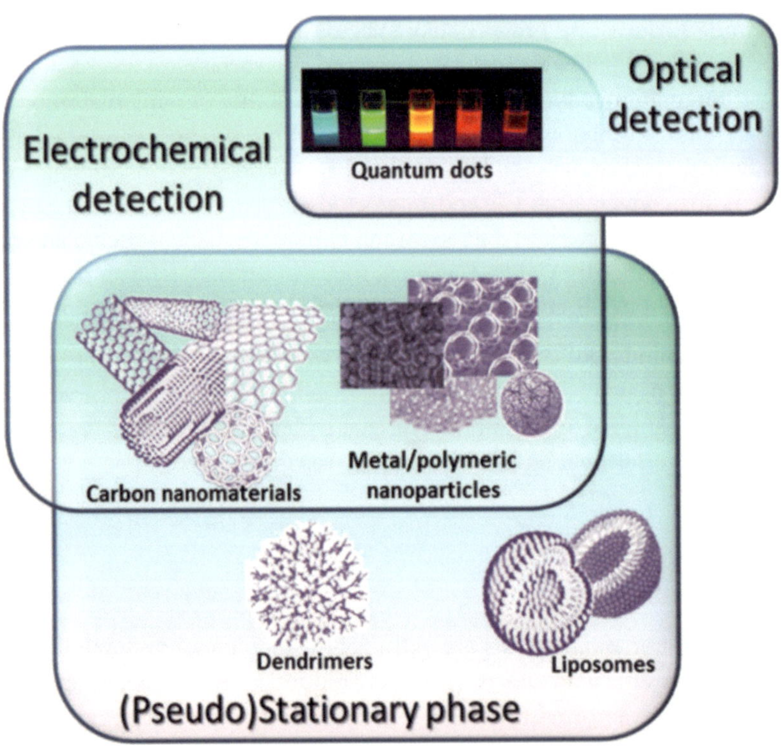

Figure 3. Summary of applications of nanomaterials in capillary electrophoresis.

http://dx.doi.org/10.5772/intechopen.72015

Acknowledgements

Financial support was provided by Grant agency of Czech Republic (GACR 16-23647Y) and project CEITEC 2020 (LQ1601) with financial support from the Ministry of Education, Youth and Sports of the Czech Republic under the National Sustainability Programme II.

Author details

Vojtech Adam[1,2] and Marketa Vaculovicova[1,2]*

*Address all correspondence to: marketa.ryvolova@seznam.cz

1 Department of Chemistry and Biochemistry, Mendel University in Brno, Brno, Czech Republic

2 Central European Institute of Technology, Brno University of Technology, Brno, Czech Republic

References

[1] Wagner FE, Haslbeck S, Stievano L, Calogero S, Pankhurst QA, Martinek P. Before striking gold in gold-ruby glass. Nature. 2000;**407**:691-692

[2] Yadugiri VT, Malhotra R. 'Plenty of room'—fifty years after the Feynman lecture. Current Science. 2010;**99**:900-907

[3] Whitesides GM. Nanoscience, nanotechnology, and chemistry. Small. 2005;**1**:172-179

[4] Macka M, Haddad PR. Determination of metal ions by capillary electrophoresis. Electrophoresis. 1997;**18**:2482-2501

[5] Haddad PR, Doble P, Macka M. Developments in sample preparation and separation techniques for the determination of inorganic ions by ion chromatography and capillary electrophoresis. Journal of Chromatography A. 1999;**856**:145-177

[6] Timerbaev AR, Shpigun OA. Recent progress in capillary electrophoresis of metal ions. Electrophoresis. 2000;**21**:4179-4191

[7] Timerbaev AR. Recent advances and trends in capillary electrophoresis of inorganic ions. Electrophoresis. 2002;**23**:3884-3906

[8] Timerbaev AR. Capillary electrophoresis of inorganic ions: An update. Electrophoresis. 2004;**25**:4008-4031

[9] DeBacker BL, Nagels LJ. Potentiometric detection for capillary electrophoresis: Determination of organic acids. Analytical Chemistry. 1996;**68**:4441-4445

[10] Soga T, Ross GA. Simultaneous determination of inorganic anions, organic acids, amino acids and carbohydrates by capillary electrophoresis. Journal of Chromatography A. 1999;**837**:231-239

[11] Soga T, Imaizumi M. Capillary electrophoresis method for the analysis of inorganic anions, organic acids, amino acids, nucleotides, carbohydrates and other anionic compounds. Electrophoresis. 2001;**22**:3418-3425

[12] Huang YF, Huang CC, CC H, Chang HT. Capillary electrophoresis-based separation techniques for the analysis of proteins. Electrophoresis. 2006;**27**:3503-3522

[13] Haselberg R, de Jong GJ, Somsen GW. Capillary electrophoresis-mass spectrometry for the analysis of intact proteins. Journal of Chromatography A. 2007;**1159**:81-109

[14] Dolnik V. Capillary electrophoresis of proteins 2005-2007. Electrophoresis. 2008;**29**: 143-156

[15] Heller C. Principles of DNA separation with capillary electrophoresis. Electrophoresis. 2001;**22**:629-643

[16] Righetti PG, Gelfi C, D'Acunto MR. Recent progress in DNA analysis by capillary electrophoresis. Electrophoresis. 2002;**23**:1361-1374

[17] Kleparnik K, Bocek P. DNA diagnostics by capillary electrophoresis. Chemical Reviews. 2007;**107**:5279-5317

[18] Venter JC, Adams MD, Myers EW, Li PW, Mural RJ, Sutton GG, Smith HO, Yandell M, Evans CA, Holt RA, Gocayne JD, Amanatides P, Ballew RM, Huson DH, Wortman JR, Zhang Q, Kodira CD, Zheng XQH, Chen L, Skupski M, Subramanian G, Thomas PD, Zhang JH, Miklos GLG, Nelson C, Broder S, Clark AG, Nadeau C, McKusick VA, Zinder N, Levine AJ, Roberts RJ, Simon M, Slayman C, Hunkapiller M, Bolanos R, Delcher A, Dew I, Fasulo D, Flanigan M, Florea L, Halpern A, Hannenhalli S, Kravitz S, Levy S, Mobarry C, Reinert K, Remington K, Abu-Threideh J, Beasley E, Biddick K, Bonazzi V, Brandon R, Cargill M, Chandramouliswaran I, Charlab R, Chaturvedi K, Deng ZM, Di Francesco V, Dunn P, Eilbeck K, Evangelista C, Gabrielian AE, Gan W, Ge WM, Gong FC, Gu ZP, Guan P, Heiman TJ, Higgins ME, Ji RR, Ke ZX, Ketchum KA, Lai ZW, Lei YD, Li ZY, Li JY, Liang Y, Lin XY, Lu F, Merkulov GV, Milshina N, Moore HM, Naik AK, Narayan VA, Neelam B, Nusskern D, Rusch DB, Salzberg S, Shao W, Shue BX, Sun JT, Wang ZY, Wang AH, Wang X, Wang J, Wei MH, Wides R, Xiao CL, Yan CH, Yao A, Ye J, Zhan M, Zhang WQ, Zhang HY, Zhao Q, Zheng LS, Zhong F, Zhong WY, Zhu SPC, Zhao SY, Gilbert D, Baumhueter S, Spier G, Carter C, Cravchik A, Woodage T, Ali F, An HJ, Awe A, Baldwin D, Baden H, Barnstead M, Barrow I, Beeson K, Busam D, Carver A, Center A, Cheng ML, Curry L, Danaher S, Davenport L, Desilets R, Dietz S, Dodson K, Doup L, Ferriera S, Garg N, Gluecksmann A, Hart B, Haynes J, Haynes C, Heiner C, Hladun S, Hostin D, Houck J, Howland T, Ibegwam C, Johnson J, Kalush F, Kline L, Koduru S, Love A, Mann F, May D, McCawley S, McIntosh T, McMullen I, Moy M, Moy L, Murphy B, Nelson K, Pfannkoch C, Pratts E, Puri V, Qureshi H, Reardon M, Rodriguez R, Rogers YH, Romblad D, Ruhfel B, Scott R, Sitter C, Smallwood M, Stewart E, Strong R, Suh E, Thomas R, Tint NN, Tse S,

Vech C, Wang G, Wetter J, Williams S, Williams M, Windsor S, Winn-Deen E, Wolfe K, Zaveri J, Zaveri K, Abril JF, Guigo R, Campbell MJ, Sjolander KV, Karlak B, Kejariwal A, Mi HY, Lazareva B, Hatton T, Narechania A, Diemer K, Muruganujan A, Guo N, Sato S, Bafna V, Istrail S, Lippert R, Schwartz R, Walenz B, Yooseph S, Allen D, Basu A, Baxendale J, lick L, Caminha M, Carnes-Stine J, Caulk P, Chiang YH, Coyne M, Dahlke C, Mays AD, Dombroski M, Donnelly M, Ely D, Esparham S, Fosler C, Gire H, Glanowski S, Glasser K, Glodek A, Gorokhov M, Graham K, Gropman B, Harris M, Heil J, Henderson S, Hoover J, Jennings D, Jordan C, Jordan J, Kasha J, Kagan L, Kraft C, Levitsky A, Lewis M, Liu XJ, Lopez J, Ma D, Majoros W, McDaniel J, Murphy S, Newman M, Nguyen T, Nguyen N, Nodell M, Pan S, Peck J, Peterson M, Rowe W, Sanders R, Scott J, Simpson M, Smith T, Sprague A, Stockwell T, Turner R, Venter E, Wang M, Wen MY, Wu D, Wu M, Xia A, Zandieh A, Zhu XH. The sequence of the human genome. Science. 2001;**291**:1304

[19] Manz A, Graber N, Widmer HM. Miniaturizes total chemical-analysis systems—A novel concept for chemical sensing. Sensors and Actuators B—Chemical. 1990;**1**:244-248

[20] Dolnik V, Liu SR, Jovanovich S. Capillary electrophoresis on microchip. Electrophoresis. 2000;**21**:41-54

[21] Dolnik V, Liu SR. Applications of capillary electrophoresis on microchip. Journal of Separation Science. 2005;**28**:1994-2009

[22] Li SFY, Kricka LJ. Clinical analysis by microchip capillary electrophoresis. Clinical Chemistry. 2006;**52**:37-45

[23] Breadmore MC. Capillary and microchip electrophoresis: Challenging the common conceptions. Journal of Chromatography A. 2012;**1221**:42-55

[24] Osbourn DM, Weiss DJ, Lunte CE. On-line preconcentration methods for capillary electrophoresis. Electrophoresis. 2000;**21**:2768-2779

[25] Shihabi ZK. Transient pseudo-isotachophoresis for sample concentration in capillary electrophoresis. Electrophoresis. 2002;**23**:1612-1617

[26] Timerbaev AR, Hirokawa T. Recent advances of transient isotachophoresis-capillary electrophoresis in the analysis of small ions from high-conductivity matrices. Electrophoresis. 2006;**27**:323-340

[27] Beckers JL, Bocek P. Sample stacking in capillary zone electrophoresis: Principles, advantages and limitations. Electrophoresis. 2000;**21**:2747-2767

[28] Quirino JP, Terabe S. Sample stacking of cationic and anionic analytes in capillary electrophoresis. Journal of Chromatography A. 2000;**902**:119-135

[29] Shihabi ZK. Stacking in capillary zone electrophoresis. Journal of Chromatography A. 2000;**902**:107-117

[30] Quirino JP, Terabe S. Approaching a million-fold sensitivity increase in capillary electrophoresis with direct ultraviolet detection: Cation-selective exhaustive injection and sweeping. Analytical Chemistry. 2000;**72**:1023-1030

[31] Quirino JP, Kim JB, Terabe S. Sweeping: Concentration mechanism and applications to high-sensitivity analysis in capillary electrophoresis. Journal of Chromatography A. 2002;**965**:357-373

[32] Aranas AT, Guidote AM, Quirino JP. Sweeping and new on-line sample preconcentration techniques in capillary electrophoresis. Analytical and Bioanalytical Chemistry. 2009;**394**:175-185

[33] Britz-McKibbin P, Chen DDY. Selective focusing of catecholamines and weakly acidic compounds by capillary electrophoresis using a dynamic pH junction. Analytical Chemistry. 2000;**72**:1242-1252

[34] Kazarian AA, Hilder EF, Breadmore MC. Online sample pre-concentration via dynamic pH junction in capillary and microchip electrophoresis. Journal of Separation Science. 2011;**34**:2800-2821

[35] Adam V, Vaculovicova M. Nanomaterials for sample pretreatment prior to capillary electrophoretic analysis. The Analyst. 2017;**142**:849-857

[36] Adam V, Vaculovicova M. Capillary electrophoresis and nanomaterials—Part I: Capillary electrophoresis of nanomaterials. Electrophoresis. 2017:1-16

[37] Rodriguez MA, Armstrong DW. Separation and analysis of colloidal/nano-particles including microorganisms by capillary electrophoresis: A fundamental review. Journal of Chromatography B-Analytical Technologies in the Biomedical and Life Sciences. 2004;**800**:7-25

[38] Surugau N, Urban PL. Electrophoretic methods for separation of nanoparticles. Journal of Separation Science. 2009;**32**:1889-1906

[39] Pyell U. Characterization of nanoparticles by capillary electromigration separation techniques. Electrophoresis. 2010;**31**:814-831

[40] Lopez-Lorente AI, Simonet BM, Valcarcel M. Electrophoretic methods for the analysis of nanoparticles. Trac—Trends in Analytical Chemistry. 2011;**30**:58-71

[41] Aleksenko SS, Matczuk M, Timerbaev AR. Characterization of interactions of metal-containing nanoparticles with biomolecules by CE: An update (2012-2016). Electrophoresis. 2017;**38**:1661-1668

[42] Salafi T, Zeming KK, Zhang Y. Advancements in microfluidics for nanoparticle separation. Lab on a Chip. 2017;**17**:11-33

[43] Shi XY, Banyai I, Lesniak WG, Islam MT, Orszagh I, Balogh P, Baker JR, Balogh LP. Capillary electrophoresis of polycationic poly(amidoamine) dendrimers. Electrophoresis. 2005;**26**:2949-2959

[44] Liu FK. Analysis and applications of nanoparticles in the separation sciences: A case of gold nanoparticles. Journal of Chromatography A. 2009;**1216**:9034-9047

[45] CS W, Liu FK, Ko FH. Potential role of gold nanoparticles for improved analytical methods: An introduction to characterizations and applications. Analytical and Bioanalytical Chemistry. 2011;**399**:103-118

[46] Sang FM, Huang XY, Ren JC. Characterization and separation of semiconductor quantum dots and their conjugates by capillary electrophoresis. Electrophoresis. 2014;**35**:793-803

[47] Stanisavljevic M, Vaculovicova M, Kizek R, Adam V. Capillary electrophoresis of quantum dots: Minireview. Electrophoresis. 2014;**35**:1929-1937

[48] Lin QH, Cheng YQ, Dong YN, Zhu Y, Pan JZ, Fang Q. High-speed separation of proteins by sodium dodecyl sulfate-capillary gel electrophoresis with partial translational spontaneous sample injection. Electrophoresis. 2011;**32**:2898-2903

[49] Wang W, Ma LH, Yao FZ, Lin XL, Xu KX. High-speed separation and detection of amino acids in laver using a short capillary electrophoresis system. Electrophoresis. 2015;**36**:335-340

[50] Terabe S. Capillary separation: Micellar electrokinetic chromatography. In: Annual Review of Analytical Chemistry. Palo Alto: Annual Reviews; 2009. pp. 99-120

[51] Duan AH, Xie SM, Yuan LM. Nanoparticles as stationary and pseudo-stationary phases in chromatographic and electrochromatographic separations. Trac—Trends in Analytical Chemistry. 2011;**30**:484-491

[52] Gonzalez-Curbelo MA, Varela-Martinez DA, Socas-Rodriguez B, Hernandez-Borges J. Recent applications of nanomaterials in capillary electrophoresis. Electrophoresis. 2017;**38**:2431-2446

[53] Hajba L, Guttman A. Recent advances in column coatings for capillary electrophoresis of proteins. Trac—Trends in Analytical Chemistry. 2017;**90**:38-44

[54] Ravelo-Perez LM, Herrera-Herrera AV, Hernandez-Borges J, Rodriguez-Delgado MA. Carbon nanotubes: Solid-phase extraction. Journal of Chromatography A. 2010;**1217**: 2618-2641

[55] Peng Y, Xie Y, Luo J, Nie L, Chen Y, Chen LN, SH D, Zhang ZP. Molecularly imprinted polymer layer-coated silica nanoparticles toward dispersive solid-phase extraction of trace sulfonylurea herbicides from soil and crop samples. Analytica Chimica Acta. 2010;**674**:190-200

[56] Guihen E. Recent highlights in electro-driven separations-selected applications of alkylthiol gold nanoparticles in capillary electrophoresis and capillary electro-chromatography. Electrophoresis. 2017;**38**:2184-2192

[57] Chigome S, Darko G, Torto N. Electrospun nanofibers as sorbent material for solid phase extraction. The Analyst. 2011;**136**:2879-2889

[58] Wang D, Zhang ZM, Luo L, Li TM, Zhang L, Chen GN. ZnO nanorod array solid phase micro-extraction fiber coating: Fabrication and extraction capability. Nanotechnology. 2009;**20**

[59] Wang D, Wang QT, Zhang ZM, Chen GN. ZnO nanorod array polydimethylsiloxane composite solid phase micro-extraction fiber coating: Fabrication and extraction capability. The Analyst. 2012;**137**:476-480

[60] Adam V, Vaculovicova M. CE and nanomaterials—Part II: Nanomaterials in CE. Electrophoresis. 2017:1-26

[61] Swinney K, Bornhop DJ. Detection in capillary electrophoresis. Electrophoresis. 2000; **21**:1239-1250

[62] Bai Y, FY D, Yang YY, Liu HW. In-capillary non-covalent labeling and determination of tomato systemin with quantum dots in capillary electrophoresis with laser-induced fluorescence detection. Journal of Separation Science. 2011;**34**:2893-2900

[63] Chen QD, Zhao WF, Fung YS. Determination of acrylamide in potato crisps by capillary electrophoresis with quantum dot-mediated LIF detection. Electrophoresis. 2011;**32**:1252-1257

[64] Guo DS, Chen GH, Tong MZ, CQ W, Fang R, Yi LX. Determination of five preservatives in food by capillary electrophoresis with quantum dot indirect laser induced fluorescence. Chinese Journal of Analytical Chemistry. 2012;**40**:1379-1384

[65] Qiu L, Bi YH, Wang CL, Li JY, Guo PL, Li JC, He WJ, Wang JH, Jiang PJ. Protein a detection based on quantum dots-antibody bioprobe using fluorescence coupled capillary electrophoresis. International Journal of Molecular Sciences. 2014;**15**:1804-1811

[66] Chen GH, Sun J, Dai YJ, Dong M. Determination of nicotinyl pesticide residues in vegetables by micellar electrokinetic capillary chromatography with quantum dot indirect laser-induced fluorescence. Electrophoresis. 2012;**33**:2192-2196

[67] Tang TT, Deng JJ, Zhang M, Shi GY, Zhou TS. Quantum dot-DNA aptamer conjugates coupled with capillary electrophoresis: A universal strategy for ratiometric detection of organophosphorus pesticides. Talanta. 2016;**146**:55-61

[68] Meng HL, Chen GH, Guo X, Chen P, Cai QH, Tian YF. Determination of five quinolone antibiotic residues in foods by micellar electrokinetic capillary chromatography with quantum dot indirect laser-induced fluorescence. Analytical and Bioanalytical Chemistry. 2014;**406**:3201-3208

[69] Chen QD, Fung YS. Capillary electrophoresis with immobilized quantum dot fluorescence detection for rapid determination of organophosphorus pesticides in vegetables. Electrophoresis. 2010;**31**:3107-3114

[70] Yao J, Li L, Li PF, Yang M. Quantum dots: From fluorescence to chemiluminescence, bioluminescence, electrochemiluminescence, and electrochemistry. Nanoscale. 2017;**9**:13364-13383

[71] Giokas DL, Vlessidis AG, Tsogas GZ, Evmiridis NP. Nanoparticle-assisted chemiluminescence and its applications in analytical chemistry. Trac—Trends in Analytical Chemistry. 2010;**29**:1113-1126

[72] Erdem A. Nanomaterial-based electrochemical DNA sensing strategies. Talanta. 2007; **74**:318-325

[73] Mao X, Liu GD. Nanomaterial based electrochemical DNA biosensors and bioassays. Journal of Biomedical Nanotechnology. 2008;**4**:419-431

[74] Liu XG, Cheng ZQ, Fan H, Ai SY, Han RX. Electrochemical detection of avian influenza virus H5N1 gene sequence using a DNA aptamer immobilized onto a hybrid nanomaterial-modified electrode. Electrochimica Acta. 2011;**56**:6266-6270

[75] Rusling JF, Sotzing G, Papadimitrakopoulosa F. Designing nanomaterial-enhanced electrochemical immunosensors for cancer biomarker proteins. Bioelectrochemistry. 2009;**76**:189-194

[76] Rusling JF. Nanomaterials-based electrochemical immunosensors for proteins. Chemical Record. 2012;**12**:164-176

[77] de Andres F, Zougagh M, Castaneda G, Rios A. Determination of zearalenone and its metabolites in urine samples by liquid chromatography with electrochemical detection using a carbon nanotube-modified electrode. Journal of Chromatography A. 2008;**1212**: 54-60

[78] Songa EA, Waryo T, Jahed N, Baker PGL, Kgarebe BV, Iwuoha EI. Electrochemical Nanobiosensor for glyphosate herbicide and its metabolite. Electroanalysis. 2009;**21**: 671-674

[79] Aragay G, Merkoci A. Nanomaterials application in electrochemical detection of heavy metals. Electrochimica Acta. 2012;**84**:49-61

[80] Luo XL, Morrin A, Killard AJ, Smyth MR. Application of nanoparticles in electrochemical sensors and biosensors. Electroanalysis. 2006;**18**:319-326

[81] Pumera M, Escarpa A. Nanomaterials as electrochemical detectors in microfluidics and CE: Fundamentals, designs, and applications. Electrophoresis. 2009;**30**:3315-3323

[82] Garcia-Carmona L, Martin A, Sierra T, Gonzalez MC, Escarpa A. Electrochemical detectors based on carbon and metallic nanostructures in capillary and microchip electrophoresis. Electrophoresis. 2017;**38**:80-94

Chapter 18

Nanofluids as Novel Alternative Smart Fluids for Reservoir Wettability Alteration

Lezorgia Nekabari Nwidee, Ahmed Barifcani,
Mohammad Sarmadivaleh and Stefan Iglauer

Additional information is available at the end of the chapter

http://dx.doi.org/10.5772/intechopen.72267

Abstract

This chapter presents an account of two metal oxide nanoparticles (zirconium and nickel oxide) on basis of their structure, morphology, crystallinity phases, and their wetting effect on solid-liquid interface. As a preliminary step to sound understanding of process mechanisms; wettability, nanoparticles, and their relations thereof were scrutinized. To investigate the nanofluids wetting inclinations, complex mixtures of the nanoparticles and NaCl brine (ZrO_2/NaCl; NiO/NaCl) were formulated and their technical feasibility as wetting agents tested via contact angle measurement. The result shows that the nanoparticles exhibit different structural and morphological features and capable of addressing reservoir wettability challenges owing to favorable adsorption behavior on the surface of the calcite which facilitated the wetting changes quantified by contact angle. We believe this study will significantly impact the understanding of wetting at solid-liquid interface which is crucial for recovery process optimization.

Keywords: nanoparticles, calcite, zirconium oxide, nickel oxide, carbonate, wettability, EOR

1. Overview of nanomaterials

In the past half century, industrial processes in general have experienced a transition in material applications owing to a shift from conventional bulk materials toward nanoscale materials. This has driven innovative applications in wide-ranging areas of science and technology globally, thus yielding a proliferating interest and investment in nanoscience and nanotechnology fields. The increase possibilities for the manipulation of matter in nanometer-scale have primarily led to this growth with nanomaterials at the leading edge of this fast-developing field. The potentials for direct control of systems at the same scale as nature such as DNA, cells, mitochondria and even reservoir rock pores can yield effective approaches in a wide variety of industrial processes such as the production of chemicals, materials, and energy [1]. Although nanomaterial appears to be a recent development owing to the current tremendous research growth and diverse applications, this material is not completely new as it has a rather shocking protracted history. The knowledge of the materials commenced as early as the 1950's by Richard Feynman who proposed that fabrications of materials and devices can be performed at atomic scales. Then in the 1980s, the term nanotechnology became even popular as established by Drexler Eric K. The current applications of these materials are not an exclusive result of modern research or laboratory synthesis, or even circumscribed to man-made materials. These materials have long been in existence with traceable applications in the old days. For instance, natural asbestos nanofibers and metal nanoparticles were used several decades ago for the control reinforcement of ceramic matrix and as color pigments in glass and luster technology respectively [2, 3].

Ever since, novel studies of nanoscale fundamentals and principles, design, characterization, production, and application of these materials [4–6] have evolved and remained intriguing and ground-breaking. Since 2000, the nanotechnology industry has experienced a growing trend and the funding of nanotechnology research has also been on the rise (**Figure 1**). For example; in 2013, the global market for nanotechnology was estimated at $22.9 billion, by 2014, the estimate had grown to about $26 billion with a further projected growth of about $64.2 billion by 2019 [7]. This innovative development involves the nanometer (nm) length scale manipulations of the structure of matter, where a nanometer represents a billionth of a meter, a distance that is equivalent to 2–20 atoms positioned next to one another. Nanomaterial has increasingly gained attention for a variety of processes such as electronic cooling and space applications, transportation, biomedicine, cooling of high-power laser diodes and submarines, and heating of buildings [8, 9]. Its wide applicability also extends to the field of environmental protection. This material exhibits great potentials as pollution reduction agent and improves the quality of air, water, and soil [10] and it is also currently being used as novel tools for oil and gas operations. Nanomaterials especially nanoparticles can be used alone

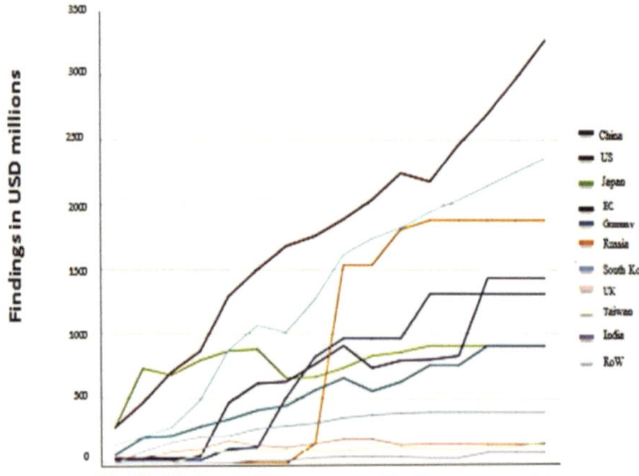

Figure 1. Worldwide nanotechnology research and funding by year [7].

or manipulated for the creation of larger scale materials to facilitate innovative applications. The material allow for clear-cut design and manipulation of atoms and molecules and its industrial applications are cost-effective and efficient [11, 12]. Nanoparticle cut across diverse fields of science and engineering and shows great potentials as effective approach for novel applications with high technological prospects and environmental friendliness. A recent drift in its application is its usage for resolving reservoir engineering challenges.

1.1. Nanoparticles

Nanoparticles are nano-sized structures with dimensions in the range of 1 to 100 nm. These materials exhibit unique properties with better potentials than bulk materials. Nanoparticles application enables the creation of new composites with unique properties which allows for innovative technological advancements. Unarguably, the effectiveness of this material cut across diverse industries such as energy, manufacturing, medicine, electronics, oil and gas industries etc., however, the understanding of the material is still very limited in EOR processes especially in wetting evaluations.

Nanoparticles are categorized as magnetic (iron, cobalt etc. and the oxides) [13–16], metallic (gold, silver, copper, and Platinum) [17, 18] or metal oxides (oxides of aluminum, zinc, silicon, magnesium, zirconium, cerium, titanium) [8, 19–23]. Among these categories, the metal oxides are the most commonly used nanoparticles in EOR [24–28] as the material offers special unique structures, compositions, physical, and chemical properties. These materials display efficient thermal conductivity effect [8], great stability [19] and excellent saline-alkaline tolerance [23]. However, its effectiveness depends greatly on the preparation methods, dispersant and subsequent applicability. Generally, nanoparticles can be prepared via chemical, biological or physical methods, although the preparation method is mainly based on the nature of the material and associated chemical reactions [29], and the applications thereof.

1.2. Nanofluid production and stability control

Nanofluids are colloidal suspensions of solid nanoparticles or nanofibers. These solid-liquid composite materials are typically two-phase systems, consisting of a carrier medium and solid phase. The carrier liquids are often water, polymer solution, oil, ethylene glycol and sodium chloride brine. The solid phases are nanoparticles of chemically stable metals and oxides usually within the range of 1-100 nm [30–39]. Nanofluid has continuously attracted great attention for various processes owing to its great thermal properties at low volume fractions of less than 1%. To achieve the same functionality with conventional suspensions of well-dispersed particles, high concentrations that are greater than 10% of particles is often a requirement. Such high concentrations increase the issues of rheology and stability which has remained a deterrent to the extensive use of conventional slurries as heat transfer fluids [35]. Nanofluid production can be achieved via chemical or physical synthesis. The chemical synthesis involves the use of methods such as thermal spraying, chemical vapor deposition, spray pyrolysis, chemical precipitation, or micro-emulsions. Whereas, the physical synthesis involves inert-gas-condensation technique and mechanical grinding approach. During its production, a one-step or two-step approach can be adopted. The one-step approach allows for the production of nanoparticle and its dispersion in a fluid in a single combined process, which is suitable for nanofluids with high-conductivity metals contents. Whereas with the two-step method, production process is performed in two separate steps, firstly, the nanoparticle is produced, and then the produced nanoparticle is dispersed in a fluid – this is considered an effective strategy for commercial use [8, 9].

In the face of the diverse functionalities of the nanoparticle, a major impediment in the manipulation and application of nanoparticle is the colloidal stability control. Nanoparticle tends to agglomerate when in suspensions irrespective of its small size. This has remained an issue with the production and utilization of nanoparticle based fluid as this behavior impacts the overall fluid stability. Such agglomeration can impede fluid flow characteristics in porous media as well as in flow based cooling applications [35]. Since nanofluids are typically produced in small quantities at laboratory scale, there is high potentials of yielding sufficiently well dispersed, homogenized, and stable fluids. However, homogeneous nanoparticle dispersion is often a challenge owing to agglomerating or clustering tendencies. The agglomeration inclination or clustering behavior of nanoparticle is dependent on the nanoparticles properties, particle concentrations, production methods, nature of dispersants, fluid homogeneity and stability. The nanofluids stability is vital for process efficiency as instability can influence the particles functionalities. Fedele et al. [19] reported stability evaluation of nanoparticle via a comparison of various preparation methods such as ball milling, sonication, and high-pressure homogenization. The ball milling method produced the least stable fluids when compared to sonication and homogenization, which produced better stable fluid. The use of magnetic stirrer or ball milling method has been shown to be rather insufficient for stable nanofluid formulation [23]. Similarly, Roustaei and Bagherzadeh [20] reported sonication and homogenization [21–23] as the most efficient methods.

An approach to ensure the stability and homogeneity of nanofluids aside the use of additives or stabilizers, is the uniform dispersibility of the particles in the solution. Attaining high-performance heat transfer nanofluids require efficient dispersion of the particles in the base fluid [8, 12, 40, 41] and ensuring an approximately monodispersed or non-agglomerated nanoparticle in liquids during the production of the suspensions. Thus, the fluid stability and excellent particle dispersion in base fluids can be significantly improved by using appropriate dispersants, suitable fluid production methods, and surface treated nanoparticles. Suitable dispersal and fluid production methods are vital to achieving desirable properties and uniform distribution of the particles in the system which can further prevent issues of agglomeration and can also improve the mechanical properties such as strength and ductility of the system [42]. Careful consideration should also be given to the concentration of the nanoparticles, as high particle concentration volume can propel high particle agglomeration.

2. Wettability

Wetting applies to several practical processes and a variety of industries such as energy, marine, manufacturing and materials. Wetting processes often involve the interaction of solids such as porous material, suspensions, or fibers, and liquids - water, ink, dye or lubricants. Typical indications of solid-liquid wetting can be illustrated using standard scenarios to correlate the dynamics such as: (1) the interaction between porous soil and water - where the water wets the solid components of the soil; (2) enhanced oil recovery processes - where the process permeates water into oil-wet porous media. However, the mineral floatation in these processes is often based on the selective wetting characteristics of the mineral particles [43]. Two key mechanisms [44] governs wettability alteration of surfaces - cleaning and coating. Cleaning involves the use of surface-active agents to desorb surfaces e.g. surfactants induced wettability alteration, where cationic surfactants can desorb the hydrophobic layer on a surface while changing the surface toward hydrophilic condition. Whereas coating involves covering a hydrophobic surface with a hydrophilic material, e.g. hydrophilic zirconium nanoparticle can adsorb on hydrophobic rock surface and form nanotextures capable of coating the hydrophobic surfaces. At large scales, wetting or non-wetting plays an essential role during oil recovery [45]. Inadequate formation wetting can prohibit efficient hydrocarbon flow, which in turn hinders the oil, gas or water movement or distribution through the pore spaces, as such fluids may appear to have flowed whereas its distribution through the pore spaces is hindered owing to poor rock wetting. Assuming a system contains only three phases (solid, liquid, and vapor), for any two of these phases to be in contact, a transitional area of molecular dimensions occurs owing to the compositional alteration of the system that leads to phase changes. For example, if a non-volatile molecular smooth solid is in contact with an inert gas, it is expected that the system will exhibit a transition region thickness of about a molecule, this would cause a change from solid molecules to gas molecules. Whereas, if similar trend occurs on an irregular surface, the transition region would reflect the physical non-uniformity of the surface and a concentration profile of the region would indicate the existence or non-existence of the solid phase. Similar concentration profile phenomenon holds for solid-liquid systems; however, the related specifics are dependent on the solubility of the solid in the liquid or the solubility of the liquid in the solid [46, 47]. In a typical solid-liquid-vapor three-phase system [48], the system would exhibit a completely dry behavior if there is an intrusion of a macroscopic vapor layer between the solid and the liquid; a partial wetting behavior if the droplet is bounded by microscopic thin film that is adsorbed on the surface of the solid; and complete wetting occurs due to macroscopic adsorbed thick wetting layer.

With respect to petroleum reservoir rocks, wetting is ascribed to the measurement of the reservoir rocks affinity for water or oil in a typical rock-fluid-oil system. An understanding of the wetting preference of rock is vital, as it unveils the mechanisms behind fluid flow in porous media, soil decontamination process evaluations, and ultimately promotes recovery efficiency. Reservoirs exhibit different wetting inclinations: water-wet, oil-wet, or intermediate-wet (**Figure 2**). Reservoir rocks considered as water-wet have high affinity for water and water predominantly occupies the tiny rock pores as well as the surface of the formation rock. Whereas, oil-wet reservoir rocks have high affinity for oil and such oil mainly occupies the tiny rock pores and the formation rock surface. For example; in controlled laboratory experiments involving the manipulations of cores or rock samples, the samples are usually cleaned and modified to a preferential wetting state. If such samples have high affinity for water or were originally water wet, then saturated to a suitable oil-wet state, the rock surface becomes even more oil-wet or hydrophobic upon exposure to oil under efficient and favorable treatment conditions. However, this does not influence the actual wetting affinity of the rock. The actual wetting affinity can be affirmed by exposing such hydrophobic rock to a water imbibition test. The water imbibing potential of the rock can be used to predict its wetting affinity.

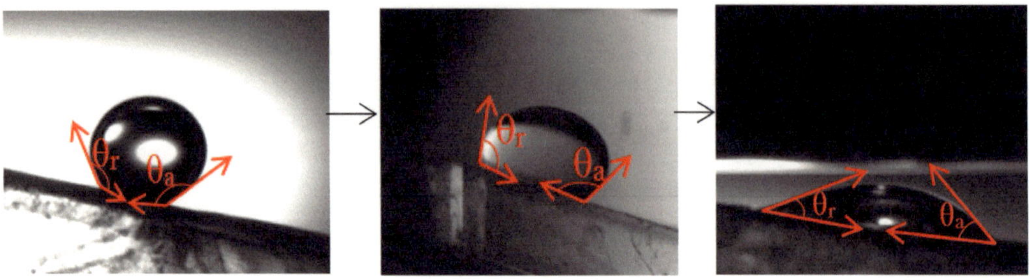

Figure 2. Rock surface wetting transition from hydrophobic to hydrophilic. Left: Oil-wet (105–180°); Center: Intermediate-wet (75–105°); right: Water-wet (0–75°).

Ideally, if the rock has high affinity for water then the oil will be displaced from the surface of the rock by water. Whereas if a rock with high affinity for oil is saturated with water, and then the rock is placed in an oil-wet environment, the oil will displace the water from the rock surface while efficiently imbibing into the rock pores. In the absence of an actual inclination for water or oil, the formation rock is considered intermediate wetting. Aside from these key-wetting preferences, there also exists fractional wetting where the formation rock exhibit different wetting inclination in different sections of the rock [49–53].

2.1. Carbonate porous media and wettability challenges

Reservoir rocks considered as porous media are formations with an interconnected network of pores or voids characterized by the rock's porosity, and physical and textural properties that exhibit a dependency on the formations constituents. Oil recovery exhibits great dependency on the formation wettability as it controls the fluid displacements of the wetting and non-wetting phase at the pore scale. Reservoir wettability is a prime factor for determining the microscopic displacement efficiency in the swept regions of a waterflood. Originally, reservoirs were strongly water-wet, and the formation traps were initially filled with water, thus the surface of reservoir rocks had high affinity for water in the presence of oil. However, overtime, oil migrated into such formations forming firm adsorbed layers of heavy hydrocarbons that poise several challenges and cannot be altered via gasoline or mere solvent applications. The oil migration and accumulation cause water to be retained in the rock pore spaces due to capillary pressure, while the rock pores surface become covered by oil, and water existed on such surfaces in the form of films. Such effect is primarily due to the rock surface wettability changes owing to the ease of invasion by a wetting fluid, which causes lithological variances. The level of oil migration also determines the formations wetting state. For instance, if the migrated oil is negligible, the possibilities are that the neighboring formations will be more oil wetting, while the tight regions of the formation would exhibit more water-wet behavior [49–55]. This behavior is more prevalent in carbonate rocks owing to the oil-wet character of this formation, which is still poorly understood.

It is well established that heterogeneous carbonate rocks are more prevalent globally. Carbonate rocks exhibit complex microstructures [56–58] and its complex nature impacts reservoir wetting preference. These formations are predominantly naturally fractured, and exhibits diverse wetting conditions; intermediate-wet or oil-wet behavior with as high as over 80% oil-wetness. The oil-wet nature of carbonate formation is due to its surface charges, which tend to attract negatively charged carboxylic acids compounds in crude oils [59–61]. Ideally, the formations positive surface charges attract crude oil acidic components. Carbonate reservoirs are problematic, as the complex wetting characteristics of this reservoir make the production capacity quite different in comparison to other conventional formations. Typically, an enormous capacity of the original oil in place is left stranded in this formation after primary and secondary oil recovery approaches are employed. Such approaches have been implemented for several decades, however, the fraction of recoverable oil from this reservoir is less than two-thirds [39, 62]. Fractured reservoir with enormous oil resources in its matrix requires advanced approaches for efficient recovery. Although water flooding enhances productivity in this reservoir by imbibing water from the formation fractures into the rock matrix, while enhancing oil flow out of the matrix through the fractures to the production well, this is mainly achievable if the capillary driving force is robust and efficient as it influences recovery efficiency, and the interaction between the matrix and the fracture is required for oil recovery from the formation matrix. Capillary forces have a significant effect on recovery capacities, however, its impact is greatly dependent on the nature of the reservoir, whether it is fractured or non-fractured. For non-fractured reservoirs, the presence of strong capillary forces during water flooding traps oil, however, the residual oil saturation becomes relatively high. Hence, the need for a reduction of the oil-water interfacial tension forces in order to remobilize residual oil in such formations. Whereas, for fractured reservoirs such as carbonate, the key driving force for efficient oil displacement in this formation is the spontaneous imbibition of water [63, 64]. Capillary effect and wettability are the underlying mechanisms in this case. This effect is attainable if the formation rock is hydrophilic [65]. Strong capillary effect occurs if the matrix is sufficiently water-wet and the fracture network holds enough water. Ideally, this is not the case with carbonate reservoir, as it is characterized by complex microstructures and poor rock wettability (intermediate-wet or oil-wet). This behavior impedes productivity, hence, harnessing substantial oil from this formation is rather unsatisfactory.

Understanding wettability in oil-wet carbonate reservoir is challenging owing to the complex nature of crude oil, and its characterization is even more difficult. Crude oil contains polar compounds which are normally surface-active and capable of altering reservoir rock surfaces when adsorbed [55, 66–69]. Among such polar compounds, asphaltenes and resins have the most polar oil fractions with high surface activity. Asphaltenes are known for their propensity to self-associate in solution, with high surface adsorption behavior. Surface wettability alteration is mainly caused by the asphaltenes through interaction of its polar functional component with the surface polar sites, which leads to operational problems, and such interaction poses even more complexities [66, 70–75]. Since the distribution of the oil in the reservoir is dependent on the degree of the reservoir rock wettability, it is, therefore, imperative to quantify the balance of forces existing at the line of contacts between the rock, oil, and water. Typically, if the oil and water are in contact with the rock surface, either of these fluids will exhibit displacement tendencies of the other or attain equilibrium as the fluids create an angle of contact with the rock. However, such interactions can be affected by factors such as the mineralogy of the rock surface, brine composition, pore roughness and the nature of the oil.

Contact angle and spontaneous imbibition tests are key approaches for assessing formation wettability among other approaches such as relative permeability, capillary pressure/displacement capillary pressure or USBM [49–52, 76]. However, there exists a remarkable variation in the test methods, which is primarily based on how much of the rock surface is exposed to the wetting phase or wetted by water. In contact angle tests, only the outer surface area of the sample is exposed to a drop of water without consideration for the inner surface of the rock, whereas, in the spontaneous imbibition tests, the whole sample is exposed to the wetting phase (**Figure 3**). Thus, the inner surface area of the rock can be accounted for upon displacement of the non-wetting phase (oil) by the wetting phase. For example, in an oil-wet carbonate rock, for oil to be displaced by the wetting phase (water/brine solutions etc.) the capillary barriers must be overcome. If the wetting phase penetrates the rock pores, two key possibilities exist; (a) rock wettability change; (b) the presence of a positive capillary due to the wettability change. Such scenarios can enhance recovery especially if the formation rock is hydrophilic. **Figure 3b** (i-ii) shows a typical case of an oil-wet rock or core sample placed in an imbibition cell containing NaCl brine solution (the wetting phase). The brine imbibes into the rock or core pores and pushes out the oil in the cores. Such expelled oil sticks on the rock surface while been collected at the top of the cell simultaneously for estimation of the recoverable oil. Formation rock with a considerable water-wet condition exhibits high potentials for allowing water into the tight rock matrix pores. Thus, more water-wet rocks allow higher rates in spontaneous imbibition with possibilities of improving recovery. However, maintaining water wetness of formation rocks depends on the extent to which the water film on the rock surface is stable. The presence of unstable water films can lead to oil migration to the rock surface (like the behavior observed in **Figure 3b**), thus, changing the rock surface wettability. With respect to a typical crude oil system, such behavior would lead to adsorption of polar compounds on the solid surface which in turn changes the wetting properties of the solid [51]. Usually the brine present in a typical carbonate reservoir exhibits a somewhat basic pH (7–8), very high concentration of Ca^{2+}, and a very small amount of CO_3^{2-}, thus the rock-water interface becomes positively charged [77]. The carboxylic materials present in the crude oil acts as surface-active materials, and partial dissociation of the acidic group leads to negatively charged oil-water interface. This behavior causes instability of the initial water film between the oil and the rock and the oil comes in contact with the rock yielding mixed-wet characteristics. Several wetting studies have been conducted with crude oil used to alter originally water-wet surfaces to oil-wet in different systems [78–89]. For example; Standnes and

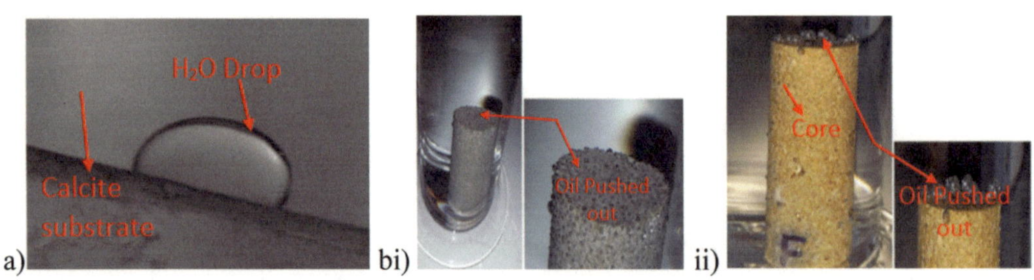

Figure 3. Contact angle versus spontaneous imbibition (a) contact angle: Outer surface area wetted by a drop of water on calcite sample; (b) imbibition: The whole carbonate core sample is wetted by brine (NaCl) in an imbibition cell at different temperature conditions during an imbibition experiment. As the brine imbibes into the cores, the oil is pushed out of the cores to the rock surface and collected at the top of the cell (i) ambient temperature (22 ± 1°C); (ii) 50 ± 1°C showing oil droplets that have been pushed out of the cores on the outer surface of the rocks.

Austad [79, 80] performed a wettability test on chalk cores and calcite mineral surfaces altered by crude oil to a sufficient oil-wet state using surfactant as the surface-active agents via spontaneous imbibition. The authors reported that the cationic surfactant changed the wettability of the chalk by desorbing the organic carboxylates from the chalk surface leading to an increased oil recovery of about 70% from the chalk. Buckley and Lord [83] altered mica surface to oil-wet using series of crude oil through atomic force microscopy (AFM), and found that the oils that produce the thickest coatings exhibited the highest water-advancing angles.

3. Nanoparticles and its effect on solid-liquid interface

Based on the wettability issues in carbonate formations mentioned above, here, we investigated two specific metal oxide nanoparticle types of interest; zirconium oxide and nickel oxide. Firstly, their structure, morphology, and crystallinity phases were examined. Then the wetting inclinations of the nanoparticles were further tested to ascertain their solid-liquid interface behavior on basis of wettability owing to the growing interest in understanding reservoir rock wetting.

3.1. Nanoparticle characterization

Nickel oxide (NiO) is a metal oxide nanoparticle in the form of dark gray (**Figure 4**) crystalline solid. This material has good chemical stability, excellent electrical properties, large exciton binding energy, and a stable wide band gap >3 eV [90–94]. NiO is considered a p-type semiconductor metal oxide particle, thus, a candidate for p-type transparent conducting films [92, 95, 96]. This material also exhibits good optical and magnetic properties; anodic electrochromism properties, excellent durability, large spin optical density, and displays strong insulating property [92, 97–100]. On the basis of reactivity, NiO surface that is considered imperfect acts as a useful oxidation catalyst, although, a perfect NiO is weakly reactive. The perfect surface inertness of NiO is in accordance with the non-metallic properties of the material bulk system [92]. NiO is suitable for usage in electrochemical super-capacitors, dye-sensitized photo cathodes applications and smart windows applications [92, 97–100]. Other processes where uniform size, well-dispersed NiO nanoparticles are also suitable are in heterogeneous catalytic processes, design of ceramics, magnetic applications, fabrication of gas sensors, films, and cathodes of alkaline batteries [101–109]. Although NiO appears to be suitable for a wide variety of processes, its property and functionality depend on the pore morphology, pore matrix-interface and process application. For instance, a very high specific surface area is required for this material in catalytic applications, whereas a rather dense material is required for cathodic applications [96].

Zirconium Oxide (ZrO_2) is a metal oxide nanoparticle in the form of white (**Figure 4**) crystalline solid. ZrO_2 has high refractive index, high melting point of 2680°C, wide region of low absorption from the near-UV > 240 nm to mid-IR range < 8 mm and high resistance against oxidation [110, 111]. This material is also characterized by high breakdown field, good thermal stability, large band gap >5ev, and high-dielectric constant >20 [112–115], thus, the material has been considered a potential challenger of other nanoparticles. In a recent report, it was established that ZrO_2 exhibits superior chemical and thermal stability than alumina and silica nanoparticles [112].

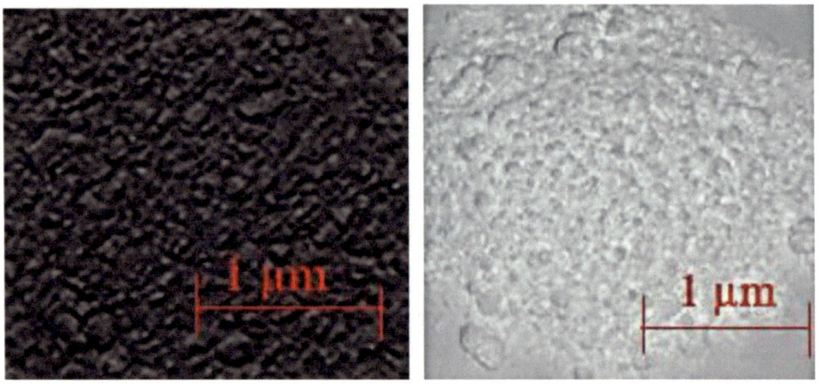

Figure 4. Nanoparticles in powder form (NiO-dark gray green color; ZrO_2-white color).

Similarly, Gopalan et al. [116] earlier reported that silica nanoparticles exhibit limited chemical and physical stability, as such, ZrO_2 nanoparticle was considered as a better alternative and also more chemically stable than γ-alumina or silica. ZrO_2 has an extraordinary high catalytic effect and it is the only metal oxide nanoparticle with four chemical properties on the surface: acidic/basic and reducing/oxidizing properties [117]. ZrO_2 has attracted attention in a wide variety of processes, as the material displays superior mechanical strength, high temperature resistance, high flexural strength, hardness, and low corrosion potential. As such, it can act as a catalyst, refractory, and insulator in transistors in fuel cells, electronic devices, and oxygen sensors, and also suitable for broadband interference filters, laser mirrors, and ionic conductors [118–121].

With respect to wettability, the recent relevance of nanoparticles in wetting processes is mainly due to the particles excellent range of physical and chemical properties as reported earlier. The materials surface and interface properties play an essential role in their overall behavior, whether during preparation or applications. However, there is a lack of understanding of nanoparticles wetting on basis of the solid-liquid interactions, especially, whether strongly hydrophobic surfaces can be rendered hydrophilic, as this is vital for EOR, soil decontamination, and carbon geo-sequestration process efficiency. Since successful oil recovery from fractured carbonate reservoirs show dependency on wettability [122], it is, therefore, necessary to establish an understanding of ZrO_2 and NiO nanoparticles properties, specifically, on the key areas that facilitate their process efficiency and subsequent influence on wetting.

4. Experimental procedure

4.1. Materials

4.1.1. Rock samples

Iceland spar calcite crystals from Ward Science as a representative of carbonate formation as calcite is a predominant mineral constituent of carbonate [38, 58] as also evident in the spectrum analysis of the calcite fraction of a carbonate rock (**Figure 5**; **Table 1**).

a) b)

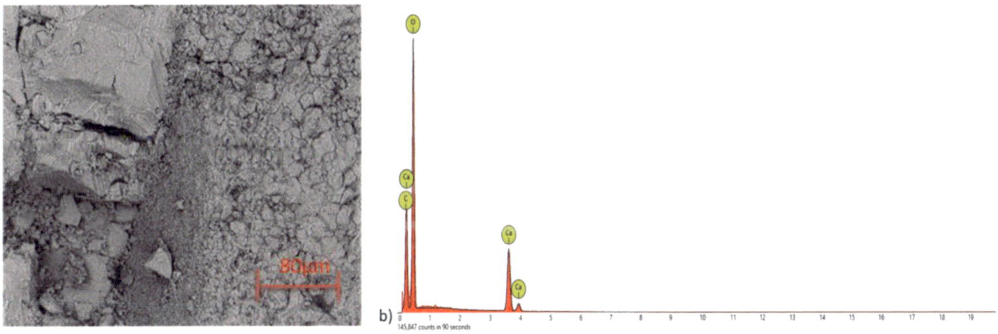

Figure 5. Spectrum analysis of the calcite fraction of the carbonate rock.

Element	Symbol	Atomic concentration (%)	Weight concentration (%)
Calcium	Ca	62.76	57.24
Carbon	C	10.78	24.64
Oxygen	O	26.46	18.12

Table 1. Carbonate rock.

4.1.2. Nanoparticles

Zirconium oxide (Purity: 99.5 wt. %; density: 5.89 g/mL at 25°C (lit.)) and nickel oxide (Purity 99.5 wt. %; density: 6.67 g/mL at 25°C) nanoparticles from Sigma Aldrich were used in this study (**Table 2**).

4.1.3. Oil phase

Dodecyltriethoxysilane obtained from Sigma Aldrich (($C_{18}H_{40}O_3Si$) - purity ≥99.0 mol. %; boiling point: 538.4 k; Density: 875 kg/m^3 - **Figure 6**) was used for altering samples to sufficiently stable oil-wet state. Toluene (purity 99.9 mol. %) obtained from Sigma Aldrich was used as model oil.

Sample	Concentration (Wt. %)	Chemical formula	Molecular weight (g/mol)	Form	Color	Particle Size (nm)
Zirconium Oxide	(0.005–0.05)	ZrO_2	123.22	Nano powder	White	< 50
Nickel Oxide	(0.005–0.05)	NiO	74.69	Nano powder	Dark gray	< 50

Table 2. Properties of nanoparticles.

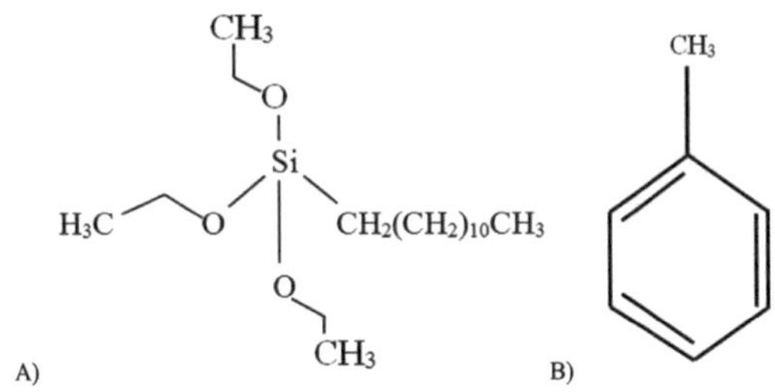

Figure 6. Structure of silane and model oil (A) Dodecyltriethoxysilane; (B) toluene.

4.1.4. Aqueous phase

Sodium Chloride (purity ≥99.5 mol. %) from Rowe Scientific and ultrapure de-ionized water from David Gray was used. The sodium chloride was dissolved in deionized water to achieve desired concentrations using a 220 V/50 Hz magnetic stirrer.

4.2. Sample preparation

Sample preparation which accounts for sample treatments and techniques used, is an important requirement in wetting analysis. Proper sample preparation, as well as, adequate sample quality and cleanliness are essential to eliminate any chances of methodical inaccuracies.

4.2.1. Calcite cleaning

The mineral crystals (calcite) originally hydrophilic were cleaned with analytical reagent grade acetone and methanol (Rowe Scientific Pty. Ltd), and de-ionized water (David Gray & Co. Ltd). This was done to remove surface fragments and inorganic contaminants. Subsequently, the samples were exposed to air plasma for 15 mins [123–125] to remove any residual organic contaminants.

4.2.2. Aging

The clean samples were modified to oil-wet by aging in the oil phase (dodecyltriethoxysilane) for 12 h at 90°C. Samples were then separated from the oil phase, cleaned with methanol, and deionized water to remove excess silane from the surface of the rock and dried.

4.2.3. Nanofluid formulation

ZrO_2 and NiO nanoparticles (Concentration - 0.005 - 0.05 wt. %) were mixed with a fixed amount of dispersals. To ensure adequate particle dispersal in the base fluid and the fluid uniformity, all fluids were formulated using high frequency ultrasonic homogenizer (a 300VT ultrasonic homogenizer and a titanium micro tip of 9.5 mm diameter) as also reported in literature [21–23, 126]. The formulations were kept in a cool place away from heat and light and the nanofluids were subjected to visual monitoring for a fixed period to ensure clear and stable solutions.

4.3. Wettability quantification

Wettability quantification was achieved via contact angle measurement and mechanistic approaches.

4.3.1. Contact Angle

Contact angle (θ) was used as the deterministic tool for wettability assessments. The aged samples were exposed to the formulated ZrO_2 and NiO nanofluids for a fixed period of one hour (1 h). Then the substrates were removed from the nanofluids and dried. A water droplet was dispensed on the modified calcite substrate and a high-performance microscopic camera (Basler scA 640–70 fm, pixel size =7.4 μm; frame rate = 71 fps; Fujinon CCTV lens: HF35HA-1B; 1:1.6/35 mm) was used to capture the water drop dispensing process. The advancing and receding contact angles were measured using a tilting stage [127] for water contact angle in air. Further analysis of the drop was done using Image J software and the standard deviation was ±3 based on replicate measurements.

4.3.2. Mechanistic quantification

Mechanistic investigation of samples was achieved by X-ray powder diffraction, atomic force microscope and scanning electron microscope.

4.3.3. X-ray diffraction (XRD)

Samples were prepared by placing the fine powders in a sample holder that has been well smeared on a glass slide for measurement in a powdered x-ray diffractometer. Diffraction arises through constructive interference due to the illumination of periodic structures of a given spacing with the light of a similar wavelength [128]. The X-ray diffraction patterns of the nanoparticle samples were recorded using powder diffractometer D8 advance (Bruker AXS, Germany), with a copper K alpha radiation source at 40 kV and 40 mA with a LynxEye detector.

4.3.4. Scanning electron microscope (SEM)

The surface morphology of the treated and untreated samples was characterized by scanning electron microscopy - Zeiss Neon 40EsB FIBSEM with an Oxford Instruments x-act Inca SDD x-ray detector and Inca software, and scanning transmission electron microscopy - Tescan Mira3 FESEM instrument. High electron beam was used to scan over the surface of the sample for improved surface characterization. Several researchers have also reported the effectiveness of SEM for morphological evaluations [129–134].

5. Results and discussion

5.1. Scanning electron microscope

The scanning electron microscope images show the micrograph of ZrO_2 and NiO nanoparticles (**Figure 7**). The ZrO_2 nanoparticle exhibits a distribution of sphere-like shaped particles in the range of 25–40 nm, while the NiO nanoparticle displays hexagonal-like shaped particles in the range of 10–20 nm. The nanoparticles exhibited approximately uniform size distribution and high trends of finely dispersed particles in the bulk state, thus, an indication of low particle-particle agglomeration inclinations. This behavior can be attributed to the intrinsic properties of metal oxide (superior stability) and preparation method (homogenization - Section 1).

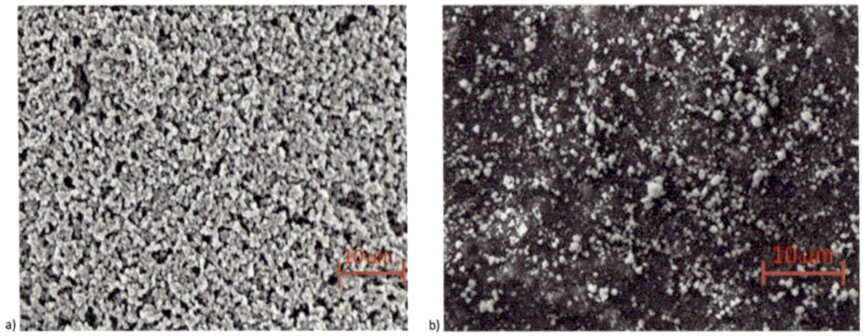

Figure 7. SEM images of: (a) ZrO_2; (b) NiO nanoparticles.

5.2. Particle size and surface area

Particle size plays a vital role in the nanoparticle characterization as their physical and chemical properties greatly depend on the particle size. The small particle size of nanoparticles yields important features such as surface area. Nanoparticle size and surface area are interrelated; as the nanoparticle size becomes negligible, the particle surface area increases. Bulk materials as opposed to nanoparticle exhibit larger particle sizes (nanoparticle avarage diameter: <100 nm; microparticle >1 μm) with tons of atoms on the inside of the particle and limited atoms at the surface, whereas, with nanoparticles more atoms are predominantly on the outer surface of the particles. Such high surface area enables the bonding of other materials on the particle surface and lead to the generation of even much stronger materials that promote better interaction with neighboring atoms or ions. Ascertaining the nanoparticles size is essential as it affects particulate materials properties and can act as an approach to determine the quality and performance of these materials. The particle size of the ZrO_2 and NiO nanoparticles was investigated to obtain more accurate and precise size of the particles. **Figure 8** shows that the ZrO_2 particle size was 21–35 nm and NiO was in the range of 10–12 nm. The average particle diameter for ZrO_2 ~ 28 nm and the average value for NiO was ~ 12 nm.

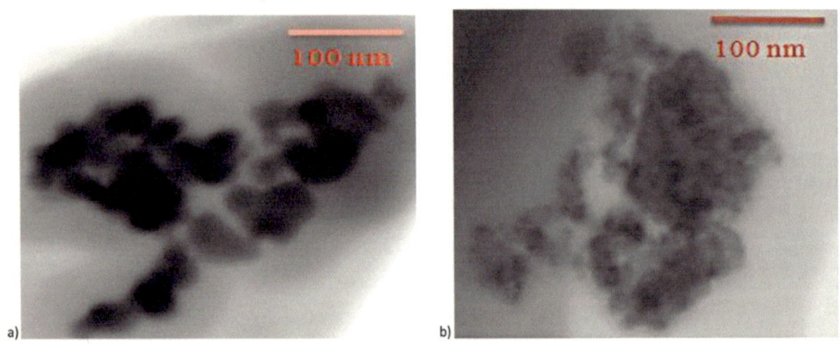

Figure 8. Particle size morphology of a) ZrO_2; b) NiO nanoparticles (<50 nm).

5.3. X-ray diffraction analysis

To better understand the crystallographic nature of ZrO_2 and NiO nanoparticles, the materials were further characterized using X-ray Diffraction (XRD). The XRD peaks usually exhibit different patterns and positioning. The pattern of the XRD of a specific sample is mainly dependent on the different arrangements of the atoms. The unit cell dimensions and angles determine the positions of the peaks. Whereas, the types and positions of the atoms within the unit cell determine the intensities of the peaks [128]. **Figure 9** shows the typical XRD patterns of the ZrO_2 (in red) and NiO (black) nanoparticles. The scan parameters used for phase identification were 2theta scan range (degree): 7.5–90; Step size (degree): 0.015; Time/step: 0.7 s and total scan time of approximately 1 hr. The XRD pattern indicates the crystallographic structure of the nanoparticles. Strong and sharp diffraction peaks at 2θ values were observed in both cases (ZrO_2 - in red; NiO - in black), especially, for NiO with precise peaks - 37, 43.5, 63 etc., while the ZrO_2 diffraction Peaks value were 28.5, 31.5, 34.5, 50.5 etc. The peaks are indications of the pure phase formation of the ZrO_2 and NiO nanoparticles. The crystallographic phases of the nanoparticles were also identified. Pure ZrO_2 exhibits a tetragonal phase (**Figure 10** (i)), and the pure NiO exhibits a cubic phase (**Figure 10** (ii)), consistent with literature [92, 135].

5.4. Contact angle

An understanding of surface chemistry is imperative for evaluating wetting behavior as porous media wetting are influenced by the rock surface morphology, as well as, the chemical compositions. Surface chemistry modifications of materials facilitate short-ranged chemical interactions. This phenomenon is predominantly governed by the surface and interfacial interactions, which act over the scale of molecules, and electrostatic surface forces that determine the extent to which

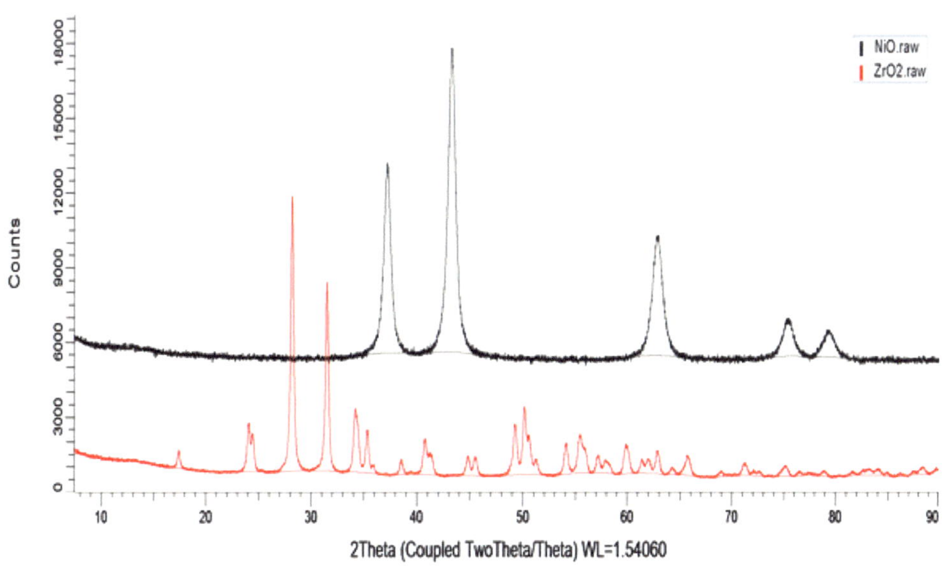

Figure 9. XRD patterns of the ZrO_2 and NiO nanoparticles.

a fluid can wet a surface [48]. Quantification of wettability of solid surfaces was also performed to ascertain the effects of ZrO_2 and NiO nanoparticle on wettability alteration of carbonate rocks. Contact angle tests were conducted at solid-liquid-air interface to ascertain the wetting variances prior and after nano-modifications. Water-advancing and receding contact angles were measured of which the advancing contact angles better defines wettability since its relevant to waterflooding [51]. The understanding of contact angle is complex as it exhibits a dependency on the solid-liquid interaction and the structure of the solid or mineralogy of the rock sample.

Prior to the contact angle measurement, calcite substrates were cleaned and aged based on procedure 4.2.2. The nanofluids were prepared using various nanoparticle concentration (0.005–0.05 wt. %) and fixed NaCl brine concentration (7 wt. %) as dispersing agents based on procedure 4.2.2. The samples were immersed in the nanofluids and subsequently contact

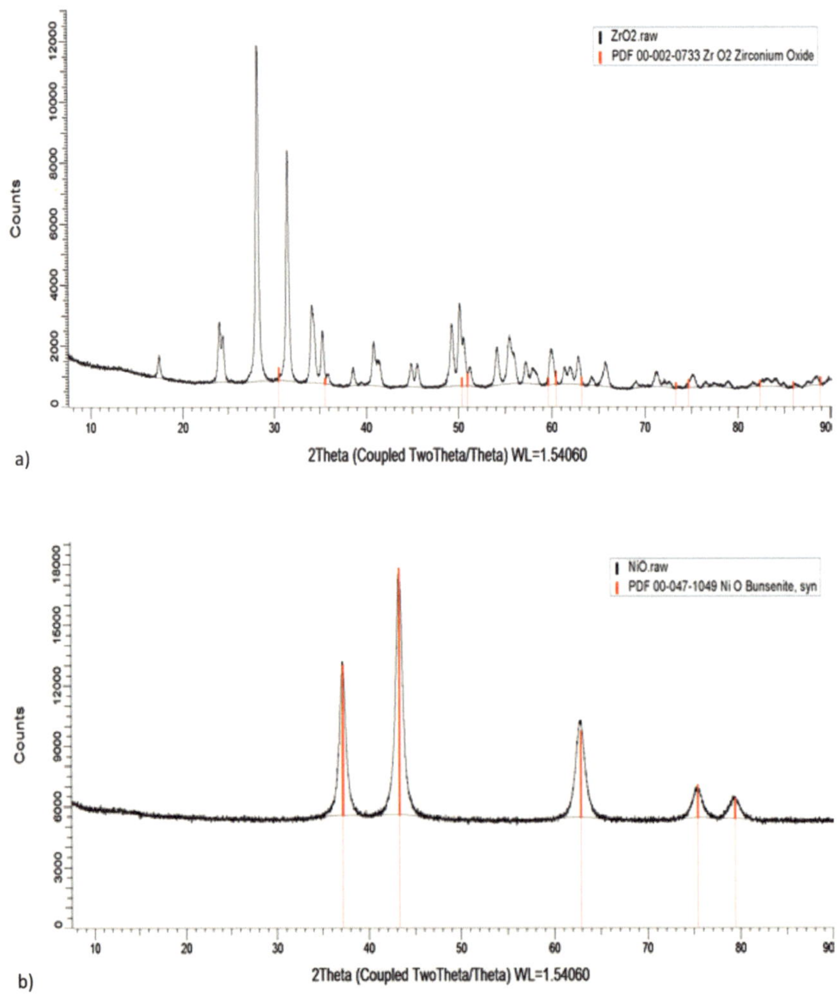

Figure 10. (a) Crystallographic phases of the ZrO_2 nanoparticles exhibiting tetragonal phase. (b). Crystallographic phase of the NiO nanoparticles exhibiting cubic phase.

http://dx.doi.org/10.5772/intechopen.72267

angles were measured. **Figures 12** and **13** (SEM) show nano-modified calcites, as compared to the fresh calcite without any nano-modification (**Figure 11**). The ZrO_2 nanoparticle exhibits a spherical-like shape and more uniformly adsorbed behavior on the calcite surface than NiO, while the NiO exhibits a hexagonal-like shape which is consistent with the earlier observation in Section 4.31. **Figure 15** shows the contact angle measurement as a function of nanoparticle concentration for ZrO_2 and NiO nanoparticle respectively in air. It is expected that the exposure of the calcite to a different environment would change its surface property. The calcite upon contact with the ZrO_2 and NiO nanofluids, film-like deposits of the nanofluids were adsorbed on the calcite surface after exposure to air. Such film thickness can vary between a fraction to numerous fractions of a molecule. However, the level of thickness is dependent on the affinity of the molecules to the substrate and the corresponding distance to the bulk critical point. Moreso, the nano-films on the rock surface may appear thinner than others, which is dependent on the nanoparticle type, and their optical and electrical properties. ZrO_2 nano-films are relatively thicker than those of other nanoparticles owing to its material properties such as large band gap >5ev and high-dielectric constant >20 (Petit and Monot, 2015). This may have formed better-adsorbed ZrO_2 nano-layers on the oil-wet calcite surface when compared to the NiO nanoparticles.

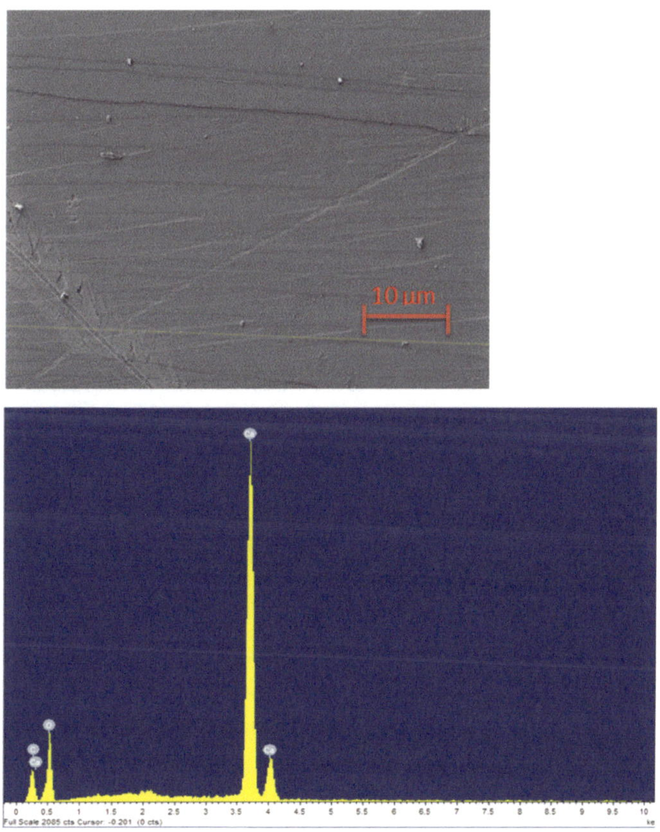

Figure 11. SEM image of pure calcite surface before nano-modification.

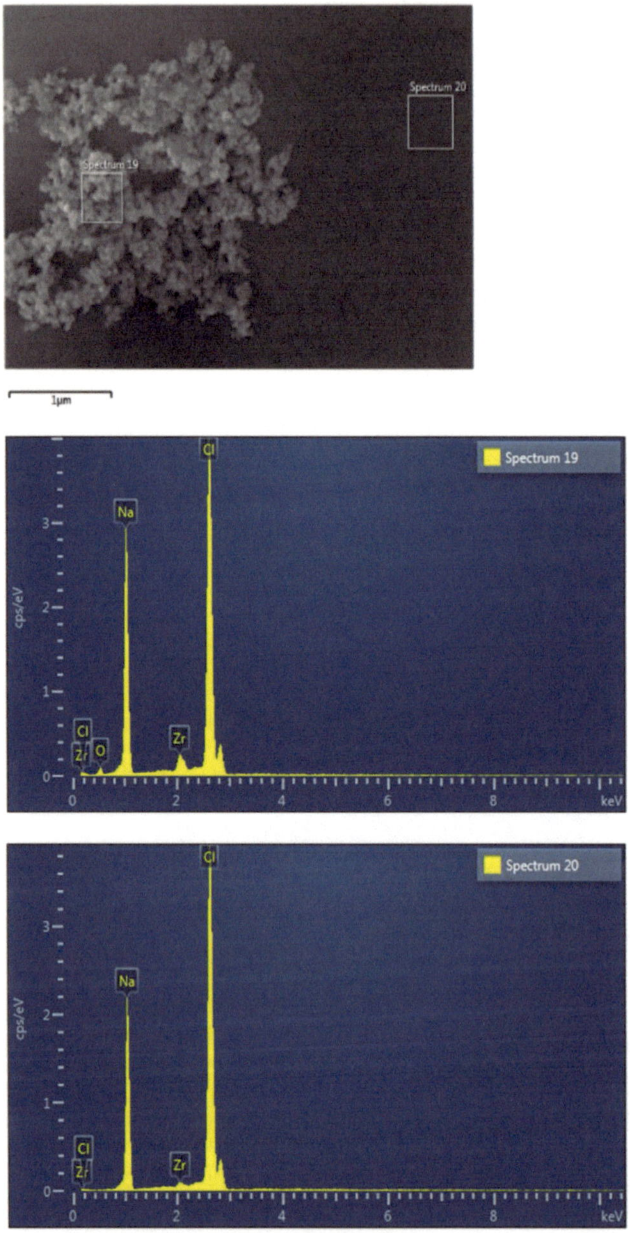

Figure 12. SEM image of calcite surface after ZrO_2 nano-modification.

As the dispensed water droplet comes in contact with the surface of the nano-coated calcite, the contact angle decreases owing to the presence of adsorbed nano-layers. This behavior is attributed to the favorable interaction of the nanoparticles with the dispersing fluid (NaCl brine) and high chemical affinity to the calcite. Such solid-liquid interaction at the interface is mainly due to electrostatic interactions. The presence of the nanoparticle increased the surface activity of the brine (NaCl), thereby modifying the calcite surface wetting propensity upon contact. The nanoparticles in suspension act as a coating mechanism by self-structuring into layered NPs and changes the entropy of the system. The particles hydrophilic nature facilitates their adsorption on the rock surface in form of a wedge film which in turn displaces the oil on the surface of the rock, yielding a hydrophilic state. Ideally, if wettability is preferentially altered to favorable water-wet condition and the IFT is ultralow, the forces that retain oil in a fractured reservoir can be overcome as capillarity is diminished through the ultralow IFTs.

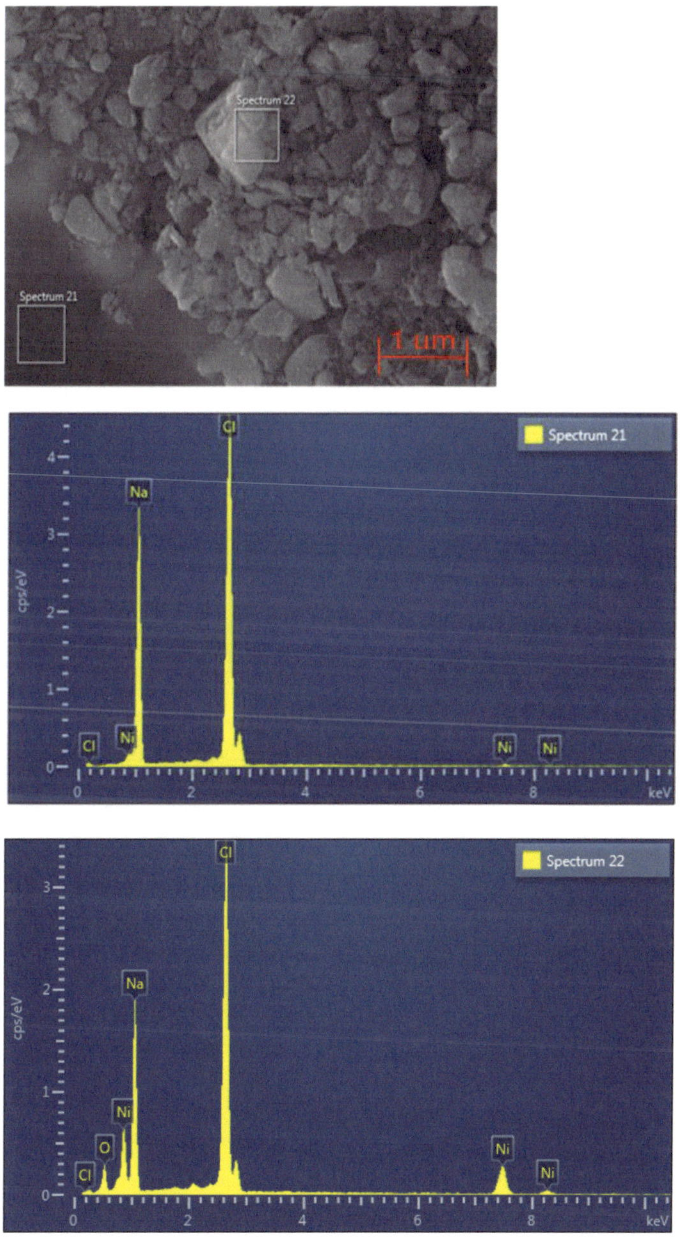

Figure 13. SEM image of calcite surface after NiO nano-modification.

Figure 14 shows the image representation of the transition phase toward water-wet in air from 88° θ_a to 48° θ_a (NiO/NaCl) and 38° θ_a (ZrO_2/NaCl). A decrease in contact angle was observed for all the systems tested with an increase in the nanoparticle concentration (**Figure 15**) consistent with literature [36–38, 136–139]. Calcite substrates coated with ZrO_2/NaCl fluids demonstrated better wetting propensities than the NiO/NaCl system. The efficiency of the systems is due to efficient surface adsorption of the particles on the pore walls of the rock, which invariably rendered the rock surface sufficiently water wet upon contact.

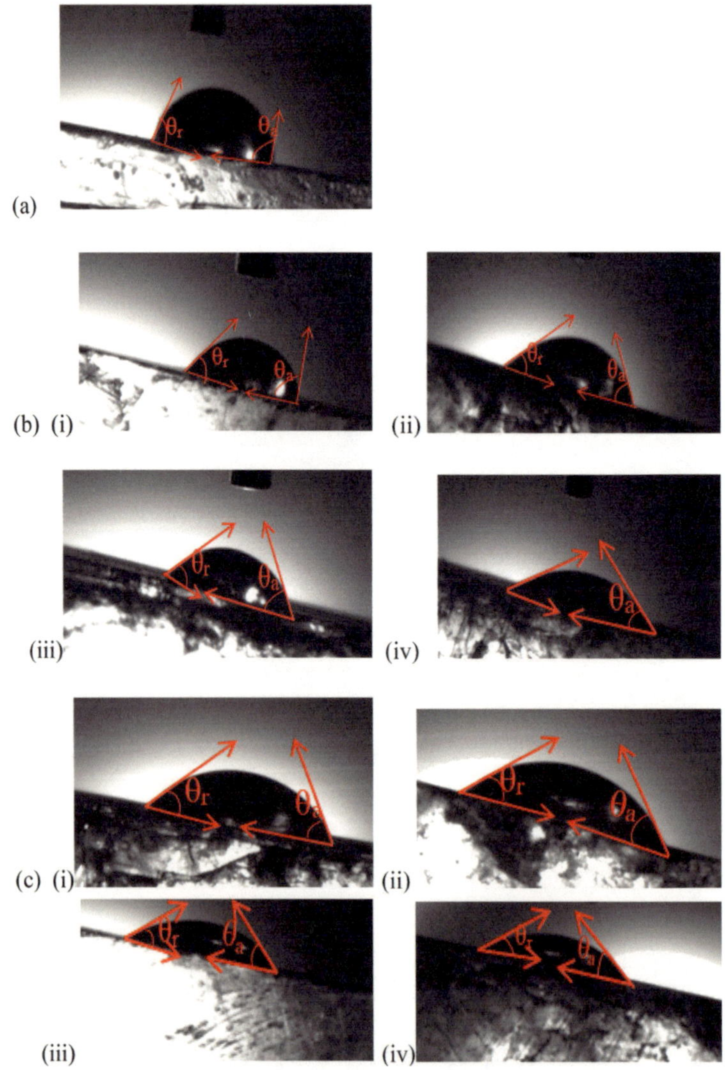

Figure 14. Contact angle images showing variation with increase in nanoparticle concentration (a) unmodified (high θ-indicating an intermediate-wet state—88° θ_a); (B, C) nano-modified—(B) NiO/NaCl modified (NiO concentration—0.005–0.05); (C) ZrO_2/NaCl nano-modified (ZrO_2 concentration—0.005-0.05) (B and C indicates low θ which represents strong interaction with the rock surface, and inclination to wet; I–IV); see graphical representation for θ values.

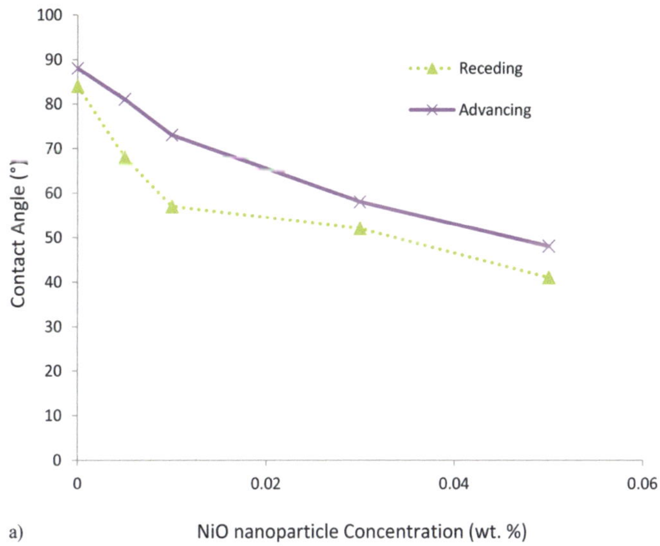

a)

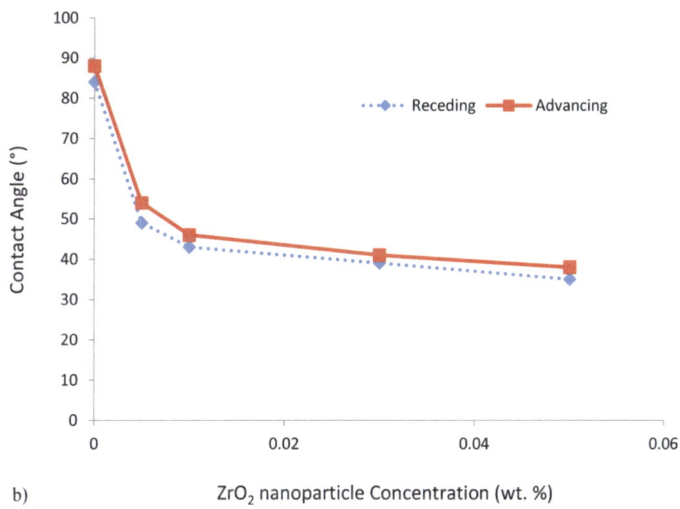

b)

Figure 15. Receding and advancing water contact angles in air (1 h exposure time) (a) NiO; (b) ZrO_2 at ambient condition.

6. Conclusions

The interfacial behavior of nickel oxide and zirconium oxide nanoparticles at solid-liquid interface was studied on their propensity to alter oil-wet surfaces toward water-wet conditions. The ZrO_2 and NiO nanoparticles exhibited very different structural and morphological features, as well as crystallinity phases. The nanoparticles exhibited particles size in ranges below 50 nm (< 50 nm) and different shape patterns. The ZrO_2 nanoparticles are sphere-like shaped particles while the NiO displayed hexagonal-like shaped particles (**Figures 7**, **12** and **13**). The XRD crystallographic structure and phase identification shows the tetragonal phase of the ZrO_2, whereas, the NiO nanoparticle has a cubic phase orientation. The nanoparticles also displayed favorable adsorption behavior on the calcite surface as evident in the SEM images, which facilitated the wetting change quantified by contact angle measurement, however, the ZrO_2 based systems exhibited more uniform surface distribution and better wetting than NiO. Thus, nanoparticles are considered efficient modifiers for wettability alteration of surfaces toward a suitable hydrophilic condition.

Author details

Lezorgia Nekabari Nwidee[1]*, Ahmed Barifcani[1,2], Mohammad Sarmadivaleh[1] and Stefan Iglauer[1]

*Address all correspondence to: l.nwidee@postgrad.curtin.edu.au

1 Department of Petroleum Engineering, Curtin University, Perth, Western Australia, Australia

2 Department of Chemical Engineering, Curtin University, Perth, Western Australia, Australia

References

[1] Campelo JM, Luna D, Luque R, Marinas JM, Romero AA. Sustainable preparation of supported metal nanoparticles and their applications in catalysis. ChemSusChem. 2009; **2**(1):18-45

[2] Colomban P. The use of metal nanoparticles to produce yellow, red and iridescent colour, from bronze age to present times in lustre pottery and glass: Solid state chemistry, spectroscopy and nanostructure. Journal of Nano Research. 2009;**8**:109-132 Trans Tech Publications

[3] Heiligtag FJ, Niederberger M. The fascinating world of nanoparticle research. Materials Today. 2013;**16**(7):262-271

[4] Kaminsky R, Radke CJ. Asphaltenes, water films and wettability reversal. Society of Petroleum Engineers Journal. 1997;**2**(4):458-493

[5] Kamyshny AS, Magdassi S. Aqueous dispersions of metallic nanoparticles. Colloidal Interfacial Aspects. 2010:747-778

[6] Edmunds N, Chhina H. Economic optimum operating pressure for SAGD projects in Alberta. Journal of Canadian Petroleum Technology. 2001;**40**(12):13-17

[7] Nanotek. Pioneering Expansion in the world of Nanotechnology. 19th International Conference on Nanotechnology and Expo. Atlanta, Georgia, USA. 2017. Retrieved on 15 April, 2017 from http://nanotechnologyexpo.conferenceseries.com/

[8] Yu W, France DM, Routbort JL, Choi SU. Review and comparison of nanofluid thermal conductivity and heat transfer enhancements. Heat Transfer Engineering. 2008;**29**(5):432-460

[9] Ramakoteswaa RN, Gahane L, Ranganayakulu SV. Synthesis, applications and challenges of nanofluids-review. IOSR Journal of Applied Physics. 2014:21-28

[10] Biswas P, Wu CY. Control of toxic metal emissions from combustors using sorbents: A review. Journal of the Air & Waste Management Association. 1998;**48**(2):113-127

[11] Serrano E, Rus G, Garcia-Martinez J. Nanotechnology for sustainable energy. Renewable and Sustainable Energy Reviews. 2009;**13**:2373-2384

[12] Khalil M, Jan BM, Tong CW, Berawi MA. Advanced Nanomaterials in oil and gas industry: Design, application, and challenges. Applied Energy. 2017;**191**:287-310

[13] Jun YW, Choi JS, Cheon J. Heterostructured magnetic nanoparticles: Their versatility and high-performance capabilities. Chemical Communications. 2007;**12**:1203-1214

[14] Pastrana-Martínez LM, Pereira N, Lima R, Faria JL, Gomes HT, Silva AMT. Degradation of diphenhydramine by photo-fenton using magnetically recoverable iron oxide nanoparticles as catalyst. Chemical Engineering Journal. 2015;**261**:45-52

[15] Avendano C, Lee SS, Escalera G, Colvin V. Magnetic characterization of nanoparticles designed for use as contrast agents for downhole measurements. In: SPE 157123 presented at the SPE international oilfield nanotechnology conference and exhibition, Noordwijk, The Netherlands. 12-14 June 2012

[16] Morrow L, Potter DK, Barron AR. Detection of magnetic nanoparticles against proppant and shale reservoir rocks. Journal of Experimental Nanoscience. 2015;**10**(13):1028-1041

[17] Hyne JB, Greidanus JW, Tyrer JD, Verona D, Rizek C, Clark PD, Clarke RA, Koo J. Aquathermolysis of heavy oils. The Second International Conference on Heavy Crude and Tar Sands, in Caracas, Venezuela, Feb 7-17, 1982

[18] White RJ, Luque R, Budarin VL, Clark JH, Macquarrie DJ. Supported metal nanoparticles on porous materials, methods and applications. Chemical Society Reviews. 2009;**38**(2):481-494

[19] Fedele L, Colla L, Bobbo S, Barison S, Agresti F. Experimental stability analysis of different water-based nanofluids. Nanoscale Research Letters. 2011;**6**(1):300

[20] Roustaei A, Bagherzadeh H. Experimental investigation of SiO_2 nanoparticles on enhanced oil recovery of carbonate reservoirs. Journal of Petroleum Exploration and Production Technology. 2015;**5**(1):27-33

[21] Tjong SC. Novel nanoparticle-reinforced metal matrix composites with enhanced mechanical properties. Advanced Engineering Materials. 2007;**9**(8):639-652

[22] Yang Y, Lan J, Li X. Study on bulk aluminum matrix nano-composite fabricated by ultrasonic dispersion of nano-sized SiC particles in molten aluminum alloy. Materials Science and Engineering A. 2004;**380**(1):378-383

[23] Mao H, Qiu Z, Shen Z, Huang W. Hydrophobic associated polymer based silica nanoparticles composite with core-shell structure as a filtrate reducer for drilling fluid at utra-high temperature. Journal of Petroleum Science and Engineering. 2015;**129**:1-14

[24] Salem RAM, Hannora AEA. Comparative investigation of nanoparticle effects for improved oil recovery-experimental work. In: Proceedings of the SPE Kuwait Oil and Gas Show and Conference, Mishref, Kuwait, 11-14 October 2015

[25] Metin CO, Baran JR, Nguyen QP. Adsorption of surface functionalized silica nanoparticles onto mineral surfaces and decane/water interface. Journal of Nanoparticle Research. 2012;**14**(11):1246

[26] Hendraningrat L, Torsæter O. Metal oxide-based nanoparticles: Revealing their potential to enhance oil recovery in different wettability systems. Applied Nanoscience. 2015;**5**(2):181-199

[27] Ahmadi MA, Shadizadeh SR. Induced effect of adding nano silica on adsorption of a natural surfactant onto sandstone rock: Experimental and theoretical study. Journal of Petroleum Science and Engineering. 2013;**112**:239-247

[28] Bayat AE, Junin R, Shamshirband S, Chong WT. Transport and retention of engineered Al_2O_3, TiO_2, and SiO_2 nanoparticles through various sedimentary rocks. Scientific Reports. 2015;**5**:14264

[29] Saravanan P, Gopalan R, Chandrasekaran V. Synthesis and characterization of nanomaterials. Defence Science Journal. 2008;**58**(4):504-516

[30] Rudyak VY. Viscosity of nanofluids-why it is not described by the classical theories. Advances in Nanoparticles. 2013;**2**(3):266-279

[31] Choi S. Enhancing thermal conductivity of fluids with nanoparticles. In: Siginer DA, Wang HP, editors. Developments Applications of Non-Newtonian Flows. New York: 231, ASME; 1995. pp. 99-105

[32] Yu W, Xie H. A review on nanofluids: Preparation, stability mechanisms and applications. Journal of Nanomaterials. 2012:1-17

[33] Al-Anssari S, Barifcani A, Wang S, Maxim L, Iglauer S. Wettability alteration of oil-wet carbonate by silica nanofluid. Journal of Colloid and Interface Science. 2016;**461**:435-442

[34] Al-Anssari S, Wang S, Barifcani A, Iglauer S. Oil-water interfacial tensions of silica nanoparticle-surfactant formulations. Tenside, Surfactants, Detergents;**54**:6

[35] Keblinski P, Prasher R, Eapen J. Thermal conductance of nanofluids: Is the controversy over? Journal of Nanoparticle Research. 2008;**10**(7):1089-1097

[36] Nwidee LN, Al-Anssari S, Barifcani A, Sarmadivaleh M, Iglauer S. Nanofluids for enhanced oil recovery processes: wettability alteration using zirconium oxide. Offshore Technology Conference (OTC-26573-MS). Kuala Lumpur, Malaysia, 22-25 March 2016a

[37] Nwidee LN, Al-Anssari S, Barifcani A, Sarmadivaleh M, Lebedev M, Iglauer S. Nanoparticles influence on wetting behaviour of fractured limestone formation. Journal of Petroleum Science and Engineering. 2017a;**149**:782-788

[38] Nwidee LN, Al-Anssari S, Barifcani A, Sarmadivaleh M, Lebedev M, Iglauer S. Wettability alteration of oil-wet limestone using surfactant-nanoparticle formulation. Journal of Colloid and Interface Science. 2017b;**504**:334-345

[39] Nwidee LN, Theophilus S, Barifcani A, Sarmadivaleh M, Iglauer S. EOR processes, opportunities and technological advancements. In: Chemical Enhanced Oil Recovery (cEOR)-a Practical Overview. InTech; 2016

[40] Wang XQ, Mujumdar AS. A review on nanofluids-part I: Theoretical and numerical investigations. Brazilian Journal of Chemical Engineering. 2008;**25**(4):613-630

[41] Yang Y, Lan J, Li X. Study on bulk aluminum matrix nano-composite fabricated by ultrasonic dispersion of nano-sized SiC particles in molten aluminum alloy. Materials Science and Engineering A. 2004;**380**(1):378-383

[42] Guo D, Xie G, Luo J. Mechanical properties of nanoparticles: Basics and applications. Journal of Physics D: Applied Physics. 2014;**47**(1):013001

[43] De Gennes PG. Wetting: Statistics and dynamics. Reviews of Modern Physics. 1985; **57**(3):827-863

[44] Mohammed M, Babadagli T. Wettability alteration: A comprehensive review of materials/methods and testing the selected ones on heavy-oil containing oil-wet systems. Advances in Colloid and Interface Science. 2015;**220**:54-77

[45] Bertrand E, Bonn D, Broseta D, Dobbs H, Indekeu JO, Meunier J, Ragil K, Shahidzadeh N. Wetting of alkanes on water. Journal of Petroleum Science and Engineering. 2002; **33**(1):217-222

[46] Myers D. Surfaces, Interfaces and Colloids. Wiley-Vch: New York; 1990

[47] Craig FF. The Reservoir Engineering Aspects of Waterflooding. SPE, Richardson, Texas: Monograph Series; 1971

[48] Bonn D, Eggers J, Indekeu J, Meunier J, Rolley E. Wetting and spreading. Reviews of Modern Physics. 2009;**81**:739-805

[49] Anderson WG. Wettability literature survey - part 1: Rock/oil/brine interactions and the effects of core handling on wettability. Journal of Petroleum Techology. 1986; **38**(10):1125-1144

[50] Anderson WG. Wettability literature survey part 2: Wettability measurement. Journal of Petroleum Techology. 1986;**38**:1246-1262

[51] Anderson WG. Wettability literature survey-part 3: The effects of wettability on the electrical properties of porous media. Journal of Petroleum Technology. 1986;**38**(12):1371-1378

[52] Morrow NR. Wettability and its effect on oil recovery. Journal of Petroleum Technology. 1990;**42**(12):1-476

[53] Abdallah W, Buckley J, Carnegie A, Edwards J, Herold B, Fordham E, Graue A, Habashy T, Seleznev N, Signer C, Hussain H. Fundamentals of wettability. Oilfield Review. 2007:44-61

[54] Nutting PG. Some physical and chemical properties of reservoir rocks bearing on the accumulations and discharge of oil. In: Wrather WE, Lahee FH, editors. Problems of Petroleum Geology. American Association of Petroleum Geologists; 1934. pp. 825-832

[55] Benner FC, Bartel FE. The effect of polar impurities upon capillary and surface phenomena in petroleum production. Drilling and production practice. American Petroleum Institute, New York, New York ,1 January 1941

[56] Moore CH. Carbonate Reservoirs Porosity Evolution and Diagenesis in a Sequence Stratigraphic Framework. Amsterdam: Elsevier; 2001

[57] Tucker ME, Wright VP. Carbonate sedimentology. John Wiley and Sons; 2009

[58] Lebedev M, Wilson ME, Mikhaltsevitch V. An experimental study of solid matrix weakening in water-saturated Savonnières limestone. Geophysical Prospecting. 2014; **62**(6):1253-1265

[59] Treiber LE, Archer DL, Owens WW. A laboratory evaluation of the wettability of fifty oil-producing reservoirs. Society of Petroleum Engineers Journal. 1972;**12**(6):531-540

[60] Chilingar GV, Yen TF. Some notes on wettability and relative permeabilities of carbonate reservoir rocks II. Journal of Energy Sources. 1983;**7**(1):67-75

[61] Buckley JS, Liu Y, Monsterleet S. Mechanisms of wetting alteration by crude oils. Society of Petroleum Engineering Journal. 1998;**3**(01):54-61

[62] Thomas S. Enhanced oil recovery-an overview. Oil & Gas Science and Technology. 2008;**63**(1):9-19

[63] Babadagli T. Scaling of co-current and counter-current capillary imbibition for surfactant and polymer injection in naturally fractured reservoirs. SPE/AAPG western regional meeting, Long Beach, California, 19-22 June 2000

[64] Cheraghian G, Hemmati M, Masihi M, Bazgir S. An experimental investigation of the enhanced oil recovery and improved performance of drilling fluids using titanium dioxide and fumed silica nanoparticles. Journal of Nanostructure in Chemistry. 2013;**3**(1):1-19

[65] Delshad M, Najafabadi NF, Anderson G, Pope GA, Sepehrnoori K. Modeling wettability alteration by surfactants in naturally fractured reservoirs. SPE Reservoir Evaluation & Engineering. 2009;**12**(3):361-370

[66] Buckley JS, Liu Y, Monsterleet S. Mechanisms of wetting alteration by crude oils. Society of Petroleum Engineering Journal. 1998;**3**(01):54-61

[67] Yu L, Buckley JS. Evolution of wetting alteration by adsorption from crude oil. Society of Petroleum Engineers Formation Evaluation. 1997;**12**(01):5-12

[68] Tong Z, Xie X, Morrow NR. Crude oil composition and the ability of mixed wettability in sandstones. SCA International Symposium of the Society of Core Analysts, Monterey, CA, USA, 30 September 2002

[69] Lord DL, Buckley JS. An AFM study of the morphological features that affect wetting at crude oil-water-mica interfaces. Colloids and Surfaces A: Physicochemical and Engineering Aspects. 2002;**206**(1):531-546

[70] Acevedo S, Escobar G, Ranaudo MA, Piñate J, Amorı′n A, Dı′az M, Silva P. Observations about the structure and dispersion of petroleum asphaltenes aggregates obtained from dialysis fractionation and characterization. Energy & Fuels. 1997;**11**(4):774-778

[71] González G, Moreira MB. The wettability of mineral surfaces containing adsorbed asphaltene. Colloids and Surfaces. 1991;**58**(3):293-302

[72] Andersen SI, Christensen SD. The critical micelle concentration of asphaltenes as measured by calorimetry. Energy & Fuels. 2000;**14**(I1):38-42

[73] Kaminsky R, Radke CJ. Asphaltenes, water films and wettability reversal. Society of Petroleum Engineers Journal. 1997;**2**(4):458-493

[74] Liu L, Buckley JS. Alteration of wetting of mica surfaces. Journal of Petroleum Science and Engineering. 1999;**24**(2):75-83

[75] Standal S, Haavik J, Blokhus AM, Skauge A. Effect of polar organic components on wettability as studied by adsorption and contact angles. Journal of Petroleum Science and Engineering. 1999;**24**(2-4):131-144

[76] Donaldson EC, Thomas RD, Lorenz PB. Wettability determination and its effects on recovery efficiency. Society of Petroleum Engineering Journal. 1969;**9**(1):13-20

[77] Zhang P, Austad T. Wettability and oil recovery from carbonates: Effects of temperature and potential determining ions. Colloids and Surfaces A: Physicochemical and Engineering Aspects. 2006;**279**(1):179-187

[78] Standnes DC, Austad T. Nontoxic low-cost amines as wettability alteration Chemicals in Carbonates. Journal of Petroleum Science and Engineering. 2003;**39**(3):431-438

[79] Standnes DC, Austad T. Wettability alteration in chalk: 1. Preparation of core material and oil properties. Journal of Petroleum Science and Engineering. 2000;**28**(3):111-121

[80] Standnes DC, Austad T. Wettability alteration in chalk: 2. Mechanism for wettability alteration from oil-wet to water-wet using surfactants. Journal of Petroleum Science and Engineering. 2000;**28**(3):123-143

[81] Standnes DC, Austad T. Wettability alteration in carbonates: Interaction between cationic surfactant and carboxylates as a key factor in wettability alteration from oil-wet to water-wet conditions. Colloids and Surfaces A: Physicochemical and Engineering Aspects. 2003;**216**(1):243-259

[82] Standnes DC, Nogaret LA, Chen HL, Austad T. An evaluation of spontaneous imbibition of water into oil-wet carbonate reservoir cores using a nonionic and a cationic surfactant. Energy and Fuels. 2002;**16**(6):1557-1564

[83] Buckley JS, Lord DL. Wettability and morphology of mica surfaces after exposure to crude oil. Journal of Petroleum Science and Engineering. 2003;**39**(3):261-273

[84] Zhang DL, Liu S, Puerto M, Miller CA, Hirasaki GJ. Wettability alteration and spontaneous imbibition in oil-wet carbonate formations. Journal of Petroleum Science and Engineering. 2006;**52**(1):213-226

[85] Seiedi O, Rahbar M, Nabipour M, Emadi MA, Ghatee MH, Ayatollahi S. Atomic force microscopy (AFM) investigation on the surfactant wettability alteration mechanism of aged mica mineral surfaces. Energy and Fuels. 2010;**25**(1):183-188

[86] Kathel P, Mohanty KK. Wettability alteration in a tight oil reservoir. Energy and Fuels. 2013;**27**(11):6460-6468

[87] Bera A, Ojha K, Kumar T, Mandal A. Mechanistic study of wettability alteration of quartz surface induced by nonionic surfactants and interaction between crude oil and quartz in the presence of sodium chloride salt. Energy and Fuels. 2012;**26**(6):3634-3643

[88] Hou BF, Wang YF, Huang Y. Mechanistic study of wettability alteration of oil-wet sandstone surface using different surfactants. Applied Surface Science. 2015;**330**:56-64

[89] Hou BF, Wang YF, Huang Y. Relationship between reservoir wettability and oil recovery by waterflooding. In: In Advances in Energy Science and Equipment Engineering, Edited by Zhou, Patty and Chen. London: Taylor Francis Group; 2015

[90] Kamal H, Elmaghraby EK, Ali SA, Abdel-Hady K. Characterization of nickel oxide films deposited at different substrate temperatures using spray pyrolysis. Journal of Crystal Growth. 2004;**262**(1):424-434

[91] Choi JM, Im S. Ultraviolet enhanced Si-photodetector using p-NiO films. Applied Surface Science. 2005;**244**(1):435-438

[92] Kunz AB. Electronic structure of NiO. Journal of Physics C: Solid State Physics. 1981; **14**(16):L455

[93] Adler D, Feinleib J. Electrical and optical properties of narrow-band materials. Physical Review B. 1970;**2**(8):3112

[94] Irwin MD, Buchholz DB, Hains AW, Chang RP, Marks TJ. P-type semiconducting nickel oxide as an efficiency-enhancing anode interfacial layer in polymer bulk-heterojunction solar cells. Proceedings of the National Academy of Sciences. 2008;**105**(8):2783-2787

[95] Sato H, Minami T, Takata S, Yamada T. Transparent conducting p-type NiO thin films prepared by magnetron sputtering. Thin Solid Films. 1993;**236**(1):27-31

[96] Bahadur J, Sen D, Mazumder S, Ramanathan S. Effect of heat treatment on pore structure in nano-crystalline NiO: A small angle neutron scattering study. Journal of Solid State Chemistry. 2008;**181**(5):1227-1235

[97] He J, Lindström H, Hagfeldt A, Lindquist SE. Dye-sensitized nanostructured p-type nickel oxide film as a photocathode for a solar cell. The Journal of Physical Chemistry B. 1999;**103**(42):8940-8943

[98] Srinivasan V, Weidner JW. An electrochemical route for making porous nickel oxide electrochemical capacitors. Journal of the Electrochemical Society. 1997;**144**(8):L210-L213

[99] Liu KC, Anderson MA. Porous nickel oxide/nickel films for electrochemical capacitors. Journal of the Electrochemical Society. 1996;**143**(1):124-130

[100] Chakrabarty S, Chatterjee K. Synthesis and characterization of nano-dimensional nickelous oxide (NiO) semiconductor. Journal of Physical Science. 2009;**13**:245-250

[101] Wu Y, He Y, Wu T, Chen T, Weng W, Wan H. Influence of some parameters on the synthesis of nanosized NiO material by modified sol-gel method. Materials Letters. 2007; **61**(14):3174-3178

[102] Thota S, Kumar J. Sol-gel synthesis and anomalous magnetic behaviour of NiO nanoparticles. Journal of Physics and Chemistry of Solids. 2007;**68**(10):1951-1964

[103] Hotovy I, Huran J, Spiess L, Romanus H, Buc D, Kosiba R. NiO-based nanostructured thin films with Pt surface modification for gas detection. Thin Solid Films. 2006;**515**(2): 658-661

[104] Min KC, Kim M, You YH, Lee SS, Lee YK, Chung TM, Kim CG, Hwang JH, An KS, Lee NS, Kim Y. NiO thin films by MOCVD of Ni (dmamb) 2 and their resistance switching phenomena. Surface and Coatings Technology. 2007;**201**(22):9252-9255

[105] Tao D, Wei F. New procedure towards size-homogeneous and well-dispersed nickel oxide nanoparticles of 30nm. Materials Letters. 2004;**58**(25):3226-3228

[106] Dooley KM, Chen SY, Ross JRH. Stable nickel-containing catalysts for the oxidative coupling of methane. Journal of Catalysis. 1994;**145**(2):402-408

[107] Yang HX, Dong QF, Hu XH, Ai XP, Li SX. Preparation and characterization of $LiNiO_2$ synthesized from Ni $(OH)_2$ and $LiOH{\cdot}H_2O$. Journal of Power Sources. 1999;**79**(2):256-261

[108] Miller EL, Rocheleau RE. Electrochemical behavior of reactively sputtered iron-doped nickel oxide. Journal of the Electrochemical Society. 1997;**144**(9):3072-3077

[109] Ichiyanagi Y, Wakabayashi N, Yamazaki J, Yamada S, Kimishima Y, Komatsu E, Tajima H. Magnetic properties of nio nanoparticles. Physica B: Condensed Matter. 2003;**329**:862-863

[110] Venkataraj S, Kappertz O, Weis H, Drese R, Jayavel R, Wuttig M. Structural and optical properties of thin zirconium oxide films prepared by reactive direct current magnetron sputtering. Journal of Applied Physics. 2002;**92**(7):3599-3607

[111] Gao P, Meng LJ, dos Santos MP, Teixeira V, Andritschky M. Characterisation of ZrO_2 films prepared by rf reactive sputtering at different O_2 concentrations in the sputtering gases. Vacuum. 2000;**56**(2):143-148

[112] Petit M, Monot J. Functionalization of Zirconium Oxide Surfaces. Chemistry of Organo-Hybrids: Synthesis and Characterization of Functional Nano-Objects. 2015. pp. 168-199

[113] Balaram N, Reddy MSP, Reddy VR, Park C. Effects of high-k zirconium oxide (ZrO_2) interlayer on the electrical and transport properties of au/n-type InP Schottky diode. Thin Solid Films. 2016;**619**:231-238

[114] Lopez CM, Suvorova NA, Irene EA, Suvorova AA, Saunders M. ZrO_2 film interfaces with Si and SiO_2. Journal of Applied Physics. 2005;**98**(3):033506

[115] Zhang HH, Ma CY, Zhang QY. Scaling behavior and structure transition of ZrO_2 films deposited by RF magnetron sputtering. Vacuum. 2009;**83**(11):1311-1316

[116] Gopalan R, Chang CH, Lin YS. Thermal stability improvement on pore and phase structure of sol-gel derived zirconia. Journal of Materials Science. 1995;**30**(12):3075-3081

[117] Tanabe K. Surface and catalytic properties of ZrO_2. Materials Chemistry and Physics. 1985;**13**(3-4):347-364

[118] Piconi C, Maccauro G. Zirconia as a ceramic biomaterial. Biomaterials. 1999;**20**(1):1-25

[119] Muhammad S, Hussain ST, Waseem M, Naeem A, Hussain J, Jan MT. Surface charge properties of zirconium dioxide. Iranian Journal of Science and Technology. 2012; **36**(A4):481

[120] Manicone PF, Iommetti PR, Raffaelli L. An overview of zirconia ceramics: Basic properties and clinical applications. Journal of Dentistry. 2007;**35**(11):819-826

[121] Pareja RR, Ibáñez RL, Martín F, Ramos-Barrado JR, Leinen D. Corrosion behaviour of zirconia barrier coatings on galvanized steel. Surface and Coatings Technology. 2006;**200**(22):6606-6610

[122] Puntervold T, Strand S, Austad T. Water flooding of carbonate reservoirs: Effects of a model base and natural crude oil bases on chalk wettability. Energy and Fuels. 2007;**21**(3):1606-1616

[123] Love JC, Estroff LA, Kriebel JK, Nuzzo RG, Whitesides GM. Self assembled monolayers of thiolates on metals as a form of nanotechnology. Chemical Reviews. 2005; **105**(4):1103-1170

[124] Iglauer S, Salamah A, Sarmadivaleh M, Liu K, Phan C. Contamination of silica surfaces: Impact on water-CO_2-quartz and glass contact angle measurements. International Journal of Greenhouse Gas Control. 2014;**22**:325-328

[125] Sarmadivaleh M, Al-Yaseri AZ, Iglauer S. Influence of temperature and Presssure on quartz-water-CO_2 contact angle and CO_2-water interfacial tension. Journal of Colloid Science. 2015;**441**:59-64

[126] Mahdi Jafari S, He Y, Bhandari B. Nano-emulsion production by sonication and microfluidization-a comparison. International Journal of Food Properties. 2006;**9**(3):475-485

[127] Lander LM, Siewierski LM, Brittain WJ, Vogler EA. A systematic comparison of contact angle methods. Langmuir. 1993;**9**(8):2237-2239

[128] Ingham B. X-ray scattering characterisation of nanoparticles. Crystallography Reviews. 2015;**21**(4):229-303

[129] Winkler K, Paszewski M, Kalwarczyk T, Kalwarczyk E, Wojciechowski T, Gorecka E, Pociecha D, Holyst R, Fialkowski M. Ionic strength-controlled deposition of charged nanoparticles on a solid substrate. The Journal of Physical Chemistry C. 2011;**115**(39): 19096-19103

[130] Nikolov A, Kondiparty K, Wasan D. Nanoparticle self-structuring in a nanofluid film spreading on a solid surface. Langmuir. 2010;**26**(11):7665-7670

[131] Ershadi M, Alaei M, Rashidi A, Ramazani A, Khosravani S. Carbonate and sandstone reservoirs wettability improvement without using surfactants for chemical enhanced oil recovery (C-EOR). Fuel. 2015;**153**:408-415

[132] Eshed M, Pol S, Gedanken A, Balasubramanian M. Zirconium nanoparticles prepared by the reduction of zirconium oxide using the rapet method. Beilstein Journal of Nanotechnology. 2011;**2**(1):198-203

[133] Son HA, Yoon KY, Lee GJ, Cho JW, Choi SK, Kim JW, Im KC, Kim HT, Lee KS, Sung WM. The potential applications in oil recovery with silica nanoparticle and polyvinyl alcohol stabilized emulsion. Journal of Petroleum Science and Engineering. 2015;**126**:152-161

[134] Zargartalebi M, Kharrat R, Barati N. Enhancement of surfactant flooding performance by the use of silica nanoparticles. Fuel. 2015;**143**:21-27

[135] Mercera PDL, Van Ommen JG, Doesburg EBM, Burggraaf AJ, Roes JRH. Stabilized tetragonal zirconium oxide as a support for catalysts evolution of the texture and structure on calcination in static air. Applied Catalysis. 1991;**78**(1):79-96

[136] Li YV, Cathles LM. Retention of silica nanoparticles on calcium carbonate sands immersed in electrolyte solutions. Journal of Colloid and Interface Science. 2014;**436**:1-8

[137] Zhang T, Murphy M, Yu H, Bagaria H, Nielson B, Bielawski C, Johnston K, Huh C, Bryant S. Investigation of nanoparticle adsorption during transport in porous media. SPE Annual Technical Exhibition, New Orleans, Louisiana, 30 September-2 October. 2013

[138] Ju B, Fan T, Li Z. Improving water injectivity and enhancing oil recovery by wettability control using nanopowders. Journal of Petroleum Science and Engineering. 2012;**86**:206-216

[139] Maghzi A, Mohammadi S, Ghazanfari MH, Kharrat R, Masihi M. Monitoring wettability alteration by silica nanoparticles during water flooding to heavy oils in five-spot systems: A pore-level investigation. Experimental Thermal and Fluid Science. 2012;**40**:168-176